T0140125

EAI/Springer Innovations in Communication and Computing

Series editor

Imrich Chlamtac, CreateNet, Trento, Italy

The impact of information technologies is creating a new world yet not fully understood. The extent and speed of economic, life style and social changes already perceived in everyday life is hard to estimate without understanding the technological driving forces behind it. This series presents contributed volumes featuring the latest research and development in the various information engineering technologies that play a key role in this process.

The range of topics, focusing primarily on communications and computing engineering include, but are not limited to, wireless networks; mobile communication; design and learning; gaming; interaction; e-health and pervasive healthcare; energy management; smart grids; internet of things; cognitive radio networks; computation; cloud computing; ubiquitous connectivity, and in mode general smart living, smart cities, Internet of Things and more. The series publishes a combination of expanded papers selected from hosted and sponsored European Alliance for Innovation (EAI) conferences that present cutting edge, global research as well as provide new perspectives on traditional related engineering fields. This content, complemented with open calls for contribution of book titles and individual chapters, together maintain Springer's and EAI's high standards of academic excellence. The audience for the books consists of researchers, industry professionals, advanced level students as well as practitioners in related fields of activity include information and communication specialists, security experts, economists, urban planners, doctors, and in general representatives in all those walks of life affected ad contributing to the information revolution.

About EAI

EAI is a grassroots member organization initiated through cooperation between businesses, public, private and government organizations to address the global challenges of Europe's future competitiveness and link the European Research community with its counterparts around the globe. EAI reaches out to hundreds of thousands of individual subscribers on all continents and collaborates with an institutional member base including Fortune 500 companies, government organizations, and educational institutions, provide a free research and innovation platform.

Through its open free membership model EAI promotes a new research and innovation culture based on collaboration, connectivity and recognition of excellence by community.

More information about this series at http://www.springer.com/series/15427

Shishir Kumar Shandilya · Smita Shandilya
Atulya K. Nagar
Editors

Advances in Nature-Inspired Computing and Applications

Editors
Shishir Kumar Shandilya
VIT Bhopal University
Bhopal, Madhya Pradesh, India

Atulya K. Nagar
Liverpool Hope University
Liverpool, UK

Smita Shandilya
Sagar Institute of Research,
    Technology and Science
Bhopal, Madhya Pradesh, India

ISSN 2522-8595   ISSN 2522-8609   (electronic)
EAI/Springer Innovations in Communication and Computing
ISBN 978-3-030-07195-0   ISBN 978-3-319-96451-5   (eBook)
https://doi.org/10.1007/978-3-319-96451-5

This Springer imprint is published by the registered company Springer Nature Switzerland AG
The registered company address is: Gewerbestrasse 11, 6330 Cham, Switzerland

To our little master 'Samarth Koshagra Shandilya'
—Shishir Kumar Shandilya, Smita Shandilya

To my lovely daughters 'Kopal' & 'Priyel'
—Atulya K. Nagar

Foreword

I had the pleasure and honor of reviewing and writing foreword for a previous book by the same editors. I want to congratulate them for another book with a similar theme. This second book is more focused on nature-inspired computing. Comparison of contents of the two books shows the progress in the field over last 2 years.

This book is an interesting mix of techniques and applications. The contents of the book probably provide a good sample: optimization, applications to computer networks, and nature-inspired computing.

Neural and genetic processes provided early inspiration to tackle problems that could not be solved using conventional mathematics. The early neural networks (perceptrons) were decidedly limited in terms of the set of problems that could be solved. The workings of a perceptron could be easily proven mathematically for linear separable problems. Most real-world problems are not linear. Yet, biological entities regularly survive in the nonlinear world. Computing scientists cautiously started borrowing some of the principles from natural processes. Looking at the model of brain gave us multilayer perceptrons. Support vector machines extended perceptrons based on mathematical optimization. Deep learning extended the concept of multilayer perceptrons based on higher level understanding of human reasoning process, especially for understanding the images. Neurocomputing is not always associated with nature-inspired computing. This book contains one chapter (Chapter "Enhanced Throughput and Accelerated Detection of Network Attacks Using a Membrane Computing Model Implemented on a GPU") that looks at P systems or membrane computing (MC) that can use neural-like processing. The Chapter "Enhanced Throughput and Accelerated Detection of Network Attacks Using a Membrane Computing Model Implemented on a GPU" in this book focuses on a different aspect of membrane computing: tissue-like MC. I enjoyed learning about these different variations of nature-inspired computing that I had either not seen before or had only looked at them peripherally.

Optimization seems to be the primary focus of many chapters in this book. Optimizing search of the solution space was one of the first and important goals of artificial intelligence. The number of possible solutions to even a simple real-world problem tends to be in billions. Evaluating each one of these solutions and identifying optimal solution is an impossible task. Nature continues to find acceptable solutions to problems that are several magnitudes more complex. Cellular automata that evolved into genetic algorithms provided excellent solutions to many optimization problems. The researchers have taken the nature-inspired computing to the next level by understanding and coding behaviors of many species. The list of early efforts included ants (ant colony optimization), fish/birds (particle swarm optimization), fireflies, bats, and bees. The list of these analogies continues to grow. I was intrigued by many more mentioned in this book. For example, Chapter "Artificial Feeding Birds (AFB): A New Metaheuristic Inspired by the Behavior of Pigeons" introduces the use of a different behavioral aspect of birds—feeding strategy. Among other applications, this approach has been used to optimize neural network training. Chapter "Application of Nature—Inspired Algorithms in Medical Image Processing" uses the optimization inspired by Lions and Monkeys. Chapter "Modified Krill Herd Algorithm for Global Numerical Optimization Problems" adds Krill herd behavior to the portfolio of species covered in this book. Readers will be able to read how the behaviors of at least eight species can help solve complex optimization problems. Having all these techniques in a single book will help readers contemplate the similarities and differences of these natural strategies to wade through the exponential search space of possible solutions.

The number of application domains in this book is almost as varied as the number of natural processes. There is an emphasis on networking applications. Graph theoretical/networking problems are often computationally expensive and can benefit greatly from heuristics from nature-inspired optimization techniques. The book also demonstrates the effectiveness of these techniques in other application domains including traffic management, numerical optimization problems, and bioinformatics.

In summary, I would like to say that the book represents a diversity of techniques for a variety of real-world applications. It will give readers sufficient but not overwhelming selection of topics. I thank the editors for an opportunity to review this wonderful collection.

Halifax, Canada

Pawan Lingras[1]
Saint Mary's University

---

[1]Pawan Lingras is a graduate of IIT Bombay with graduate studies from the University of Regina. He is currently a Professor and the Director of Computing and Data Analytics at Saint Mary's University, Halifax. He is also involved with a number of international activities, having served as a Visiting Professor at Munich University of Applied Sciences and IIT Gandhinagar, as a Research Supervisor at Institut Superieur de Gestion de Tunis, as a University Grants Commission (India) Scholar-in Residence, and as a Shastri Indo-Canadian scholar. He has delivered more than 35 invited talks at various institutions around the world. He has authored more than 200 research papers in various international journals and conferences. He has also co-authored three textbooks, and co-edited two books and eight volumes of research papers. His academic collaborations/co-authors include academics from Canada, Chile, China, Germany, India, Poland, Tunisia, the UK, and USA. His areas of interests include artificial intelligence, information retrieval, data mining, web intelligence, and intelligent transportation systems. He has served as a general co-chair, program co-chair, review committee chair, program committee member, and reviewer for various international conferences on artificial intelligence and data mining. He is also on the editorial boards of a number of international journals. His fundamental research has been supported by the Natural Science and Engineering Research Council (NSERC) of Canada for 25 years, and the applications of his research to the industry are supported by a number of funding agencies, including NSERC, the Government of Nova Scotia, NRC-IRAP, and MITACS.

Preface

Nature-inspired computing underlines the concept of learning and behaving as per the biological species to achieve the adaptability for survival by fulfilling certain objectives. Nature-inspired computing techniques have successfully been experimented on and applied to machine-learning and advanced artificial intelligence. It is still a less-explored area but has potential to change the meaning of the learning process of machines. Most of the current research is focused on methods encouraged by these concepts. It is having an especially important role in science and engineering and is becoming a backbone of almost all developing technologies.

This presented book has focused on current research while highlighting the empirical results along with theoretical concepts to provide a comprehensive reference for students, researchers, scholars, professionals, and practitioners in the field of advanced artificial intelligence, nature-inspired algorithms, and soft computing.

This book contains quality research contributions from leading scholars from all over the world with comprehensive coverage of each specific topic, highlighting recent and future trends and describing the latest advances. The book is aimed to bring together leading researchers and practitioners in this field who are working on nature-inspired computing and related techniques.

We express our heartfelt gratitude to all the authors, reviewers, and publishers, especially Mr. Jakub Tlolka, Eliska Vlckova and Ms. Lucia Zatkova for their kind support. Special thanks to Prof. Pawan Lingras for accepting our request to write the Foreword and guiding us throughout the process of this publication. We hope that this book will be beneficial to all the concerned readers.

Bhopal, India
Bhopal, India
Liverpool, UK

Shishir Kumar Shandilya
Smita Shandilya
Atulya K. Nagar

Contents

About the Editors

Dr. Shishir Kumar Shandilya, Division Head, Cyber Security & Digital Forensics, VIT Bhopal University, is a renowned academician and active researcher with proven record of teaching and research. He is a Cambridge University Certified Professional Teacher and Trainer and Senior Member of IEEE-USA. He has received "IDA Teaching Excellence Award" for distinctive use of Technology in Teaching by Indian Didactics Association, Bangalore (2016) and "Young Scientist Award" for consecutive two years (2005 and 2006) by Indian Science Congress and MP Council of Science and Technology. He has written seven books of international fame (published in USA, Denmark, and India) and published quality research papers. He is an active member of various international professional bodies.

Smita Shandilya, Associate Professor in the department of Electrical & Electronics Engineering, SIRTS, is an eminent scholar and researcher. She is Senior Member of IEEE-USA and has delivered several invited talks in national seminars of high repute. Her research interests are Power System Planning and Smart Micro Grids. She is also involved in the establishment of Energy Lab in the Institute (first in any Private Institute in M.P.), and Establishment of Training cum Incubator center in Collaboration with iEnergy Pvt. Ltd.

Atulya K. Nagar holds the Foundation Chair as Professor of Mathematical Sciences and is Dean of the Faculty of Science at Liverpool Hope University, UK; he has been the Head of Department of Mathematics and Computer Science which he established at the University. He is an internationally respected scholar working at the cutting edge of theoretical computer science, applied mathematical analysis, operations research, and systems engineering. He received a prestigious Commonwealth Fellowship for pursuing his doctorate (D.Phil.) in Applied Non-Linear Mathematics, which he earned from the University of York (UK) in 1996; and he holds B.Sc. (Hons.), M.Sc., and M.Phil. (with Distinction) in Mathematical Physics from the MDS University of Ajmer, India. His research expertise spans both Applied Mathematics and Computational Methods for non-linear, complex, and intractable problems arising in Science, Engineering, and Industry. In problems like these, the effect is known, but the cause is not. In this approach of mathematics, also known as "Inverse Problems", sophisticated mathematical modeling and computational algorithms are required to understand such behavior.

Automatic Generation of Cyber Architectures Optimized for Security, Cost, and Mission Performance: A Nature-Inspired Approach

Neal Wagner, Cem Ş. Şahin, Jaime Pena and William W. Streilein

Abstract Network segmentation refers to the practice of partitioning a computer network into multiple segments and restricting communications between segments to inhibit a cyberattacker's ability to move and spread infection. While segmentation is widely recommended by cybersecurity experts, there is no clear guidance on what segmentation architectures are best to maximize a network's security posture. Additionally, the security gained by segmentation does not come without cost. Segmentation architectures require resources to implement and may also cause degradation of mission performance. Network administrators currently rely on judgment to construct segmentation architectures that maximize security while minimizing resource cost and mission degradation. This chapter proposes an automated method for generating segmentation architectures optimized for security, cost, and mission performance. The method employs a hybrid approach that combines nature-inspired optimization with cyber risk modeling and simulation to construct candidate architectures, evaluate them, and intelligently search the space of possible architectures to hone in on effective ones. We implement the method in a prototype decision system and demonstrate the system via a case study on a representative network environment under cyberattack.

This material is based upon the work supported by the Assistant Secretary of Defense for Research and Engineering under Air Force Contract No. FA8721-05-C-0002 and/or FA8702-15-D-0001. Any opinions, findings, conclusions, or recommendations expressed in this material are those of the author(s) and do not necessarily reflect the views of the Assistant Secretary of Defense for Research and Engineering.

N. Wagner (✉) · C. Ş. Şahin · J. Pena · W. W. Streilein
MIT Lincoln Laboratory, Lexington, MA, USA
e-mail: neal.wagner@ll.mit.edu

C. Ş. Şahin
e-mail: cem.sahin@ll.mit.edu

J. Pena
e-mail: jdpena@ll.mit.edu

W. W. Streilein
e-mail: wws@ll.mit.edu

© Springer International Publishing AG, part of Springer Nature 2019
S. K. Shandilya et al. (eds.), *Advances in Nature-Inspired Computing and Applications*, EAI/Springer Innovations in Communication and Computing, https://doi.org/10.1007/978-3-319-96451-5_1

1 Introduction

Network segmentation (NS) is a defensive mitigation technique designed to reduce the damage due to cyberattack. Its goal is to limit attacker access to and movement within a network by partitioning the network into multiple segments or enclaves and restricting communications between enclaves and between enclaves and the Internet. This partitioning is typically implemented by the use of firewalls, network egress and ingress filters, application-level filters, and/or physical (hardware) infrastructure [8]. NS is widely regarded as critical for network security [9, 16, 20, 25] but is poorly understood with only vague guidance (e.g., [20, 23]) on how to apply it. For even small networks, many different NS architectures are possible and the number of possibilities grows exponentially with network size.

The problem is compounded by the fact that security does not come without cost. Segmentation architectures require resources to implement and maintain and also may cause degradation of mission performance. Network administrators must select architectures that maximize security posture while minimizing resource cost and mission degradation. Currently, administrators are forced to rely on judgment to balance tradeoffs between security, cost, and mission performance. A further compounding factor is that for many enterprise networks, the problem of mission mapping has not been solved [11, 26]. Mission mapping refers to the mapping of an organization's mission onto the cyber assets (e.g., devices/servers, software applications, communication protocols, etc.) used to execute it. Because many organizations do not know exactly what cyber assets are being used to support their mission and how they are being used, they cannot reasonably estimate the mission degradation that may result due to a given NS architecture.

This chapter proposes an automated method for constructing NS architectures that are optimized for security, cost, and mission performance. The method employs a hybrid approach that combines nature-inspired optimization with cyber risk modeling and simulation to construct candidate architectures, evaluate them, and intelligently search the space of possible architectures to generate efficacious ones. The proposed method is implemented in a prototype decision system and demonstrated via a case study on a representative network environment under cyberattack. Our work addresses an important gap in the area of cybersecurity decision support (CSDS): the need for systems that leverage data-driven methods to generate optimal/near-optimal security decisions.

The field of CSDS is still quite young with only a handful of studies to date. Two recent studies seek to aggregate input from subject matter experts (SMEs) to address cyber threats: in [10] SME assessments are used to forecast threats and recommend security measures while in [17] SME rankings of cyberattacks and relevant security components are aggregated to provide security assessments of computer systems during the system design phase.

Another study details a cyber infrastructure to facilitate and secure individual-based decision-making and negotiation for Internet-of-Things devices with respect to applications in health care [12]. In [18], a Bayesian Belief Network, cyber vulnerability assessment, and expected loss computation are combined to compute appropriate premiums for cyber insurance products while [4] utilizes game theory and combinatorial optimization to evaluate cybersecurity investment strategies. Finally, in [13], a decision support system to assist cyber defenders protecting mobile ad-hoc networks (MANETs) is developed to remediate malicious intrusions and reduce network energy costs.

In this chapter, we build upon the work of [27] to develop an automated decision system to generate NS architectures optimized for multiple objectives. In [27], a semiautomated system for the NS decision problem is developed that optimizes for one objective only, namely security risk. Here, we present a fully automated implementation of the system that includes optimization with respect to the three critical, and often conflicting, objectives: security, cost, and mission performance. The contributions of this chapter are the following.

- We provide a method to automatically generate effective NS architectures and realize this method in a fully automated decision system. The system utilizes a novel hybrid algorithm that combines nature-inspired optimization with cyber risk modeling and simulation.
- The system outputs NS architectures that are optimized for security, cost, and mission performance.
- We provide a mission performance model that allows for approximated measurement of mission degradation due to a given segmentation architecture when the mapping of cyber assets onto organizational mission is not available.

The rest of this chapter is organized as follows: Sect. 2 provides an overview of the network segmentation defensive mitigation, Sect. 3 describes the decision system's overall design, Sect. 4 details the security, cost, and mission performance models used to evaluate candidate architectures, Sect. 5 presents a case study that demonstrates the system for a representative network environment under cyberattack, and Sect. 6 concludes.

2 Network Segmentation Defensive Mitigation

Network segmentation (NS), as described above, is a cyber defensive mitigation meant to inhibit an attacker's ability to move and spread infection throughout the network. NS accomplishes this by partitioning the network into multiple enclaves and restricting communications between enclaves and between enclaves and the Internet. This partitioning is implemented by the use of firewalls, network egress and ingress filters, application-level filters, and/or physical (hardware) infrastructure [8]. Figure 1a, b provides example NS architectures.

Figure 1a gives an example architecture in which no partitioning is utilized. Here, the entire network (i.e., network devices, servers, routers, switches, etc.) is contained within a single enclave in which all network devices are allowed direct communication with all other network devices. Additionally, the enclave is connected to the

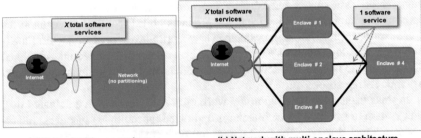

Fig. 1 Two example network segmentation architectures: **a** a network with no partitioning and **b** a network partitioned into four enclaves, three that allow direct communications to the Internet and a fourth enclave that allows direct communications to the other three network enclaves but not to the Internet

Internet, that is, it allows direct communication to the Internet where cyberattackers preside. Communications between the network enclave and the Internet occur via one or more software services (e.g., web browsers, email, ssh, etc.) running on the enclave. In the figure, X software services allow this communication.

In Fig. 1a, a cyber attacker may exploit vulnerabilities present in one or more of the software services to penetrate the network enclave. Common strategies for penetration include sending phishing emails with infected attachments or links to infected websites to entice a network user to open the attachment or browse to the infected website and, thus allow the attacker to infect her device and gain a foothold in the network. Once inside, the attacker can easily move and spread infection to other network devices because the infected device is allowed to directly communicate with all other network devices.

The example given by Fig. 1b is an NS architecture that is partitioned into four enclaves, three that are allowed direct communication with the Internet (Enclaves #1–3 in the figure) and a fourth that is allowed direct communications to the other three enclaves but not to the Internet (Enclave #4 in the figure). Note also that Enclaves #1–3 do not have direct communications with each other. Communications from Enclaves #1–3 to the Internet occur via the same set of X software services (split up over the three enclaves) as given in Fig. 1a. An additional software service provides communication from these Internet-facing enclaves to Enclave #4.

For the architecture of Fig. 1b, an attacker may also be able to penetrate the network and gain access to one of Enclaves #1–3. However, upon penetration, the attacker's ability to spread infection to other network devices is hampered by the communication restrictions. In order for the attacker to infect a device in Enclave #4, she must exploit an additional software service, the one that allows communication with that enclave. To spread to other Internet-facing enclaves, she must also exploit additional services. The restrictions created by the architecture serve to slow down the rate at which the attacker can spread infection and allows the defender more time to patch vulnerable software services and/or cleanse infected enclave devices.

An NS architecture selected by the defender consists of the following components.

- A set of network enclaves. An enclave is defined as a group of devices with homogeneous reachability.
- Software services that allow communications between enclaves and between enclaves and the Internet.
- The rate at which software services are patched.
- The rate at which enclaves are cleansed. Enclave cleansing is a process by which an enclave's devices are cleansed to remove infection and restored to their original state. The result of enclave cleansing is to dis-entrench an attacker who has penetrated.

3 Automatic Generation of NS Architectures: Decision System Design

Extending the work presented in [27], we have designed an automatic method for generating NS architectures that are optimized for security, cost, and mission performance. The proposed method is realized as a fully automated decision system. The system inputs parameters that characterize the network environment and cyber threat and outputs an effective architecture for that environment.

Figure 2 gives a graphical depiction of the decision system. In the figure, the system is represented by the gray box. System inputs (blue and red inflowing arrows to the left of the figure) are used to characterize the network environment including its security posture and existing cyber threat. System output (blue arrow to the right of the figure) is an optimized NS architecture for the given network environment. The

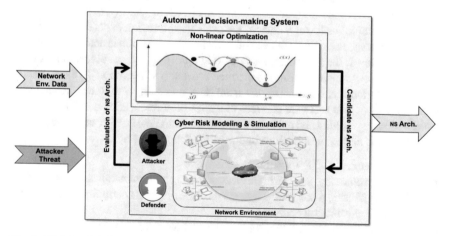

Fig. 2 High-level design of the decision system to automatically generate NS architectures optimized for security, cost, and mission performance

decision system is comprised of two algorithmic components, nonlinear optimization, and cyber risk modeling and simulation (mod/sim) (depicted by the yellow and green boxes inside the gray box in the figure). The optimization component suggests candidate NS architectures while the mod/sim component evaluates these architectures. These two components run iteratively: the optimization component suggests a candidate architecture, it is evaluated via the mod/sim component, and the evaluation is then fed back to the optimization component where it is utilized to guide its search to construct newer, more promising candidate architectures. Together, these components work to search the space of possible NS architectures and hone in on effective architectures.

As discussed in Sect. 1, the number of possible NS architectures grows exponentially with network size and thus deterministic search methods are intractable for all but the smallest networks. Here, we employ soft computing algorithms inspired by nature to execute the function of the optimization component. The current version of the system utilizes a Simulated Annealing (SA) algorithm to explore the space of NS architectures.

SA is an adaptation of the Metropolis–Hastings Monte Carlo method for approximating the global optimum of a given function. It is inspired by the process of annealing in metallurgy [2]. Generally, the algorithm starts with an initial solution, generates a new solution that is a neighbor to it in the decision problem search space, evaluates both the old and new solutions against a given objective function, probabilistically accepts the new solution in place of the old solution, and then repeats these steps (using the currently accepted solution) for some number of iterations. The algorithm mimics the annealing process by initially having a relatively higher probability of accepting a new solution that is inferior to the currently accepted solution and progressively lowering that probability at each iteration. Note that if a new solution is superior to the currently accepted solution, it is always accepted.

Algorithm 1 specifies the SA algorithm for the NS decision system. In the algorithm, the objective is to minimize the combined risk to security, cost, and mission performance, and thus lower values of the evaluation function *eval* (lines 8, 9 of the algorithm) are superior. Computation of the evaluation function is detailed in Sect. 4.

In Algorithm 1, the function Generate-Neighbor (line 7 of the algorithm) is used to generate a new candidate solution architecture that is a "neighbor" to the currently accepted solution architecture. Recall from Sect. 2 that an NS architecture includes a set of enclaves, the software services that allow communications, and the rates at which software patching and enclave cleansing occur. A new architecture is generated by altering the current architecture in one of the following ways:

- Increasing or decreasing the number of enclaves.
- Adding or removing software services allowing communication between two enclaves or altering a software service to change one of its endpoints (i.e., change its source or destination enclave).
- Changing the software patching rate or the enclave cleansing rate.

Algorithm 1 Optimize-NS-Architecture(s_0, k_{max})

1: {s_0: initial segmentation architecture, k_{max}: max. no. of iterations}
2: $s \leftarrow s_0$ {Accept s_0 as current solution}
3: $s_{best} \leftarrow s$ {Save the best solution found so far}
4: $k \leftarrow k_{max}$
5: **repeat**
6:      $T \leftarrow \frac{k}{k_{max}}$ {Set temperature T}
7:      $s_{new} \leftarrow$ Generate-Neighbor(s)
8:      $eval(s)$ {Evaluate s, s_{new}}
9:      $eval(s_{new})$
10:     **if** $eval(s_{new}) < eval(s)$ **then**
11:         $s \leftarrow s_{new}$ {Accept s_{new} if superior to s (lower risk)}
12:         **if** $eval(s_{new}) < eval(s_{best})$ **then**
13:             $s_{best} \leftarrow s_{new}$ {Save the best solution found so far}
14:         **end if**
15:     **else**
16:         $r \leftarrow$ random value $\in [0, 1]$ {Probabilistically accept inferior s_{new}}
17:         **if** $r \leq e^{\frac{(1-eval(s_{new})-eval(s))}{T}} \over e^{(1/T)-1}$ **then**
18:             $s \leftarrow s_{new}$
19:         **end if**
20:     **end if**
21:     $k \leftarrow k - 1$ {Decrement k to reduce T}
22: **until** $k = 0$
23: **return** s_{best}

Figure 3a, b illustrates operations that increase/decrease the number of enclaves. Figure 3a depicts the Split Enclaves operation. Here, the number of enclaves is increased by splitting an existing enclave into two or more enclaves. Figure 3b depicts the Merge Enclaves operation: the number of enclaves is decreased by merging two or more enclaves into a single enclave.

The operations described above are used to generate new NS architectures that are then evaluated by the mod/sim component. This combination of generation and evaluation serves to drive the decision system to explore promising areas of the search space and automatically construct effective architectures. The decision system is implemented using a combination of Java 1.8, Scala 2.11, Python 2.7, and ECMAScript 7.

Figure 4 shows a screen capture of the user interface of the prototype system. The plot to the left of the figure gives the fitness value of the current solution architecture over several iterations of the SA algorithm and the three graphs to the right of the figure give graphical representations of the starting, current, and best-so-far architectures generated during the run. Note that Fig. 4 provides an illustrative example of the prototype system interface; results shown do not correspond to experiments discussed in this paper. Section 4 below details the system's fitness function while Sect. 5 describes architectures generated during experimentation and their graphical representations.

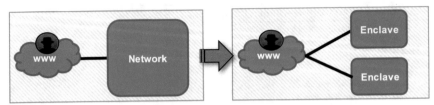

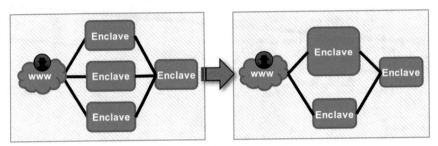

(a) Split Enclave operation

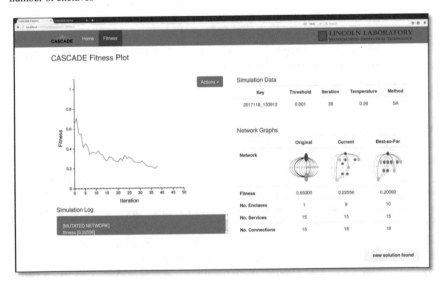

(b) Merge Enclaves operation

Fig. 3 Generating a new NS architecture by changing the number of enclaves: **a** the Split Enclaves operation to increase the number of enclaves and **b** the Merge Enclaves operation to decrease the number of enclaves

Fig. 4 Screen capture of the user interface of the prototype decision system

4 NS Architecture Evaluation Model

The decision system described in the previous section requires an evaluation function to measure the effectiveness of generated solution architectures. As discussed in Sect. 1, we wish to generate architectures that are optimized for security, cost, and mission performance. The following sections detail the security, cost, and mission performance models used for evaluation.

4.1 Measuring Security Risk

We utilize a continuous-time Markov chain model (CTM) developed in [24, 28] to measure the security inherent to an NS architecture. The CTM characterizes security risk for a given network environment under threat from external attackers whose goal is to penetrate and spread throughout the network. The model captures changes in network state due to the arrival of new software vulnerabilities, patches, exploits, and the communications allowed within a given NS architecture. Arrivals of these security-related events are modeled as Poisson processes, and thus transitions between states are characterized by sampling rates in which intervals between event samples are exponentially distributed with a given mean. Here, we use Poisson processes to capture attackers arriving to exploit vulnerabilities and defenders arriving to remediate vulnerabilities and/or cleanse infected enclaves as used in [14]. We will refer to these sampling rates as simply Poisson rates.

The model consists of three entities: (i) a network environment, (ii) one or more network enclaves (i.e., groups of devices with homogeneous reachability), and (iii) one or more software services. A network environment is characterized by a set of enclaves and communication pathways that connect these enclaves. Communication pathways represent *functional information flows* (FIF) which include physical connections (i.e., by a physical line), transitive connections (e.g., a server enclave being able to communicate with the Internet even though no direct line exists), and more complex flows (e.g., an email is sent from the Internet that arrives at a DMZ enclave, is pulled by an exchange server, and is finally downloaded and read by a user in a LAN enclave). A FIF is modeled only as a communication pathway from a source enclave to a destination enclave. Intermediate enclaves between the source and destination are not modeled. Note that the Internet is also modeled as a single enclave in which we assume the attacker presides and, thus, is always compromised.

A FIF connecting two enclaves is enabled by one or more services (i.e., software applications) running on the destination enclave that are subject to vulnerabilities that may be exploited by an attacker and patches that remediate these vulnerabilities. Figure 5a, b depict the Markov process states for an individual service and enclave, respectively.

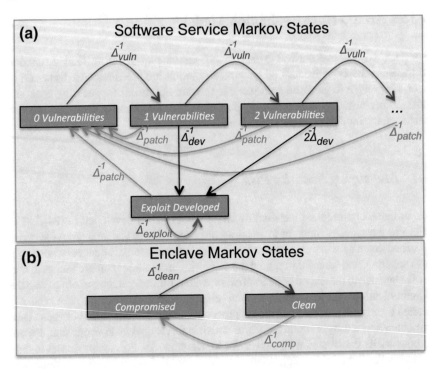

Fig. 5 **a** Markov process states for a software service represent how many vulnerabilities are present and whether an exploit for any of the vulnerabilities has been developed since it was last patched. **b** Markov states for an enclave represent whether or not the enclave has been penetrated by an attacker since it was last cleansed

As given in Fig. 5a, a service is characterized by states reflecting its current number of vulnerabilities $(0, 1, 2, \ldots)$ and whether or not attackers have developed an exploit for one or more of these vulnerabilities. To capture transitions between states we utilize three Poisson rates: the vulnerability arrival rate Δ_{vuln}^{-1}, the exploit development rate Δ_{dev}^{-1}, and the patch rate Δ_{patch}^{-1}. A service starts out in the 0-vulnerability state, transitions to higher vulnerability states with the arrival of vulnerabilities, transitions to the exploit developed state upon the development of 1 or more exploits for its existing vulnerabilities, and finally, transitions back to the 0-vulnerability state upon a patching event. While the service is in the exploit developed state, it can be exploited by the attacker. Exploit events are captured by the exploit deployment rate $\Delta_{exploit}^{-1}$.

Figure 5b shows the Markov states for an enclave. An enclave is characterized by two states, a clean state and a compromised state. An enclave starts out in the clean state, transitions to a compromised state with the arrival of exploit events for one or more of its software services, and then transitions back to a clean state upon the arrival of an enclave cleansing event.

The enclave cleansing rate Δ_{clean}^{-1} is a directly specified parameter while the enclave compromise rate Δ_{comp}^{-1} is specified as a function of the Markov processes governing the states of one or more services running on the enclave. Services running on an enclave are independent and Δ_{comp}^{-1} is computed as

$$\Delta_{comp}^{-1} = \sum_{i=1}^{N} \Delta_{comp}^{-1}(s_i) \tag{1}$$

where s_i represents ith service, $i \in [1, N]$, running on a given enclave E, N is the total number of available services on E and $\Delta_{comp}^{-1}(s_i)$ is the compromise rate of service s_i. Note that the compromise rate for a single service $\Delta_{comp}^{-1}(s_i)$ is captured via the Markov process model depicted in Fig. 5a.

An event-based simulation is used to compute the security risk for a given NS architecture. The simulation computes, for each enclave in a given architecture, the expected probability that the enclave is compromised (i.e., penetrated by the attacker) at any moment. A simulation run is executed by the following steps. (1) A given NS architecture is instantiated with its enclaves, their services, and the communications topology that specifies which services allow communication between enclaves and between enclaves and the Internet. (2) Events are generated via the above-described Poisson rates. A run is terminated when it reaches a specified maximum number of simulated time units.

The security risk for an NS architecture as a whole is measured as the expected probability of enclave penetration by the attacker over all enclaves,

$$Sec(env, s) = \frac{1}{|encls(s)|} \sum_{e \in encls(s)} P_{penetrate}(e) \tag{2}$$

where env is a network environment under cyber threat, s is a segmentation architecture, $encls(s)$ is the set of all enclaves in s, e represents a single enclave and varies over all enclaves of $encls(s)$, $P_{penetrate}(e)$ is the probability of attacker penetration for enclave e, and $Sec(env, s)$ is the expected probability of enclave penetration by the attacker for environment env and segmentation architecture s. This measure is normalized to $[0, 1]$ where lower values mean lower security risk.

4.2   Measuring Cost

Cost is characterized as an information technology (IT) maintenance cost: greater segmentation (i.e., more enclaves) incurs a higher IT cost to maintain. We utilize a normalized exponential function to capture the cost increase as the number of enclaves increases. The cost function is given by

$$C(env, s) = \frac{e^{(N \times k)/M} - 1}{e^k - 1} \tag{3}$$

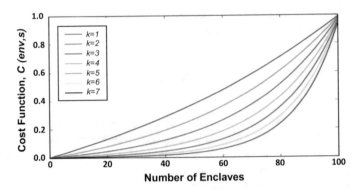

Fig. 6 IT maintenance cost function (Eq. 3) for $M = 100$ and $k = 1, \ldots, 7$

where *env* is a network environment under cyber threat, *s* is a segmentation architecture, N is the total number of enclaves in *s*, M is the maximum number of enclaves that can be supported by the IT maintenance process, k is a steepness constant, and $C(env, s)$ is the IT maintenance cost for a given environment *env* and segmentation architecture *s*. This function is normalized to [0, 1] where lower values mean lower cost.

Figure 6 gives a graphical depiction of the cost function for maximum number of enclaves, $M = 100$ and several values of the steepness constant k. The cost function of Eq. 3 is designed to be flexible: it can characterize a linear increase in cost with increases in the total number of enclaves, N (as shown by the blue line in the plot when $k = 1$) and it can also characterize an exponential increase in cost with increases in N (shown by the black line in the plot when $k = 7$).

4.3 Measuring Mission Performance

As discussed in Sect. 1, the security gained by deploying an NS architecture may have detrimental impacts on mission performance. Because many organizations have not solved the mission mapping problem, that is they do not know exactly what cyber assets are being used to execute the mission and how they are being used, they cannot easily measure and predict the negative effects to mission performance that a given NS architecture can cause. Administrators and network architects are thus put in the difficult position of having to use subjective judgment to select a NS architecture that maximizes security posture but does not degrade mission performance to an unacceptable level.

Our goal is to construct a model to measure the *mission degradation*, the fraction of mission performance that is negatively impacted, due to a given NS architecture. The model should provide viable estimation of mission degradation for networks environments in which knowledge of mission dependencies to cyber assets is incomplete or unavailable. Towards this goal, we focus on two components of NS architecture that the defender must select: the software patching and enclave cleansing rates.

The idea is to capture mission degradation when complete knowledge of mission execution is not known. Here, we assume cleansing devices in an enclave or patching software services results in a cost to mission performance, that is the mission is degraded to some fraction of its optimum performance. When an enclave is cleansed, its devices are unavailable for some amount of time and this lack of availability negatively impacts the mission. When a software service is patched, it is unavailable for some amount of time which can also negatively impact the mission, albeit potentially to a lesser degree as patching usually takes less time than cleansing and does not necessarily require devices to be completely unavailable during the patching process. However, there exists another form of mission degradation due to software patching: that of *software dysfunctionality*. New versions of software may not function exactly as older versions due to upgrades, new features, or new software bugs introduced. The altered functionality of a software application after it has been patched can cause mission operations that depend on it to execute less quickly, less accurately, and/or less completely. Regardless of what mission is being executed or whether or not that mission has been mapped to the cyber assets that support it, the base assumption is that more frequent enclave cleansing and/or software patching by the defender results in more mission degradation.

$$DA' = w_{clean} \cdot \left(\frac{\Delta^{-1}_{clean}}{\Delta^{-1}_{cleanMax}} \right) + \left(w_{patchTime} + w_{patchDisf} \right) \cdot \left(\frac{\Delta^{-1}_{patch}}{\Delta^{-1}_{patchMax}} \right) \qquad (4)$$

$$DA = max \left[\left(\frac{\Delta^{-1}_{clean}}{\Delta^{-1}_{cleanMax}} \right), \left(\frac{\Delta^{-1}_{patch}}{\Delta^{-1}_{patchMax}} \right), DA' \right] \qquad (5)$$

Equations 4 and 5 specify a mathematical model to capture the defender's activity with respect to cleansing and patching. Equation 4 characterizes a defender's activity as a weighted average of the ratios of the rates of cleansing and patching to their maximum possible rates, respectively. In Eq. 4, Δ^{-1}_{clean} and Δ^{-1}_{patch} represent the defender rates for cleansing and patching, respectively, $\Delta^{-1}_{cleanMax}$ and $\Delta^{-1}_{patchMax}$ represent the maximum possible rates of cleansing and patching, respectively, and w_{clean}, $w_{patchTime}$, and $w_{patchDisf}$ are weights representing the relative impact that cleansing and patching have to mission degradation, respectively. Note that there are two weights associated with the impact of patching, $w_{patchTime}$, and $w_{patchDisf}$. $w_{patchTime}$ reflects the impact that patching has on mission-related cyber asset availability and $w_{patchDisf}$ represents the impact that patching has on mission-related cyber asset functionality. DA' gives the measure of cleansing and patching activity by the defender as a weighted average where $w_{clean} + w_{patchTime} + w_{patchDisf} = 1.0$ This measure is normalized to [0, 1] where higher values mean more defender activity, a value of 1 means the maximum amount of defender activity, and a value of 0 means no defender activity (i.e., the defender never cleanses or patches).

Equation 5 computes the finalized measure of defender activity, DA, as a maximum of the individual ratios of the cleansing and patching rates to their respective maximum possible rates and DA' computed in Eq. 4. We utilize the maximum function because it is possible that one activity (either cleansing or patching) when executed at the maximum possible rate may effectively cause maximum activity and thus maximum mission degradation in and of itself. For example, if the maximum possible rate of enclave cleansing is once per day and the defender chooses that rate, she may completely shut down mission operations regardless of what the patching rate is.

We utilize a normalized exponential function similar to the one given in Eq. 3 to capture the increase in mission degradation as the defender activity in cleansing and patching increases. The mission degradation function is given by

$$MD(env, s) = \frac{e^{(DA \times k)/DA_{max}} - 1}{e^k - 1} \tag{6}$$

where env is a network environment under cyber threat, s is a segmentation architecture, DA is the defender activity component of s given by Eq. 5, k is a steepness constant similar to that given in Eq. 3, and DA_{max} is the maximum possible defender activity for cleansing and patching and always has a value of 1.0. This function is normalized to $[0, 1]$ where lower values mean lower mission degradation.

4.4   Combined Measure of Security, Cost, and Mission Performance

The final evaluation function computes a combined measure that represents the overall risk to security, cost, and mission performance and consists of the three component measures described above. The combined risk measure is given by

$$R(env, s) = w_1 \cdot Sec(env, s) + w_2 \cdot C(env, s) + w_3 \cdot MD(env, s) \tag{7}$$

where $Sec(env, s)$ is the security risk component (Eq. 2), $C(env, s)$ is the cost component (Eq. 3), $MD(env, s)$ is the mission degradation component (Eq. 6), and w_1, w_2, and w_3 are weights in $[0, 1]$ representing the relative importance of the security, cost, mission performance component measures, respectively, subject to the constraint $w_1 + w_2 + w_3 = 1.0$. This final risk measure is normalized to $[0, 1]$ where lower values mean lower combined risk. Overall, the objective of the decision system is, for a given network environment under cyber threat env, to generate a segmentation architecture s, that minimizes $R(env, s)$ of Eq. 7.

5 Experiments

We demonstrate the system via a case study on a representative network environment under cyberattack. The aim is to use the system to improve an initial segmentation architecture to minimize $R(env, s)$ of Eq. 7 for multiple scenarios in which the relative importance of security, cost, and mission performance vary.

Figure 7 shows the initial architecture, which represents an unsegmented network, that is a network with only a single enclave such that all network devices can communicate directly with all other network devices. In this single enclave network, direct communications are allowed from enclave to the Internet where cyberattackers preside. Communications from the network to the Internet are made through 15 software services (applications).

The network environment is specified by parameters described in Sect. 4. We leverage real vulnerability, patch, and exploit data to characterize a representative software service and its associated expected rates of vulnerability arrival, patching, and exploit development using the process given in [28, 29]. Below is a summary of this process.

5.1 Network Environment Parameters

Within the CTM (Sect. 4), a software service requires specification of multiple rate parameters including (i) vulnerability arrival rate (Δ_{vuln}^{-1}), (ii) patch arrival rate (Δ_{patch}^{-1}), (iii) exploit development rate (Δ_{dev}^{-1}), and (iv) exploit occurrence rate ($\Delta_{exploit}^{-1}$). Additionally, each enclave requires specification of the enclave cleansing rate ($\Delta_{cleanse}^{-1}$). Here, the goal is to characterize a representative service.

We utilize data from the National Vulnerability Database [21] as well as results from large-scale vulnerability studies [5–7, 19]. These studies define a *vulnerability lifecycle* that captures the state of a vulnerability over time. We use this to

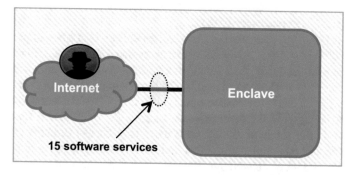

Fig. 7 Initial segmentation architecture to be improved by the decision system

Fig. 8 Time dependencies between vulnerability lifecycle phases (purple arrows represent transitions between phases)

characterize vulnerability phases including vulnerability disclosure (when the vulnerability becomes known), exploit development (when an exploit for the vulnerability is developed), exploit deployment (when the exploit is used), and patching (when a patch for the vulnerability becomes available). Figure 8 shows time dependencies between these phases. From the figure, vulnerability disclosure kicks off two processes in parallel: exploit development and patch arrival. Once an exploit has been developed for the vulnerability (and before a patch has arrived), exploits that may result in compromises can now occur for that service. Patching ends the lifecycle by rendering its exploit(s) ineffective. We use data collected with respect to these phases to compute rates for a representative service.

Δ_{vuln}^{-1}: To characterize the vulnerability arrival rate of a representative service, we average the most common services given in [7] as shown below:

$$\Delta_{vuln}^{-1} = \frac{\sum_{i \in N} \frac{V_i}{T}}{|N|} \tag{8}$$

where i represents the most vulnerable application for a vendor from the top vendor list derived by [7]. V_i is the weighted sum of vulnerabilities for application i over time period T where weights are given by each vulnerability's severity score [3]. N is a set containing the most vulnerable applications from top vendor list derived by [7] which collects vulnerability data over a 7-year period (2000–2007) and groups them by vendors (e.g., vulnerabilities for Microsoft products, for Adobe products, etc.). Note that Eq. 8 considers the set of all known vulnerabilities for a given software service s over a given time period, T. Using Eq. 8, we compute the expected vulnerability arrival rate, Δ_{vuln}^{-1}, as one every 65 days.

Δ_{patch}^{-1}: [6] analyzes data on nearly 15 K vulnerabilities over a 5-year period (from 2001 to 2006) and derives vulnerability discovery dates and patch availability dates from public sources. We fit their results to a Poisson distribution and then compute a weighted average of these fitted results. Our computation yields an expected patch availability rate of one every 25 days (after vulnerability arrival). Recall from Sect. 3 that one of the choices the defender must make when specifying a segmentation architecture is the patching rate. The defender may choose to patch at a slower or faster rate and this choice will affect both the security posture and the mission performance of a given network environment. The above-computed rate represents the fastest rate at which a defender may choose to patch, that is the defender cannot apply patches to software services of a given network environment before those patches are released to the public.

Δ_{dev}^{-1}: [6] executes a similar process to derive exploit availability (exploit development) dates. This study derives exploit development rates ranging from ≈ 8 days before disclosure to ≈ 2 days after disclosure. Additionally, [7] reports that for $\approx 90\%$ of vulnerabilities collected, exploits are available within 10 days of their corresponding disclosure dates. For our experiments, we use the midpoint of these bounds, an expected rate of exploit development as one every 5 days (after vulnerability arrival).

$\Delta_{exploit}^{-1}$: Incident reports are generally difficult to come by because organizations do not like to share data on detected compromise events within their networks. Additionally, different kinds of exploits can be executed in varying amounts of time. Some exploits take advantage of vulnerabilities in server software (for example, the Shellshock vulnerability [22]) while others utilize phishing emails to entice users to download and open malicious file attachments. Exploits that target server vulnerabilities can be executed at any time by the attacker, but exploits that rely on phishing require the victim to trigger the exploit (e.g., by opening a malicious email attachment). We use a conservative estimation for our experiments, an expected rate of exploit deployment as one every 5 days (after exploit development).

The final rate to consider is the enclave cleansing rate, $\Delta_{cleanse}^{-1}$. Here, we consider the maximum rate that the defender may choose to cleanse. As discussed in Sect. 4, enclave cleansing usually results in greater device downtime than software patching, and thus cleansing may have greater detrimental effects on mission performance than patching does. For these experiments, we assume that the maximum rate of enclave cleansing is one every 7 days. For many network environments and organizational missions, cleansing that frequently would likely make it impossible to execute mission operations.

The settings for network environment parameters are given in Table 1. Parameters Δ_{vuln}^{-1}, Δ_{dev}^{-1}, and $\Delta_{exploit}^{-1}$ represent the vulnerability arrival, exploit development, and exploit deployment rates, respectively, and remain static over all experiments. Parameters $\Delta_{patchMax}^{-1}$ and $\Delta_{cleanMax}^{-1}$ represent the maximum possible rates of patch-

Table 1 Network environment parameter settings

Parameter	Description	Setting
Δ_{vuln}^{-1}	Vulnerability arrival rate	1 every 65 days
$\Delta_{patchMax}^{-1}$	Max. patch rate	1 every 25 days
$\Delta_{patchInit}^{-1}$	Initial patch rate	1 every 30 days
Δ_{dev}^{-1}	Exploit development rate	1 every 5 days
$\Delta_{exploit}^{-1}$	Exploit deployment rate	1 every 5 days
$\Delta_{cleanMax}^{-1}$	Max. enclave cleanse rate	1 every 7 days
$\Delta_{cleanInit}^{-1}$	Initial enclave cleanse rate	1 every 365 days

ing and enclave cleansing, respectively (also static over all experiments). Parameter $\Delta^{-1}_{patchInit}$ represents the initial setting for patching rate selected by the defender. Note that the patching rate Δ^{-1}_{patch} specified by a given architecture can vary over the course of a single experiment as the defender may choose slower or faster rates for patching (subject to the constraint given by $\Delta^{-1}_{patchMax}$).

Parameter $\Delta^{-1}_{cleanInit}$ in Table 1 represents the initial setting for cleansing rate selected by the defender. Note that like the patch rate, the cleansing rate Δ^{-1}_{clean} can vary over the course of a single experiment as the defender may choose slower or faster cleansing rates subject to the constraint given by $\Delta^{-1}_{cleanMax}$. We utilize a study that sheds light on the activities of a real cyberattacker to specify $\Delta^{-1}_{cleanInit}$. In [15], a nation-state-sponsored cyber attacker that was able to compromise hundreds of enterprise networks across many countries is documented. The study investigates the time that the attacker, code-named APT1 (Advanced Persistent Threat 1), was able to persist inside an enterprise network before being removed/cleansed from the environment. The study reports persistence times that are up to 4 years with an average persistence time of approximately 1 year, and we use this average persistence time to set $\Delta^{-1}_{cleanInit}$.

Recall from Sect. 4 that the mission performance component measure requires the setting of weights w_{clean}, $w_{patchTime}$, and $w_{patchDisf}$ of Eq. 4 representing the relative impact that cleansing and patching have to mission degradation, respectively. For these experiments, we use $w_{clean} = 0.7$, $w_{patchTime} = 0.15$, and $w_{patchDisf} = 0.15$ to represent a network environment in which enclave cleansing causes a significant degradation of mission performance while patching causes smaller performance degradation with respect to both downtime and dis-functionality. Both the mission performance and cost component measures (Eqs. 6 and 3, respectively) also require setting the steepness constant k to capture the increase in these measures as defender cleansing/patching rates and the number of enclaves increase, respectively. For these experiments, we select $k = 1.0$ for the mission performance measure, which characterizes a nearly linear rise in mission degradation when cleansing and patching rates are increased. For the cost measure, we select $k = 6.0$ with maximum number of enclaves $M = 25$, which characterizes an exponential rise in cost as the number of enclaves increases to the maximum. With this setting, there are small increases in cost when the number of enclaves is less than 10 but quickly growing cost when the number of enclaves is greater.

5.2 Search Parameters

As mentioned above, we aim to explore how varying the relative importance of the three component objectives, security, cost, and mission performance, affect the choice of segmentation architecture by the defender. We execute several experiments in which the weights of Eq. 7, w_1, w_2, and w_3 are varied. We vary w_1, the security component measure weight, from [0,1] in increments of 0.1. For each value of w_1, we explore all possible combinations of values for w_2 (cost component measure weight) and w_3 (mission performance component weight) at increments of 0.1 subject to the constraint that $w_1 + w_2 + w_3 = 1.0$. We thus execute 66 total experiments overall with each experiment starting with the initial architecture given by Fig. 7 and consisting of 50 iterations of the decision engine.

We constrain the search to explore architectures with at least three software services allowing direct communication from a network enclave to the Internet. This constraint is used to prevent the system from generating architectures that are disconnected from the Internet as we assume that most enterprise networks require communication with external networks via the Internet. The output of each experiment is the best architecture found, that is the one with the minimum value of $R(env, s)$ from Eq. 7.

5.3   Results

Figure 9 gives the results of the 66 experiments executed. Security and mission performance weightings are plotted against the normalized fitness of the best architecture found by the system where lower fitness values represent better solutions (i.e., less risk) with respect to the weightings used.[1] Note that the weightings for cost are also varied as described above, but these weights are not plotted in the figure for clarity, although they can easily be inferred from the weight values of the other two component measures. In the figure, the surface of the result space is colored using a spectrum from pink (higher combined risk) to turquoise (lower combined risk).

From Fig. 9, we see that better fitness outputs occur when the security weight is 0.0 signifying scenarios in which all importance is placed on cost and/or mission performance and no importance is given to security. This intuitively makes sense as it is relatively easy to specify an architecture with minimal cost or minimal degradation of mission performance if security is not a concern: simply do not partition the network at all (reduces cost to its minimum value) and/or reduce rates of cleansing and patching (reduces degradation of mission performance). We also see that the worst fitness output occurs when the security weight is 1.0, meaning that all importance is placed on security and none is given to cost or mission performance. This makes sense as well: as discussed in Sect. 1, the number of possible ways to partition a network and restrict communications between enclaves grows exponentially with network size and, thus, it is much more difficult to construct a partitioning that results in minimal security risk.

The surface of the result space in Fig. 9 consists of several peaks and valleys in the middle areas of the plot where weightings for security, cost, and mission performance are all greater than 0.0 and represent scenarios in which importance is given to all

---

[1] Normalized fitness is computed by R/R_{worst} where R is the fitness output of the experiment (Eq. 7) and R_{worst} is the worst fitness (i.e., largest) outputted by any of the 66 experiments executed.

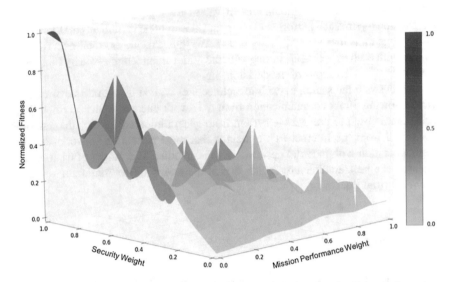

Fig. 9 Experimental results for varying weightings of security, cost, and mission performance. Security and mission performance weights are plotted against the normalized fitnesses of the best architectures found by the decision system for each experiment (cost weights are not displayed for clarity)

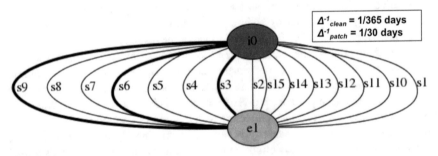

Fig. 10 Segmentation architecture generated by the system when cost weight is 1.0 and security and mission performance weights are both 0.0

three objectives. This illustrates the inherent complexity of the tradeoffs between security, cost, and mission performance and shows that different weightings given to these objective measures can result in very different architectures.

Figures 10, 11, 12 and 13 give graphical depictions of the best architectures outputted by the system in four of the 66 experiments. In the figures, the architecture is represented as a graph in which each node represents an enclave. The topmost node (labeled $i0$) represents the Internet and the other nodes represent the enclaves of the network (labeled $e1, e2, \ldots$). Lines connecting two nodes represent software services that allow communication between enclaves or between an enclave and the Internet. The enclaves are color-coded to represent their security with colors ranging

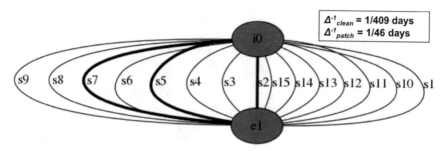

Fig. 11 Segmentation architecture generated by the system when mission performance weight is 1.0 and security and cost weights are both 0.0

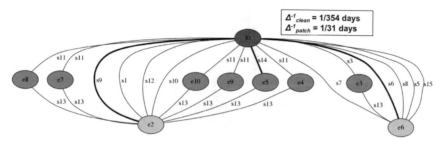

Fig. 12 Segmentation architecture generated by the system when security weight is 1.0 and cost and mission performance weights are both 0.0

from red (very insecure) to green (secure). The Internet node is always red as that is assumed to be compromised at all times. Also shown in the figures are the cleansing and patching rates Δ_{clean}^{-1} and Δ_{patch}^{-1}, respectively, associated with the architecture.

Figures 10 and 11 show the best architectures outputted by the system when the cost weight is 1.0 and when the mission performance weight is 1.0, respectively, and represent scenarios in which all importance is placed on either cost or mission performance and no importance is given to security. From the figures, these architectures do not use any partitioning at all, that is the system chooses a single enclave in which all network devices can communicate directly with all other devices. This follows as partitioning does not offer any benefit to either cost or mission performance.

When the cost weight is 1.0 (Fig. 10), the cleansing and patching rates are unchanged from their initial settings (1/365 and 1/30 days, respectively). However, when the mission performance weight is 1.0 (Fig. 11), the cleansing and patching rates are reduced from their initial settings to 1/409 and 1/46 days, respectively. This also follows as the cost measure only considers the number of enclaves present and does not consider rates of cleansing or patching while the mission performance measure does consider these rates and, thus, reduce these rates to improve mission performance. As expected, the security risk component measure (Eq. 2) for both of these architectures is high: 0.670 for the architecture of Fig. 10 and 0.810 for the

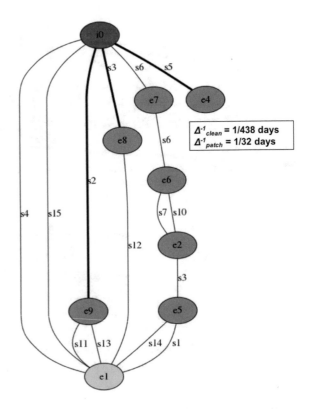

Fig. 13 Segmentation architecture generated by the system when security weight is 0.3, cost weight is 0.4, and mission performance weight is 0.3

architecture of Fig. 11. This means that for those architectures a network enclave is compromised (i.e., penetrated by the attacker) an average of 67 and 81% of the time, respectively.

Figure 12 shows the best architecture generated by the system when the security weight is 1.0, representing a scenario where all importance is given to security and no importance is given to either cost or mission performance. Here, the system partitions the network into 9 enclaves and increases the cleansing rate Δ_{clean}^{-1} to 1/351 days to achieve a security risk component measure of 0.165, a significant improvement in security compared to the architectures of Figs. 10 and 11. As expected the cost and mission performance component measures are not as good (i.e., higher) with cost component measure of 0.016 (compared to 0.0 for the architecture of Fig. 10) and mission performance degradation of 0.019 (compared to 0.013 for the architecture of Fig. 11).

In Fig. 13, the best architecture generated when the security and mission performance weights are each set to 0.3 and the cost weight is set to 0.4 is shown, representing a scenario where roughly equal importance is given to all three objectives. This architecture gives an interesting and nonintuitive solution to satisfy the three objectives. It utilizes fewer enclaves than the architecture of Fig. 12 which gives an improved cost component measure of 0.004. It also uses slower cleansing and patching rates to produce an improved mission performance degradation component measure of 0.018. With fewer enclaves and slower cleansing and patching rates, one might expect the security risk component measure to be worse (i.e., higher) but this is not the case. The architecture makes use of fewer total software services and communication pathways that rely on depth to actually *improve* the security risk component measure to 0.130 compared to the architecture generated when the security weighting is 1.0 (which has a security risk component measure of 0.165).

Because of the nonzero cost and mission performance weightings, the system is forced to explore alternative solutions beyond just increasing the total number of enclaves or increasing cleansing/patching rates to improve security. Here, the system uses depth to position some enclaves multiple hops away from the Internet. This makes it more difficult for an attacker as she must wait for new vulnerabilities to appear and utilize new exploits at each hop in order to spread to internal enclaves.

These results show that the system can construct segmentation architectures that satisfy simple scenarios when only one objective is given importance as well as complex scenarios when importance is placed on three different and conflicting objectives. The system utilizes nature-inspired search to explore effective solutions that are sometimes counterintuitive and, thus, can potentially generate architectures that are better than the ones specified by human experts. These experiments were run on a midrange business laptop (Dell Latitude E6540 with Intel Core i7 @ 2.80 GHz and 16.0 GB RAM) and took approximately 3–5 min to execute per experiment. Because the system is automated, it can potentially be used in high-performance computing environments to generate solution architectures for network environments at larger scales (e.g., 10^2 or 10^3 enclaves).

6 Conclusion

This chapter presents a nature-inspired cybersecurity decision system that automatically generates NS architectures optimized for three conflicting objectives, security, cost, and mission performance. The system makes use of two algorithmic components: nonlinear optimization and cyber risk modeling and simulation. These components work together iteratively to intelligently search the space of possible architectures and hone in on effective architectures. The fully integrated, automated system is implemented and demonstrated for a representative network environment under cyberattack. Results show that the system can generate architectures that satisfy both simple scenarios where only one objective is considered and complex scenarios where multiple objectives must be considered. Furthermore, our experiments highlight the system's ability to explore new areas of the decision search space to discover intelligent architectures that satisfy contrary objectives. Specifically, we show that the system, when faced with increasing importance weightings for cost and mission performance can find novel architectures that actually *increase* overall security as well. This illustrates the system's ability to find counterintuitive solutions that may be better than solutions constructed by human experts.

Future work is focused on testing the decision system in real or emulated network environments to validate the effectiveness generated architectures. Further work will consider alternative algorithms for the optimization component of the system such as genetic algorithms, grammatical evolution, and/or memetic algorithms. Finally, we plan to extend the decision system to solve new cybersecurity decision problems such as optimal configuration of wireless sensor networks or software-defined perimeter defense [1].

References

1. Bilger B et al (2013) Software Defined Perimeter. Technical report, Cloud Security Alliance
2. Brownlee J (2011) Clever algorithms: nature-inspired programming recipes. Lulu.com
3. CVSS (2015) Common vulnerability scoring system v3.0. https://www.first.org/cvss/specification-document
4. Fielder A et al (2016) Decision support approaches for cyber security investment. Decis Support Syst 86:13–23
5. Frei S (2009) Security econometrics the dynamics of (in) security. PhD thesis, ETH ZURICH
6. Frei S et al (2006) Large-scale vulnerability analysis. In: Proceedings of the 2006 SIGCOMM workshop on large-scale attack defense, LSAD '06. ACM, New York, pp 131–138
7. Frei S et al (2010) Modeling the security ecosystem - the dynamics of (in) security. Springer, US, pp 79–106
8. Gezelter R (1995) Computer security handbook. Wiley, New York (chap Security on the Internet)
9. Google Inc (2012) google's approach to it security. Technical report, Google
10. Grimaila M, Badiru A (2013) A hybrid dynamic decision making methodology for defensive information technology contingency measure selection in the presence of cyber threats. Oper Res 13(1):67–88
11. Guion J et al (2017) Dynamic cyber mission mapping. In: Proceedings of the industrial and systems engineering conference
12. Hakansson A, Hartung R (2014) An infrastructure for individualised and intelligent decision-making and negotiation in cyber-physical systems. Procedia Comput Sci 35:822–831
13. Huber C et al (2016) Cyber fighter associate: A decision support system for cyber agility. In: Conference on information science and systems (CISS)
14. Lippmann R et al (2012) Continuous security metrics for prevalent network threats: introduction and first four metrics. Technical report, MIT Lincoln Laboratory
15. Mandia K et al (2013) APT1: Exposing One of China's Cyber Espionage Units. FireEye Inc, Technical report
16. Microsoft, Inc (2015) Enterprise security best practices. Technical report, Microsoft. https://technet.microsoft.com/en-us/library/dd277328.aspx
17. Miller S et al (2016) Modelling cyber-security experts' decision making processes using aggregation operators. Comput Secur Vol 62(62):229–245
18. Mukhopadhyay A et al (2013) Cyber-risk decision models: to insure it or not? Decis Support Syst 56:11–26
19. Nappa A et al (2015) The attack of the clones: a study of the impact of shared code on vulnerability patching. In: 2015 IEEE symposium on security and privacy, pp 692–708
20. National Security Agency (2013) Top 10 information assurance mitigation strategies. Technical report
21. NVD (2016) The national vulnerability database. https://nvd.nist.gov/
22. Perlroth N (2014) Security experts expect 'Shellshock' software bug in bash to be significant. New York times

23. Reichenberg N (2014) Improving security via proper network segmentation. Security week
24. Riordan J et al (2016) A model of network porosity. Technical report, MIT Lincoln Laboratory
25. SANS Institute (2015) Critical security controls for effective cyber defense version 6.0. Technical report
26. Schulz AE et al (2015) Cyber network mission dependencies. Technical report TR-1189, Massachusetts Institute of Technology Lincoln Laboratory
27. Wagner N et al (2016) Towards automated cyber decision support: a case study on network segmentation for security. In: IEEE symposium on computational intelligence for cyber security
28. Wagner N et al (2017) Capturing the security effects of network segmentation via a continous-time markov chain model. In: Proceedings of the ACM spring simulation multi-conference
29. Wagner N et al (2017) Quantifying the mission impact of network-level cyber defensive mitigations. J Def Model Simul Appl Methodol Technol 14(3):201–216

Optimizing Resource Allocation of Wireless Networks with Carrier Aggregation Using Evolutionary Programming

Marcus Vinícius Gonzaga Ferreira and Flávio Henrique Teles Vieira

Abstract In this chapter, we propose a resource allocation scheme for wireless networks that aims to maximize the total data rate and attain certain Quality of Service (QoS) parameters using an Evolutionary Programming (EP) heuristic. The performance of the resource allocation algorithm is verified and compared to others in the literature using computational simulations. In these simulations, we also consider current and candidate techniques for next-generation networks such as f-OFDM (filtered-Orthogonal Frequency Division Multiplexing) and carrier aggregation in order to show that is possible to provide higher data rate in relation to 4G networks and other algorithms in the literature.

1 Introduction

In the near future, different wireless technologies will connect billions of machines, health monitors, self-driving cars, and countless other devices for which applications have not yet even been conceived [23]. To attain these demands, drastic improvements need to be made in cellular network architecture and the next-generation 5G wireless networks will have to expand capabilities of communications [15].

Globally, the cellular industry has converged to 3rd Generation Partnership Project (3GPP) Long-Term Evolution (LTE), including LTE-Advanced, as the common air interface. A cellular operator that was previously fragmented among multiple air interfaces now has one standard, resulting in huge economies of scale for infrastructure and user equipment. Thus, many researchers in the field propose that 5G will become an extension of this communications platform [20, 23].

M. V. G. Ferreira (✉) · F. H. T. Vieira
Institute of Informatics, Federal University of Goiás, Alameda Palmeiras,
Quadra D, Câmpus Samambaia, Goiânia, Goiás CEP 74690-900, Brazil
e-mail: marcusferreira@inf.ufg.br

F. H. T. Vieira
e-mail: flavio@emc.ufg.br

© Springer International Publishing AG, part of Springer Nature 2019
S. K. Shandilya et al. (eds.), *Advances in Nature-Inspired Computing and Applications*, EAI/Springer Innovations in Communication and Computing,
https://doi.org/10.1007/978-3-319-96451-5_2

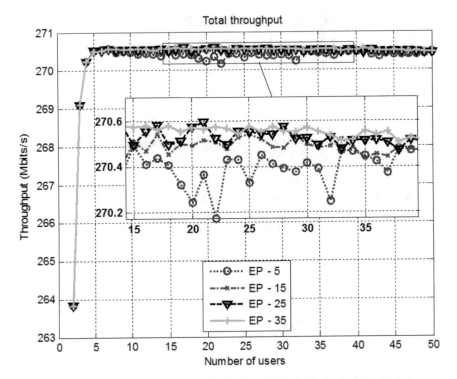

Fig. 1 Total throughput using f-OFDM for 5, 15, 25, and 30 individuals considered in EP

With focus on technologies that aim to improve data transmission to support the next-generation wireless networks, one of the topics covered in this work includes the concept of carrier aggregation, an LTE-Advanced feature that operators are deploying globally to provide higher data rate, better coverage, and lower latency [22]. This chapter also considers the recently introduced waveform framework, named f-OFDM (filtered-Orthogonal Frequency Division Multiplexing), which can enable such a spectrum slicing operation and allow for efficient coexistence of multiple sub-bands, aiming to improve the spectrum utilization and enable flexible waveform [7, 26]. The choice of f-OFDM technology is justified by the improvement of the system transfer rate by 10% when using the guard bands of LTE networks, besides supporting the asynchronous transmission of different users and being compatible with Multiple-Input Multiple-Output (MIMO) [4].

There are several proposals for resource allocation in wireless systems [5, 9, 14, 17, 25]. In [25], an algorithm is proposed to allocate resource blocks in an LTE system based on a Particle Swarm Optimization (PSO) algorithm, aiming to maximize the total bit rate of the system. In [14], it is proposed an algorithm that aims to guarantee the criterion of minimum rate required by user. In [9], it is proposed an algorithm that aims to minimize the delay criterion. These allocation schemes, in addition to others not mentioned here, have the same purpose of maximizing the data rate through

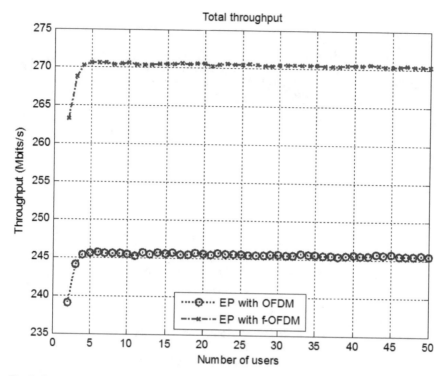

Fig. 2 Total throughput using OFDM and f-OFDM

heuristics focusing on several QoS parameters, such as delay, minimum transmission rate, etc.

In this chapter, we propose a resource allocation scheme for wireless networks, considering next-generation techniques, in order to attain the QoS parameter of minimum transmission rate. The resource allocation problem is solved by applying an Evolutionary Programming (EP) heuristic. This chapter is divided as follows: First, we present concepts regarding resource allocating problem and f-OFDM technique. Further, the EP is applied in order to decide which resource blocks to be allocated to each user, given a minimum rate requirement per user. The performance of the proposed EP based allocation scheme is validated by simulations comparing with other state-of-the-art algorithms. Finally, we present the final considerations.

2 System Model and Problem Formulation

Consider an Long-Term Evolution-Advanced (LTE-Advanced))-based cellular network with a single evolved NodeB (eNB) consisting of N User Equipments (UEs)) as $\alpha = \alpha_1, \alpha_2, \ldots, \alpha_N$. Each user $\alpha_n \in \alpha$ may have a different Carrier Aggregation

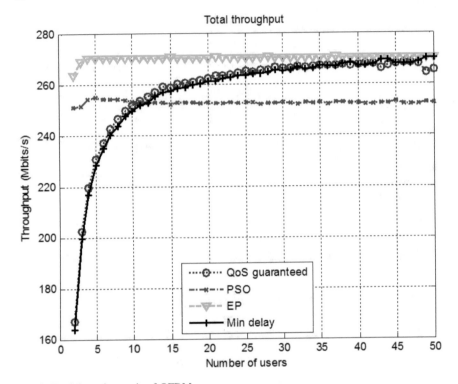

Fig. 3 Total throughput using f-OFDM

(CA) capability which is modeled as $\mu = \{\mu_n | \mu_n \in \{1, 2, 3, 4, 5\}\}_{1 \times N}$, where μ_n represents the maximum number of Component Carriers (CCs) that $\alpha_n \in \alpha$ can support (for instance, if $\alpha_n \in \alpha$ is an LTE Release 8 user, $\mu_n = 1$). All UEs in a given Transmission Time Interval (TTI), compete for $\mathcal{M}$ nonoverlapping orthogonal CCs as $\beta = \{\beta_1, \beta_2, \ldots, \beta_{\mathcal{M}}\}$ where each $\beta_m \in \beta$ has a different number of Resource Blocks (RBs) and can be written as $\gamma = \{\gamma_m | \gamma_m \in \{6, 15, 25, 50, 75, 100\}\}_{1 \times \mathcal{M}}$, where γ_m shows the number of RBs in a given β_m [22].

Assume maximum supported Modulation and Coding Scheme-index (MCS-index) of all UEs in different RBs is modeled as follows [22]:

$$K = \{k_{m,p,n} | k_{m,p,n} \in \{0, 1, \ldots, h\}\}_{M x P x N} \tag{1}$$

where $k_{m,p,n}$ represents maximum MCS-index of α_n on β_m in pth RB; value of $k_{m,p,n}$ depends on channel quality ranges between 0 and h; value of h for the uplink is 22 and for the downlink is 28 and $P = \max[\gamma]$. We also define a resource block allocation matrix as a binary matrix as follows [22]:

$$A = \{a_{m,p,n} | a_{m,p,n} \in \{0, 1\}\}_{M x P x N} \tag{2}$$

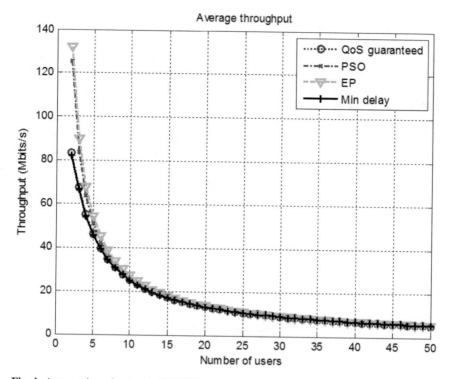

Fig. 4 Average throughput using f-OFDM

representing RB allocation map where $a_{m,p,n} = 1$ if and only if pth RB located in β_m is allocated to α_n uniquely and $a_{m,p,n} = 0$ otherwise. The resource allocation matrix A must satisfy the interference constraint defined in (3), i.e., two or more UEs cannot utilize same RB simultaneously.

$$\sum_{n=1}^{N} a_{m,p,n} \leq 1 \, for \, 1 \leq p \leq P, 1 \leq m \leq M \tag{3}$$

It is defined CC allocation matrix as a binary $\mathcal{M} \times N$ matrix [22]

$$E = \left\{ e_{m,n} | e_{m,n} = 1 \leftrightarrow \sum_{p=1}^{P} a_{m,p,n} \geq 1 \right\}_{\mathcal{M} \times N} \tag{4}$$

where $e_{m,n}$ represents whether the β_m is assigned to α_n or not. To represent the allocated MCS(s) to each α_n for a given TTI, a user rewarded matrix is defined as follows:

$$B = \left\{ b_{m,n} | b_{m,n} \in \{0, 1, \ldots, h\} \right\}_{\mathcal{M} \times N} \tag{5}$$

where $b_{m,n}$ represents the allocated MCS-index for α_n in β_m for each TTI. In [3], it is specified corresponding data rate of each MCS-index using table presentation. In this work, to depict the relationship between MCS-index and the data rate, it is used notation $r = R(b)$ where R maps each MCS-index to a corresponding data rate according to [3]; in other words, r is the achieved transmission rate for a UE on an RB with MCS b. It is defined users' data rate matrix as $R = \{r_{m,n} | r_{m,n} = R(b_{m,n})\}_{M \times N}$ as a $M \times N$ matrix where $r_{m,n}$ represents the data rate per RB for α_n on β_m [22].

Furthermore, eNB is responsible for all admission-related procedures, resource scheduling, and the link adaptation. After receiving Channel State Information (CSI) from all UEs, a resource allocation map is constructed and the MCS-index is determined accurately by eNB and then sent to each UE through control channels [22]. According to the assigned MSC-index, modulation type and coding rate of each UE in each assigned CC can be determined [3].

In this work, the scheduler is applied to maximize cell throughput while attending minimum data rate required for each user. For nth UE, the optimization problem is defined as the achievable throughput over tth TTI and can be calculated as (t is positive integer number)

$$f^{(t)} = \sum_{n=1}^{N} \sum_{m=1}^{M} \sum_{p=1}^{P} r_{m,n}^{(t)} \times a_{m,p,n}^{(t)} \tag{6}$$

where $r_{m,n}^{(t)}$ and $a_{m,p,n}^{(t)}$ are $r_{m,n}$ and $a_{m,p,n}$ in tth TTI, respectively, subject to

$$\sum_{n=1}^{N} a_{m,p,n} \leq 1, \tag{7}$$

$$\sum_{m=1}^{M} e_{m,n} \leq \mu_n, \tag{8}$$

$$b_{m,n} \leq k_{m,p,n} \tag{9}$$

$$r_{m,n} \geq r_{m,n}^{min} \tag{10}$$

for $1 \leq p \leq P$, $1 \leq m \leq M$, and $1 \leq n \leq N$. Expression (7) assures that each RB in the network is assigned to a maximum of one UE. Expression (8) guarantees that the number of assigned CC to each UE is less than its maximum aggregation capability. Expression (9) ensures that the assigned MCS-index for each UE in each CC is less than the maximum supported MCS-index in each of assigned RBs in its corresponding CC. Expression (10) ensures that the scheduler attends the minimum data rate required for each user.

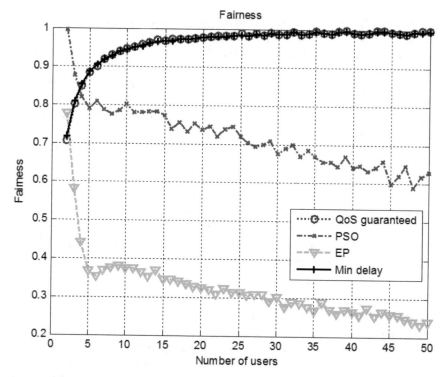

Fig. 5 Fairness index using f-OFDM

3 Filtered OFDM

To support the increased diversity of the future services to be provided by 5G networks, the system bandwidth is expected to be divided into several sub-bands, where
the frame structure and within each sub-band can be configured according to the
individual traffic type and the associated channel conditions [4, 26]. The recently
introduced waveform framework, named f-OFDM, can enable such a spectrum slicing operation and allow efficient coexistence of multiple sub-bands [26, 28].

The main idea of f-OFDM is as follows. To reduce the interference between adjacent sub-bands, the baseband OFDM signal, or even other types of waveforms, of
each sub-band is filtered by a band-limited filter to suppress its out-of-band emission.
In this way, the interference from adjacent sub-bands can be restrained to a sufficient
level. In each sub-band, tailored structure including subcarrier spacing, cyclic prefix
(CP) length, and transmission time interval (TTI) can be configured to achieve the
design target associated with each type of service. For example, the structure for
the enhanced mobile broadband (eMBB) sub-band could aim for high spectrum efficiency, while that for the ultra-reliability and low latency communication (uRLLC)
sub-band should target a low latency with larger subcarrier spacing and short TTI [26].

Another advantage of f-OFDM is to support multi-user asynchronous transmission in the uplink. With per-UE (User Equipment) filtering to suppress the out-of-band leakage, the interference between UEs becomes negligible. In LTE, the timing advance (TA) signal sent to each UE by the base station (BS) brings large signaling overhead, especially when a large number of users are present. By introducing per-UE filtering to suppress the inter-UE interference, the UEs do not need to maintain stringent synchronization with the BS (thus to exploit the orthogonally of OFDM), and the TA signaling overhead can be reduced [4].

In LTE specifications, 10% of the system bandwidth is reserved as guard band on a carrier basis, to meet the adjacent channel leakage ratio (ACLR) and spectrum mask requirement [26]. By applying f-OFDM, the baseband OFDM signal can be shaped with ultra-narrow transition region, and thus the guard band can be reutilized to transmit useful signals, leading to enhanced spectrum utilization [28].

Although the filter length may go beyond the CP length and cause inter-symbol interference (ISI), the ISI leakage to neighboring OFDM symbol is expected to be negligible for the following reasons [26, 28]:

- Since the bandwidths of both the transmit and receipt filters are generally larger than the subcarrier spacing, the main energy span of the end-to-end filter in time domain is significantly smaller than the OFDM symbol length and depending on the filter design can be made even smaller than the CP length;
- A well-designed filter has an almost flat frequency response over the entire sub-band bandwidth at both transmit and receipt. Therefore, the majority of subcarriers within the sub-band will not be impacted by transmit and receipt filtering and thus will not experience observable time-domain spreading. It is the side lobe suppression of only a few edge subcarriers that will cause a minor time-domain spreading, which is much smaller than the CP length.

4 Evolutionary Programming

Evolutionary programming was first developed by Lawrence Fogel, which focused on the use of an evolutionary process for the development of control systems using finite state machine representations [10]. Further works considered the application of evolutionary programming to control systems [24], function optimization, and system identification [11–13, 21, 27].

Evolutionary programming is a global optimization algorithm and is an instance of an evolutionary algorithm from the field of evolutionary computation, sibling of other evolutionary algorithms such as genetic algorithm and learning classifier systems [8]. It is inspired by the theory of evolution by means of natural selection. A population of a species reproduces, creating progeny with small phenotypical variation. The progeny and the parents compete based on their suitability to the environment, where the generally more fit members constitute the subsequent generation and are provided

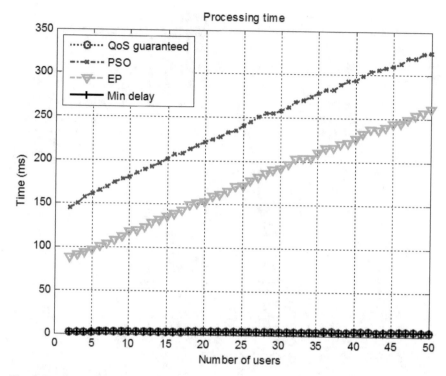

Fig. 6 Processing time using f-OFDM

with the opportunity to reproduce themselves. This process repeats, improving the adaptive fit between the species and the environment [8].

To describe the metaphor of the algorithm application, it is formulated a global minimization problem which can be formalized as a pair (S, f), where $S \subseteq R^n$ is a bounded set on R^n and $f : S \to R$ is an n-dimensional real-valued function. The problem is to find a point $x_{min} \in S$ such that $f(x_{min})$ is a global minimum on S. More specifically, it is required to find an $x_{min} \in S$ such that [27]

$$\forall x \in S : f(x_{min}) \leq f(x) \tag{11}$$

where f does not need to be continuous but it must be bounded. According to the description of Bäck and Schwefel [6], the classical EP is implemented as described in Algorithm 1.

Algorithm 1: Classical Evolutionary Programming Algorithm [27]

1. Generate the initial population of μ individuals, and set $k = 1$. Each individual is taken as a pair of real-valued vectors, $(x_i, \eta_i), \forall i \in \{1, \ldots, \mu\}$, where x_i's are objective variables and η_i's are standard deviations for Gaussian mutations (also known as strategy parameters in self-adaptive evolutionary algorithms).

2. Evaluate the fitness score for each individual $(x_i, \eta_i), \forall i \in \{1, \ldots, \mu\}$, of the population based on the objective function, $f(x_i)$.

3. Each parent $(x_i, \eta_i), i = 1, \ldots, \uparrow \mu$, creates a single offspring (x_i', η_i') by for $j = 1, \ldots, n$

$$x_i'(j) = x_i(j) + \eta_i(j)N_i(0, 1) \tag{12}$$

$$\eta_i'(j) = \eta_i(j) \exp\big(\tau' N(0, 1) + \tau N_j(0, 1)\big) \tag{13}$$

where $x_i(j), x_i'(j), \eta_i(j)$, and $\eta_i'(j)$ denote the jth component of the vectors x_i, x_i', η_i, and η_i', respectively. $N(0, 1)$ denotes a normally distributed one-dimensional random number with mean zero and standard deviation one. $N_j(0, 1)$ indicates that the random number is generated anew for each value of j. The factors τ and τ' are commonly set to $\left(\sqrt{2\sqrt{n}}\right)^{-1}$ and $\left(\sqrt{2n}\right)^{-1}$.

4. Calculate the fitness of each offspring $(x_i', \eta_i'), \forall i \in \{1, \ldots, \mu\}$.

5. Conduct pairwise comparison over the union of parents (x_i, η_i) and offspring $(x_i', \eta_i'), \forall i \in \{1, \ldots, \mu\}$. For each individual, q opponents are chosen uniformly at random from all the parents and offspring. For each comparison, if the individual's fitness is no smaller than the opponents', it receives a "win."

6. Select the μ individuals out of (x_i, η_i) and $(x_i', \eta_i'), \forall i \in \{1, \ldots, \mu\}$, that have the most wins to be parents of the next generation.

7. Stop if the halting criterion is satisfied; otherwise, $k = k+1$ and go to Step 3.

5 Resource Allocation Algorithm Using Evolutionary Programming

The optimization problem described in (6) can be solved using the classical EP by making some considerations. The classical EP does not have constraints, so it is considered the minimum bandwidth constraint (10) as a penalty function, as described in (14). The penalty function is related to the percentage of the minimum rate stipulated to each user. When the minimum rates for all users are achieved, the penalty function is zero, i.e., the constraint is attained.

$$F = \sum_{n=1}^{N} f_n^{(t)} - \left[\max\left(r_{m,n}^{\min} \right) \right]^2 \sum_{n=1}^{N} \left[\min\left(0, \frac{r_{m,n} - r_{m,n}^{\min}}{r_{m,n}^{\min}} \right) \right]^2 \qquad (14)$$

The optimization in EP algorithm is carried out by evaluating the cost of each solution through the objective function, as described in Algorithm 2. In this chapter, the solution vector of the proposed problem represents an integer index vector of resource blocks allocated to users. To implement such solution, we propose a rounding of values which represent individual (x_i, η_i) to the nearest integer.

The lowest costs are memorized and used in algorithm, in other words, the resource blocks are allocated to users. If there are still available resource blocks, these are allocated to users with better channel conditions.

Algorithm 2: Resource Allocation Algorithm Using Evolutionary Programming
1. Receive network system parameters.
2. Initialize the algorithm parameters.
3. Evaluate the cost of each solution through the objective function (14) using EP algorithm described in Algorithm 1.
4. Save the lowest cost and allocate the resource blocks according to it.
5. Allocate the remaining blocks for users with better channel conditions.
6. Calculate users bit rate.

6 Simulation Results

In this section, we present the simulation results of the resource block algorithm carried out using software MATLAB® version R2015a.

Channel conditions for each user and RBs in terms of Signal-to-Interference-plus-Noise-Ratio (SINR) were generated for each TTI according to parameters shown in Table 1.

Simulations were carried out considering parameters for downlink scenario transmission shown in Table 2.

Bit rate and MCS associated to SINR are defined with 4 bits CQI, as shown in Table 3.

The EP and PSO algorithm simulations were performed with 15 and 30 individuals, respectively, and 100 maximum iterations as stop criterion.

Simulations were performed comparing the results of the proposed EP algorithm with the following resource allocation algorithms: PSO algorithm [25], QoS guaranteed algorithm [14], and delay minimization algorithm [9]. All the algorithms studied were implemented using f-OFDM and carrier aggregation techniques.

Table 1 Simulation parameters for channel modeling [19, 1, 2]

Multipath model	Rayleigh
Multipath delay profile	ETU (Extended Typical Urban)
Path loss model	$L = 128.1 + 37.6 \log 10(R)$, R in kilometer
Lognormal shadowing	Mean 0, standard deviation 10 dB
Distance between UE and eNB	1 km
White noise power density	-174 dBm/Hz
Maximum eNB transmitter power	46 dBm
eNB antenna gain after cable loss	15 dBi
UE antenna gain	0 dBi
UE noise figure	9 dB
UE interfering margin	4 dB
UE speed	3 km/h

Table 2 Downlink simulation scenario

CA capability (number of resource blocks)	[25, 50, 50, 100, 100]
Minimum rate required per user	0.768 Mbits/s
Subframe length	1 ms
Number of TTIs simulated	1000
Modulation scheme	OFDM/f-OFDM

Table 3 CQI, modulation scheme and bit rate [18]

CQI	Modulation	Bit rate (1/1024)	Information bits per symbol
0	QPSK	0	0.00
1	QPSK	78	0.15
2	QPSK	120	0.23
3	QPSK	193	0.38
4	QPSK	308	0.60
5	QPSK	440	0.88
6	QPSK	602	1.18
7	16-QAM	378	1.48
8	16-QAM	490	1.91
9	16-QAM	616	2.41
10	64-QAM	466	2.73
11	64-QAM	567	3.32
12	64-QAM	666	3.90
13	64-QAM	772	4.52
14	64-QAM	873	5.12
15	64-QAM	948	5.55

It was simulated scenarios for different numbers of individuals in the implementation of the EP, as shown in Fig. 1. It is verified that there is a low difference in terms of total throughput when considered more than 15 individuals in simulation. Therefore, from here the simulations were performed considering a fixed number of 15 individuals.

Figure 2 shows the comparative results of the EP algorithm using OFDM and f-OFDM. It is verified that when using f-OFDM in the allocation algorithm, the performance in terms of throughput improves considerably in relation to OFDM.

The proposed EP algorithm presents the highest values of total throughput in all scenarios with different number of users considered in allocation, as can be seen in Fig. 3. This result is explained by the fact that the proposed algorithm gives a higher priority in resources allocation to users with better channel conditions, consequently guaranteeing higher throughput. QoS guaranteed and min-delay algorithm presents similar values, higher than PSO algorithm when considered more than 10 users.

Figure 4 shows that EP and PSO algorithm present similar values of average throughput if it is considered less than 10 users. By considering more than 10 users, the algorithm of this work presents similar values. The average throughput is calculated through the simple mean of total throughput as a function of the number of users considered in simulation.

In terms of fairness index [16], the QoS guaranteed and min-delay algorithms present similar values, mostly the highest values. The EP algorithm presents the lowest values, as can be seen in Fig. 5. The low performance of proposed algorithm in terms of fairness is due to the fact that it provides higher allocation priority to users with better channel conditions.

Figure 6 presents the processing time (in milliseconds) of simulation, considering that simulations were performed using a microcomputer with the following configuration: Processor Intel Core IR-3570 3.40 GHz, 8 Gb RAM, HD SATA III 7200 RPM, Windows 10 64 bits. The QoS guaranteed and min-delay algorithms present the lowest values of processing time. This fact was expected, since EP and PSO algorithms are heuristics that aim to maximize throughput, at the expense of increasing complexity. The proposed EP algorithm reaches considerably lower values when compared to the PSO algorithm, due to the reduced number of individuals considered, 15, while PSO algorithm considers 30 individuals.

7 Conclusion

We propose in this chapter a scheme of resource block allocation for next-generation wireless systems. For this, we propose to apply Evolutionary Programming to solve the resource allocation problem, defined as the maximization of system total throughput subject to the minimum transmission rate required for each user.

The results presented in the simulations show that when using f-OFDM in the allocation scheme, the performance of the proposed algorithm in terms of throughput is considerably improved in relation to OFDM. The values presented are on average 10% higher.

Simulation results show that the proposed EP algorithm outperforms the PSO, QoS guaranteed, and min-delay algorithms in terms of total and average throughput, considering all scenarios with different number of users, at the expense of increased processing time and low fairness index. This fact is justified by the characteristic of the proposed algorithm to provide higher resources allocation priority to users with better channel conditions. It is important to observe that the proposed EP algorithm presents better results in general than PSO algorithm, even considering half the number of individuals in the optimization.

Acknowledgements I would like to express my special thanks of gratitude to my parents and friends who helped me a lot in finalizing this project.

References

1. 3GPP TS 36.104 version 8.3.0 Release 8 (2008) LTE, Evolved Universal Terrestrial Radio Access (E-UTRA), Base Station (BS) radio transmission and reception
2. 3GPP TR 36.931 version 9.0.0 Release 9 (2011) LTE, Evolved Universal Terrestrial Radio Access (E-UTRA), Radio Frequency (RF) requirements for LTE Pico Node B
3. 3GPP TR 36.213 (2012) Evolved Universal Terrestrial Radio Access (E-UTRA), Physical layer procedures, 3rd generation partnership project (3GPP)
4. Abdoli J, Jia M, Ma J (2015) Filtered OFDM: a new waveform for future wireless systems. In: IEEE 16th international workshop on signal processing advances in wireless communications (SPAWC), 2015
5. Alasti M, Neekzad B, Hui J, Vannithamby R (2010) Quality of service in WiMAX and LTE networks [Topics in Wireless Communications]. IEEE Commun Mag 48(5):104–111, mai de 2010, ISSN: 0163-6804. https://doi.org/10.1109/mcom.2010.5458370
6. Back T, Schwefel H-P (1993) An overview of evolutionary algorithms for parameter optimization. Evol Comput 1(1):1–23
7. Bi M, Jia W, Li L, Miao X, Hu W (2017) Investigation of F-OFDM in 5G fronthaul networks for seamless carrier-aggregation and asynchronous transmission. Optical Fiber Communications Conference and Exhibition (OFC)
8. Brownlee J (2012) Clever algorithms: nature-inspired programming recipes. Revision 2
9. Ferreira MVG, Vieira FHT, Abrahão DC (2015) Minimizing delay in resource block allocation algorithm of LTE downlink. In: VI international workshop on telecommunications (IWT)
10. Fogel LJ (1962) Autonomous automata. Ind Res 4:14–19
11. Fogel DB (1991) System identification through simulated evolution: a machine learning approach to modeling. Needham Heights
12. Fogel DB (1992) Evolving artificial intelligence. PhD thesis, University of California, San Diego, CA, USA
13. Fogel LJ (1994) Computational intelligence: imitating life, chapter evolutionary programming in perspective: the top-down view. IEEE Press, New York, pp 135–146
14. Guan N, Zhou Y, Tian L, Sun G, Shi J (2011) QoS guaranteed resource block allocation algorithm for LTE systems. In: IEEE 7th international conference on wireless and mobile computing, networking and communications (WiMob), pp 307–312
15. Gupta A, Jha RK (2015) A survey of 5G network: architecture and emerging technologies. IEEE Access 3:1206–1232

16. Jain R, Hawe W, Chiu D (1984) A quantitative measure of fairness and discrimination for resource allocation in shared computer systems. DEC-TR-301. Accessed on 26 Sept 1984
17. Kausar R, Chen Y, Chai KK (2012) QoS aware packet scheduling with adaptive resource allocation for OFDMA based LTE-advanced networks, vol 2011. IET Conference Publications, Germany, pp 207–212
18. Kawser M, Imtiaz Bin Hamid N, Nayeemul Hasan M, Shah Alam M, Musfiqur Rahman M (2012) Downlink SNR to CQI mapping for different multiple antenna techniques in LTE, vol. 2, pp. 756–760, 2012
19. Ni M, Xu X, Mathar R (2013) A channel feedback model with robust SINR prediction for LTE systems. In: 7th European conference on antennas and propagation (EuCAP), Institute for Theoretical Information Technology, RWTH Aachen University, 2013
20. Olwal TO, Djouani K, Kurien AM (2016) A survey of resource management toward 5G radio access networks. IEEE Commun Surv Tutorials 18(3):1656–1686, Thirdquarter de 2016, ISSN: 1553-877X. https://doi.org/10.1109/comst.2016.2550765
21. Porto VW (2000) Evolutionary computation 1: basic algorithms and operations, chap 10: Evolutionary programming. IoP Press, Bristol, pp 89–102
22. Rostami S, Arshad K, Rapajic P (2015) A joint resource allocation and link adaptation algorithm with carrier aggregation for 5G LTE-advanced network. In: 22nd International Conference on Telecommunications (ICT 2015)
23. Rysavy Research/4G Americas (2015) LTE and 5G innovation: igniting mobile broadband. Accessed on Aug 2015
24. Sebald AV, Fogel DB (1990) Design of SLAYR neural networks using evolutionary programming. In: Proceedings of the 24th Asilomar Conference on Signals, Systems and Computers, pp 1020–1024
25. Su L, Wang P, Liu F (2012) Particle swarm optimization based resource block allocation algorithm for downlink LTE systems. In: 18th Asia-Pacific conference on communications (APCC), pp 970–974
26. Wu D, Zhang X, Qiu J, Gu L, Saito Y, Benjebbour A, Kishiyama Y (2016) A field trial of f-OFDM toward 5G. In: IEEE Globecom Workshops (GC Wkshps)
27. Yao X, Liu Y, Lin G (1999) Evolutionary programming made faster. IEEE Trans Evol Comput 3:82–102
28. Zhang X, Jia M, Chen L, Ma J, Qiu J (2015) Filtered-OFDM—enabler for flexible waveform in the 5th generation cellular networks. In: Proceedings of IEEE global communication conference, pp 1–6, Dec 2015

Artificial Feeding Birds (AFB): A New Metaheuristic Inspired by the Behavior of Pigeons

Jean-Baptiste Lamy

Abstract Many optimization algorithms and metaheuristics have been inspired by nature. These algorithms often permit solving a wide range of optimization problems. Most of them were inspired by exceptional or extraordinary animal behaviors. On the contrary, in this chapter, we present Artificial Feeding Birds (AFB), a new metaheuristic inspired by the very trivial behavior of birds searching for food. AFB is very simple, yet efficient, and can be easily adapted to various optimization problems. We present application to unconstrained global nonlinear optimization, with several benchmark functions and the training of Artificial Neural Networks (ANN), and to the resolution of ordering combinatorial optimization problems, with two examples: the traveling salesman problem and the optimization of rainbow boxes (a recent visualization technique for overlapping sets). We compare the results with those produced with Artificial Bee Colony (ABC), Firefly Algorithm (FA), Genetic Algorithm (GA), and Ant Colony Optimization (ACO), showing that AFB gives results equivalent or better than the other metaheuristics. Finally, we discuss the choice of inspiration sources from nature, before concluding.

1 Introduction

Many algorithms have been inspired by nature. Two well-known examples are Artificial Neural Networks (ANN) [1] and Genetic Algorithms (GA) [5]. More recently, researchers inspired themselves from the behavior of animals for inventing new optimization algorithms: social organization of insects [3] like ants [6] and honey bees [9], cohesion within a swarm in flight [18], communication by light between fireflies [22], parasitic behavior of cuckoo [24], ability of pigeons to orientate themselves spatially according to the sun position and the North pole direction [7]. These algorithms typically rely on *swarm intelligence*, i.e., they consider a population of agents that

J.-B. Lamy (✉)
LIMICS (Laboratoire d'informatique médicale et d'ingénierie des connaissances en e-santé), Université Paris 13, Sorbonne Université, Inserm, 93017 Bobigny, France
e-mail: jean-baptiste.lamy@univ-paris13.fr

© Springer International Publishing AG, part of Springer Nature 2019
S. K. Shandilya et al. (eds.), *Advances in Nature-Inspired Computing and Applications*, EAI/Springer Innovations in Communication and Computing, https://doi.org/10.1007/978-3-319-96451-5_3

interact between themselves and with their environment [8]. These agents are very simple but they can achieve complex tasks together, and in particular they can solve optimization problems [2]. These algorithms are called *metaheuristics* when they provide a top-level strategy that can be used to guide a low-level heuristic search strategy [16, 21]. Consequently, a metaheuristic is not specific to a given type of problem; it can solve very different problems, depending on the chosen low-level search strategy.

Most of the metaheuristics were actually inspired by *exceptional* or *extraordinary* animal behaviors. For instance, the ability of fireflies to emit light is exceptional: only a few animal species are able to emit light. Even the social organization of insects is quite rare: many species do not organize themselves in huge colonies (even not all bee species). However, from an evolutionary point of view, the most efficient behaviors lead to a higher chance of survival and thus, they are expected to be observed more frequently. Consequently, we might regard exceptional behaviors as poorly efficient ones (in terms of performances or capability of adaptation), and common behaviors as more efficient.

In this chapter, we propose Artificial Feeding Birds (AFB), a new metaheuristic that follows a different kind of inspiration: it has been inspired by a very *trivial* and *common* behavior that we observed on birds like pigeons, when they are searching for food on sidewalks or in a garden. Our hypothesis is that, if pigeons are so common, it is because their food search strategy is efficient, and thus it is an interesting inspiration source for algorithms. AFB presents several advantages: (1) the metaheuristic is very simple, (2) it provides good results, and (3) it is easy to adapt to new optimization problems, and in particular it makes no assumption on the solution space and does not require the computation of distances between solutions. Here, we will show the adaptability of AFB by applying it to unconstrained global nonlinear optimization, including the training of ANN, and to the resolution of ordering optimization problems, with two examples: the Traveling Salesman Problem (TSP) and the optimization of rainbow boxes [14]. Rainbow boxes are a recent information visualization technique for overlapping set that requires to solve a complex optimization problem.

The rest of the chapter is organized as follows. Section 2 presents a brief state of the art of nature-inspired optimization algorithms. Section 3 describes the behavior that we observed on pigeons and other birds searching for food. Section 4 presents the metaheuristic algorithm, and its adaptation to two problems: unconstrained global nonlinear optimization and ordering optimization. Section 5 presents various experiments performed for determining parameters values, and for testing AFB on several benchmark functions, on TSP and on rainbow boxes optimization, and comparing AFB with other metaheuristics. Finally, Sect. 6 discusses the main results, the differences between AFB and other metaheuristics, and gives some perspectives.

2 Related Works

Many optimization algorithms have been inspired by nature [23]. Many of them are based on the social behavior of insects [3] and their ability to communicate through chemical substances, moves or light. An example is the behavior of ants, which inspired Ant Colony Optimization (ACO) [6]. Ants explore their environment for searching food and they leave a track behind them using chemical substances named pheromones. These pheromones are then considered by other ants as signals indicating them the directions to follow or not to follow. Fireflies' behavior and their use of light to attract sexual mates also inspired an algorithm [22]. The Firefly Algorithm (FA) has been used subsequently for training ANN [4].

The Artificial Bee Colony (ABC) algorithm [9] has been inspired by the communication between bees through their waggle dances when they are searching nectar. When a bee has found an interesting food source, she goes back to the hives and performs the waggle dance to "recruit" other bees for bringing them to the food source. ABC considers three types of bee: workers, onlookers, and scouts. Each worker is associated with a food source, whose position corresponds to a solution of the optimization problem. Better solutions correspond to richer source food. On each cycle, each worker tries to improve her solution by trying a nearby solution, and keep this new solution if it is better than the current one. Then, she communicates to onlookers the quality of her solution. Onlookers obtain this information from workers; on each cycle, each onlooker chooses a worker to help, the choice is random but with a higher probability to choose the workers with better solutions. Then, the onlooker tries to improve the solution, in a way similar to the worker. When a solution cannot be improved after a fixed number of trials, it is abandoned and the scout bee is in charge of finding a new random food source for the worker. The ABC metaheuristic has been adapted to unconstrained global nonlinear optimization [9], the optimization of constrained problems [10], the training of ANN [11], and clustering [12].

Particle Swarm Optimization (PSO) is a technique inspired by the behavior of animals that move in a swarm (flying insects or birds, fishes) [18]. In the swarm, the move performed by each individual at a given time depends on the position of the other individuals and of the quality of the solutions associated with their positions. FA can also be seen as an improvement of PSO.

More recently, several bird-inspired algorithms were published. XS Yang et al. inspired themselves from the parasitic behavior of cuckoo [24]. H Duan et al. proposed an algorithm inspired by pigeons and their ability to orientate themselves spatially according to the sun position and the North pole direction [7]. The algorithm has been used for air robot path planning.

Fig. 1 The four types of move observed on birds when they are searching for food

3   Behaviors Observed on Birds

Our observations were carried initially on pigeons, and then in groups mixing various types of birds feeding at the same place. Pigeons are very common birds in European towns, and they are easy to observe. They feed by pecking seeds or crumbs of food on the ground. When no food is in reach, they explore their environment, using the two modes of movement at their disposal: walking and flying.

We observed that a pigeon performs four types of move when searching for food (Fig. 1): (1) walking to a new position close to his current position (because they walk slowly), (2) flying and landing at an arbitrary semi-random position, (3) flying and returning to a memorized position rich in food (such as a picnic area), and (4) flying and landing close to another pigeon. Typically, a pigeon walks for searching food (one or several move 1). After a while, if no food is found, he flies and goes to a random place (move 2), to a memorized position (move 3) or join another pigeon (move 4). Then, he begins to walk again (move 1), and so on.

This simple behavior optimizes the food search. Move 1 (walk) allows a local search. This is meaningful because there is a high probability to find food close to a position where food has already been found (e.g., if crumbs of a sandwich are present somewhere, it is probable to find other crumbs of the same sandwich nearby). Move 2 (fly to random position) allows the random exploration of space. Move 3 (return to a memorized position) allows retrieving food, or continuing to look for it in the surroundings. Move 4 (join another bird) allows benefit from the food that the other bird might have found. This leads to big groups of pigeons when an important quantity of food is available in a given place.

These observations were carried on pigeons; however, many other birds present a similar behavior, including sparrows and gooses. When several species of birds are mixed together and feed at the same place, we observed that the size of the bird has an impact on move 4 (join another bird): a big bird can join a smaller one. On the contrary, a small bird is frightened by bigger ones and does not join them. We observed this behavior in population mixing gooses and pigeons.

4 Translation in Algorithms

4.1 Metaheuristic

We designed a metaheuristic inspired by the bird feeding behavior. We consider a multi-agent system, each agent being an *artificial bird*. The position of each bird corresponds to a candidate solution for the optimization problem. Each bird also keeps in memory the best position he found, i.e., the one corresponding to the solution that minimizes the best the cost function. When the current position of a bird is better than the memorized position, the current position is memorized and the bird is considered to "have fed."

The metaheuristic performs several cycles. In each cycle, each bird performs one of the four moves described previously. For a given bird, the next move is determined as follows: if the bird has flown in the previous cycle, he walks. If the bird has eaten in the previous cycle, he walks. Otherwise, one of the four moves is randomly chosen, with different probabilities associated with each move. In addition, we considered two sizes of birds: small and big ones. Only big birds can perform move 4 and join another (small or big) bird. While this rule does not exactly match our observation, it efficiently avoids that all birds get stuck in a local minimum.

Two moves (3 and 4) are generic and independent from the optimization problem. On the contrary, the two other moves (1 and 2, i.e., walk and random fly) are problem-dependent. Therefore, we can define an optimization problem as a triplet of three functions (*cost*, *fly*, *walk*), as follows:

- *cost* : $A \rightarrow \mathbb{R}$, the cost function to minimize, where A is the admissible set of solutions for the cost function,
- *fly* : $\phi \rightarrow A$, a function that returns a random position,
- *walk* : $\mathbb{N} \rightarrow A$, a function that returns a random position close to the current position of the bird indicated by the given integer index.

The metaheuristic takes five parameters:

- n, the number of artificial birds,
- r, the ratio of small birds in the total bird population (the other being big birds),
- p_2, the probability that a bird chooses move 2,
- p_3, the probability that a bird chooses move 3, and
- p_4, the probability that a bird chooses move 4.

The probability for move 1 is thus $p_1 = 1 - p_2 - p_3 - p_4$.

The metaheuristic defines six per-bird variables, $1 \leq i \leq n$:

- $x_i \in A$, the current position of bird i,
- $f_i \in \mathbb{R}$, the value of the cost function for x_i,
- $X_i \in A$, the best position found and memorized by bird i,
- $F_i \in \mathbb{R}$, the value of the cost function for X_i,
- $s_i \in \{0, 1\}$, the size of bird i (0 is a small bird, e.g., a pigeon, and 1 a big bird, e.g., a goose), and

---

Algorithm 1 The AFB metaheuristic in pseudocode.

For $1 \leq i \leq n$:
    $x_i = X_i = fly()$
    $f_i = F_i = cost(x_i)$
    $m_i = 2$
    $s_i = 0$ if $i \leq r \times n$, 1 otherwise

Repeat:
    For $1 \leq i \leq n$:
        If $m_i \in \{2, 3, 4\}$ or $f_i = F_i$:
            $p = 1$
        Else, if $s_i = 0$:
            $p =$ random real number between $p4$ and 1
        Else:
            $p =$ random real number between 0 and 1

        If $p \geq p_2 + p_3 + p_4$:
            $m_i = 1$
            $x_i = walk(i)$
            $f_i = cost(x_i)$
        Else, if $p \geq p_3 + p_4$:
            $m_i = 2$
            $x_i = fly()$
            $f_i = cost(x_i)$
        Else, if $p \geq p_4$:
            $m_i = 3$
            $x_i = X_i$
            $f_i = F_i$
        Else:
            $m_i = 4$
            $j =$ random integer number between 1 and $n, j \neq i$
            $x_i = x_j$
            $f_i = f_j$

        If $f_i \leq F_i$:
            $X_i = x_i$
            $F_i = f_i$

    Check stopping condition

The best solution found is X_k, with $1 \leq k \leq n$ such as $F_k = min(\{F_i \mid 1 \leq i \leq n\})$

- $m_i \in \{1, 2, 3, 4\}$, the type of move performed by bird i at the previous cycle (1 walk, 2 fly to a random position, 3 fly to the memorized position, 4 fly to the position of another bird).

Algorithm 1 shows the metaheuristic. It initializes the variables, runs cycles, and finally determines the best solution found. During initialization, the position x_i of each bird is randomly defined using the $fly()$ function, the current cost f_i is computed, and m_i is set to 2 (because the random initialization is comparable to move 2).

In each cycle, for each bird i, the algorithm chooses one of the four possible moves using the previously described rules, updates m_i with the chosen move, performs the

---

Algorithm 2 $fly()$ and $walk()$ functions for optimization problems in $\mathbb{R}^d$.

Function $fly()$:
    $x' \in \mathbb{R}^d$
    For $1 \le k \le d$:
        x'_k = random real number between x_{min} and x_{max}
    Return x'

Function $walk(i)$:
    $x' \in \mathbb{R}^d, x'_k = x_{ik}$ for $1 \le k \le d$
    j = random integer number between 1 and $n, j \neq i$
    k = random integer number between 1 and d
    $\Delta = |x_{ik} - x_{jk}|$
    if $\Delta = 0$: $\Delta = 0.001$
    r = random real number between -1 and 1
    $x'_k = x'_k + r \times \Delta$
    If $x'_k < x_{min}$: $x'_k = x_{min}$
    Else, if $x'_k > x_{max}$: $x'_k = x_{max}$
    Return x'

---

move, and updates the best position if needed. Moves 1 and 2 call the $walk()$ and $fly()$ functions, respectively, and then the $cost()$ function. Moves 3 and 4 move the bird to the best memorized position or to the position of another random bird, respectively. These two moves do not test a new solution and thus do not require to call the $cost()$ function.

If the current cost f_i is lower or equal to the best memorized cost F_i, then the current position and cost are memorized. The condition "lower **or equal**" allows the modification of the memorized position in order to keep a solution that is not better than the previous one, but different; this potentially increases the diversity of the solutions memorized by the population of birds.

Finally, it is necessary to include a stopping condition in the algorithm. We suggest stopping the algorithm after a predefined number of solutions have been tested (i.e., to limit the number of calls to the $cost()$ function). As our metaheuristic does not test a new solution for each bird in each cycle, this stopping condition allows a fair comparison with other optimization algorithms that test more solutions per cycle.

At the end of the process, the best solution found is the best position memorized by the birds.

The $walk()$ and $fly()$ functions depend on the optimization problems. $fly()$ returns a random solution, and $walk()$ a new solution close to one of the given birds. Simple $walk()$ functions just modify the bird position. More sophisticated $walk()$ functions (as the two presented below) first evaluate the local density of birds at the given bird's position. The local density is roughly estimated by Δ, a partial distance computed between the walking bird and another bird chosen randomly. Then, $walk()$ modifies the solution of the bird by performing a move (or a change) that is proportional to Δ.

4.2 Adaptation to Unconstrained Global Nonlinear Optimization

In this section, we apply the AFB metaheuristic to the global optimization of a real function with d parameters, i.e., $A = \mathbb{R}^d$. The solutions are points in a space with d dimensions, whose coordinates are between x_{min} and x_{max}.

Algorithm 2 describes the $fly()$ and $walk()$ functions we propose for global nonlinear optimization. The $fly()$ function simply returns a random position. The $walk()$ function modifies a randomly chosen coordinate k of the current bird position x_i. The modification has a maximum amplitude which is Δ, the absolute value of the difference between coordinates x_{ik} and x_{jk}, where j is another randomly chosen bird index. This allows smaller amplitudes when birds are closer. This is a similar local search heuristic than the one proposed in the ABC algorithm [9]; however, our metaheuristic differs.

4.3 Adaptation to Ordering Optimization

In this section, we apply the AFB metaheuristic to ordering problems, i.e., problems in which an optimal order of the elements of a set T must be found. Ordering problems are a subcategory of combinatorial optimization problem. Algorithm 3 describes the $fly()$ and $walk()$ functions we propose for solving ordering problems. Bird positions x_i are ordered sequences of the elements in T. The $fly()$ function generates a random order. The $walk()$ function corresponds to a variant of the 2-opt local search heuristic [17], in which the sequence is opened at two points, and reconnected after reversing one of the two parts (e.g., if the sequence ABCDEF is split between B-C and E-F, the resulting sequence is ABEDCF). We modified the heuristic in order to take into account the local similarity of the bird's position with the position of another random bird j (Fig. 2). The similarity is (roughly) estimated by Δ, the number of elements in x_j between the element located in position k (the first opened edge) and the previous element in x_i. If no satisfying value can be found for Δ after 100 tries, a default random value is used.

5 Experimentations

5.1 Implementation

The algorithms described in the previous section were implemented in the Python language and executed with the PyPy2 interpreter (a version of Python integrating a Just-In-Time (JIT) compiler). Other algorithms (ABC, FA, GA, and ACO) have also been implemented in the same language, for comparison purpose.

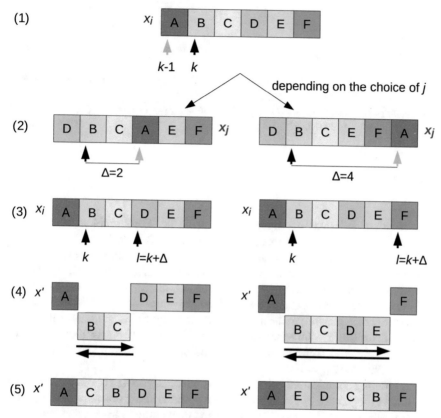

Fig. 2 Examples of the modified 2-opt local search heuristic on sequences of six elements, ABCDE-F. Notice how the order of another random bird j affects the 2-opt for bird i (the schema shows the results for two different birds j). Variable names ($i, j, k, l, \Delta, \ldots$) correspond to those in Algorithm 3

The AFB implementation in Python is available online under the GNU LGPL Open Source license, including most of the examples of the chapter:
https://bitbucket.org/jibalamy/metaheuristic_optimizer

5.2 Benchmarks and Tests

For nonlinear optimization, we selected five functions frequently used in benchmarks (Fig. 3): a five-dimensional sphere function, the Rosenbrock function, the 10-dimensional Rastrigin function, the Eggholder function (we added a constant to this function, so as its global minimum is about 0, as for the other functions), and the Himmelblau's function.

$$Sphere(x_1,...,x_n) = \sum_{i=1}^{n} x_i^2 \qquad\qquad n = 5 \text{ and } -100 < x_i < 100$$

$$Rosenbrock(x,y) = (1-x^2) + 100 \times (y-x^2)^2 \qquad -2.048 < x < 2.048 \text{ and } -2.048 < y < 2.048$$

$$Rastrigin(x_1,...,x_n) = \sum_{i=1}^{n} x_i^2 - 10 \times \cos(2\pi x_i) + 10 \qquad n = 10 \text{ and } -600 < x_i < 600$$

$$Eggholder(x,y) = -(y+47)\sin\left(\sqrt{|y+\tfrac{x}{2}+47|}\right) - x\sin\left(\sqrt{|x-(y+47)|}\right)$$
$$+959.640662720851 \qquad\qquad -512 < x,y < 512$$

$$Himmelblau(x,y) = (x^2+y-11)^2 + (x+y^2-7)^2 \qquad\qquad -6 < x,y < 6$$

Fig. 3 The five benchmark functions for experimentation

In addition, we tested the training of ANN, using the "Xor6" problem, which is also commonly used for benchmarking [4]. It consists of an ANN with two input neurons I_1 and I_2, two hidden neurons H_1 and H_2, and one output neuron O. Neurons have no bias, and there are thus six coefficients to optimize: I_1-H_1, I_1-H_2, I_2-H_1, I_2-H_2, H_1-O, and H_2-O. The four learning samples (I_1, I_2, O) are $(0, 0, 0)$, $(0, 1, 1)$, $(1, 0, 1)$, and $(1, 1, 0)$; they correspond to a logical exclusive or. The cost function takes six parameters corresponding to the six coefficients of the ANN, and returns the Mean Squared Error (MSE). We tested the Xor6 problem with two activation functions: sinus and sigmoid, leading to two functions to optimize: $Xor6_{sin}$ and $Xor6_{sig}$, with $x_{min} = -100$ and $x_{max} = 100$.

For ordering optimization, we tested two problems. The first one is the Traveling Salesman Problem (TSP). It is a well-known optimization problem in which a traveling salesman must visit a set T of towns and then return to his starting town. The objective is to find the optimal order for visiting the towns, in order to minimize the total distance of the trip. We used the FRI26 dataset, which includes 26 towns.

The second problem is the optimization of rainbow boxes. Rainbow boxes [13, 14] are an information visualization technique that we recently proposed for *overlapping sets*. We applied this technique for the visualization of drug properties [15]. Several elements and sets are to be visualized, each element can belong to several sets, and each set can contain several elements. In rainbow boxes, each element is represented by a column, and each set by a rectangular box that covers the columns corresponding to the elements belonging to the set (see example in Fig. 4). When these elements are not presented in adjacent columns, holes are present in the box. The optimization problem consists in finding the optimum column order to minimize the number of holes. Therefore, the *cost()* function computes and returns the number of holes produced by a given column order. We tested two previously presented rainbow boxes datasets [13], amino acid and histones, and we generated two sets of 100 random datasets, the first with small datasets (20 elements, 8 sets, and 30 membership relations) and the second with larger ones (30 elements, 15 sets, and 60 membership relations).

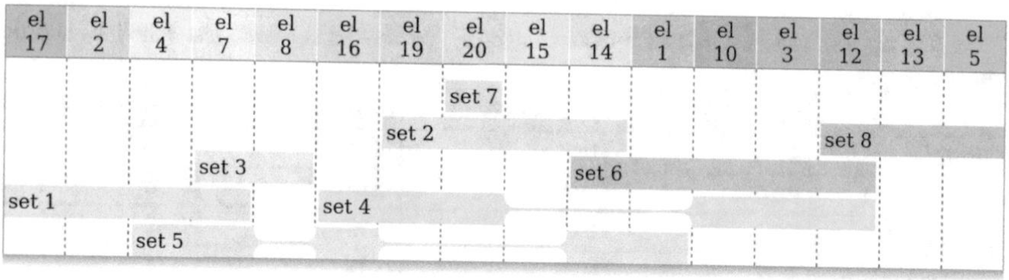

el 17	el 2	el 4	el 7	el 8	el 16	el 19	el 20	el 15	el 14	el 1	el 10	el 3	el 12	el 13	el 5

Fig. 4 Example of rainbow boxes, corresponding to one of the small random datasets ("el" stands for "element"). Here, after optimization, there are three holes, one in set four, and two in set five

5.3 Parameter Values

First, n was set to 20 semi-arbitrarily: this value gives good results and is used in several other population-based algorithms, such as ABC. Similarly, r was set to 0.75 (corresponding to 15 small birds and 5 big ones).

Finally, for studying and determining the values of the parameters p_2, p_3, and p_4, we created two instances of the AFB metaheuristic, one in charge of optimizing the parameters of the other. Table 1 shows the best parameter values found for each benchmark function. Since each parameter p_{1-4} corresponds to one of the four moves, we can see that some moves are not pertinent for optimizing some functions (However, notice that $p_1 = 0$ does not mean that the birds never walk, because they *always* walk after landing or feeding; it just means that the bird never walk after an unsuccessful walk move). However, additional tests showed that the values found for p_4 are overestimated in Table 1: when optimizing parameter values, we computed the mean of 20 runs for each set of parameter values, but there is still a certain variability in the results. A high p_4 value means that the AFB algorithm will focus more on the best solutions found, and this possibly benefits more to better runs, which are those selected during the optimization process.

The last line of Table 1 shows the parameter values that we retained. These values will be systematically used in the rest of the chapter.

5.4 Results on Unconstrained Global Nonlinear Optimization

We tested AFB on the optimization of the benchmark functions presented in Sect. 5.2, and we compared the results with those obtained with ABC and FA. The stopping condition was fixed to 40,000 tested solutions. For ABC, parameters were the following: $n = 20$ (number of bees), $limit = 100$ (a food source is considered exhausted if it cannot be improved after 100 cycles). These values correspond to the conditions used by Karaboga for the ABC algorithm: 2000 cycles for 20 bees [9]. For FA, we were unable to find a set of parameter values that performed well on all tests. We used the following values: $n = 20$ (number of fireflies), $\alpha = 0.022$, $\alpha_{fade} = 1.0$, $\beta = 0.442$, $\gamma = 3.413$

Table 1 The best values obtained for parameters p_2, p_3 and p_4 for each test function, and the retained values

	p_1 (walk)	p_2 (random fly)	p_3 (memory)	p_4 (join other)
$Sphere()$	0	0	0.64	0.36
$Rosenbrock()$	0.54	0.05	0.36	0.05
$Rastrigin()$	0	0	0.58	0.42
$Eggholder()$	0.37	0.01	0.31	0.31
$Himmelblau()$	0	0.20	0.66	0.14
$Xor6_{sin}()$	0.06	0.09	0.52	0.33
$Xor6_{sig}()$	0.20	0	0.51	0.29
Retained values	**0.25**	**0.01**	**0.67**	**0.07**

for $Rosenbrock()$, $\alpha = 0.268$, $\alpha_{fade} = 1.0$, $\beta = 0.128$, $\gamma = 9.807$ for $Eggholder()$, and $\alpha = 0.37$, $\alpha_{fade} = 0.98$, $\beta = 0.91$, $\gamma = 0$ for others. These values were obtained by running another optimization algorithm on the parameter values.

Table 2 gives the results. For ABC, they are similar to the ones published [4, 9]. AFB gives the best results for all functions but $Xor6_{sig}()$ and $Himmelblau()$, but for these two functions, AFB is close to the best results. Compared to ABC, AFB performs much better for two functions, $Rosenbrock()$ and $Eggholder()$. We explain this difference as follows. In ABC, a single coordinate of the solution is modified, and then a greedy selection is performed between the new solution and the previous one. This implies that the solution can move only in a single axis, and the move needs to improve the results to be conserved. But the $Rosenbrock()$ function has a narrow "valley" that requires diagonal moves. In AFB, walk moves also involve a single coordinate; however, several walks can be performed before returning to the best memorized position. This permits diagonal moves.

Finally, computation times were similar between AFB and ABC, but higher for FA.

5.5 Results on Ordering Problems

We compared the results obtained with AFB with those obtained with GA and ACO. The GA we implemented is the random-key algorithm proposed by Snyder et al. [19] for generalized TSP. We tested the algorithm both with and without local optimizations (2-opt and swap). We used two ACO algorithms: ACO-pants,[1] a Python ACO implementation for TSP with some TSP-specific optimizations, for TSP and the *MAX-MIN Ant System* (MMAS) [20] for rainbow boxes optimization (without any specific optimization).

---

[1]https://github.com/rhgrant10/Pants.

Algorithm 3 $fly()$ and $walk()$ functions for solving ordering problems.

Function $fly()$:
 $x' =$ sequence of the elements in T, in a random order
 Return x'

Function $walk(i)$:
 $\Delta = 0$
 Repeat maximum 100 times:
  $j =$ random integer number between 1 and $n, j \neq i$
  $k =$ random integer number between 1 and $|T|$
  $\Delta' =$ position of x_{ik} in $x_j -$ position of $x_{i(k-1)}$ in x_j
  If $1 < abs(\Delta') < |T| - 1$:
   $\Delta = \Delta'$
   Break
 If $\Delta = 0$: $\Delta =$ random integer number between 2 and $n - 1$
 $l = (k + \Delta)$ modulo $|T|$
 If $k > l$: Swap k and l
 $x' =$ clone of sequence x_i
 Reverse the order of elements between x'_k and x'_l
 Return x'

Table 2 Comparison of the results obtained when minimizing various functions with AFB and ABC. Results are the means over 250 runs, and the lower values are the best. (*) 0 results are understood at double precision (which is 1e-323)

		AFB	ABC	FA
Sphere()	Result	**5.07e-81**	6.23e-17	2.36e-32
	Std. deviation	2.88e-80	3.05e-17	1.62e-32
	Time (ms)	15	15	310
Rosenbrock()	Result	**2.64e-05**	8.73e-03	2.84e-03
	Std. deviation	8.05e-05	1.40e-02	3.56e-02
	Time (ms)	13	10	176
Rastrigin()	Result	**0 (*)**	7.94e-15	95.04
	Std. deviation	0	1.02e-13	113.30
	Time (ms)	26	28	383
Eggholder()	Result	**0 (*)**	0.48	11.3
	Std. deviation	0	2.86	18.16
	Time (ms)	17	17	185
Himmelblau()	Result	6.00e-31	5.80e-17	**2.46e-31**
	Std. deviation	3.37e-31	3.17e-17	4.04e-31
	Time (ms)	11	11	156
Xor6$_{sin}$()	Result	**1.24e-06**	9.31e-06	1.34e-04
	Std. deviation	2.78e-06	1.13e-05	8.39e-04
	Time (ms)	37	39	337
Xor6$_{sig}$()	Result	4.40e-02	**4.05e-02**	1.26e-01
	Std. deviation	3.77e-02	3.10e-02	1.75e-02
	Time (ms)	41	39	49

For TSP, the stopping condition was fixed to 40,000 tested solutions and we performed 250 runs for each algorithm. The best possible tour has a distance of 937. For rainbow boxes optimization, the stopping condition was fixed to 40,000 for large random datasets, and to 10,000 for the others. We performed 250 runs for each of the amino acid and histone datasets, and one run for each of the random datasets. The best-known results are four for amino acids and six for histones. Table 3 gives the results. AFB performed better than other algorithms. For TSP, AFB yielded a mean distance of 941.0. This represents a 0.43% error margin compared to the best possible solution. ACO performed well on rainbow boxes small datasets, but poorly on large ones.

Results obtained with GA for TSP are not on par with those published by Snyder et al. [19]. Two reasons can explain that. First, the number of tested solutions was much lower in our experiment and the GA needs to run longer, especially if local optimizations are used because they perform a lot of calls to the cost function. Second, we did not implement the TSP-specific optimizations proposed by Snyder et al. (such as considering the two tours ABCD and BCDA as identical), since we were targeting ordering problems in general.

6 General Discussion

In this chapter, we presented Artificial Feeding Birds (AFB), a new metaheuristic, inspired by the behavior of pigeons searching for food. We showed that AFB was able to solve various problems: unconstrained global nonlinear optimization, including training of neural networks, traveling salesman problem, and rainbow boxes optimization. We also showed that AFB competes well with ABC, FA, GA, and ACO algorithms.

We focused on rapid tests and rather small datasets, because our short-term goal is to optimize visualizations, such as rainbow boxes, and they often need to be produced dynamically, on the fly. In particular for TSP, the results presented here need to be confirmed on bigger datasets and with longer computation times. For rainbow boxes optimization, we previously proposed a specific heuristic algorithm [13, 14]; however, it was limited to 25 elements or less, due to computation time. Here, we showed that AFB could be used up to 30 elements, and possibly even more.

Lones [16] identified several metaheuristic aspects that are shared by many nature-inspired optimization algorithms. The following ones are found in AFB:

(a) *Neighborhood search* consists of testing new solutions that are close (or similar) to the previously tested solution. In AFB, the *walk* move performs a local search, by producing a position that is close to the current position.

(b) *variable neighborhood search* is similar to neighborhood search, but considers moves with variable steps, depending, e.g., on the position of other agents. In AFB, the two *walk*() functions we proposed perform variable neighborhood search: the step of the walk move is proportional to Δ, which depends on the position of another bird.

Table 3 Comparison of the results obtained on ordering problems. Lower values are the best

			AFB	GA	GA+local op	ACO
TSP	*(26 towns)*	Result	**941.0**	1082.0	983.3	968.21
		Std. deviation	7.1	66.2	36.1	13.6
		Time (ms)	133	574	103	4203
Rainbow boxes	*Amino acid dataset*	Result	**4.04**	7.16	6.46	4.82
		Std. deviation	0.20	1.24	1.42	0.69
		Time (ms)	205	219	263	648
	Histone dataset	Result	**6**	6.39	6.53	6.27
		Std. deviation	0	0.73	0.82	0.5
		Time (ms)	550	512	557	623
	Small random datasets	Result	**3.85**	5.39	4.92	3.87
		Std. deviation	1.08	1.16	1.25	1.02
		Time (ms)	184	194	208	658
	Large random datasets	Result	**12.59**	16.63	15.59	21.21
		Std. deviation	1.90	1.94	1.58	1.53
		Time (ms)	1052	3496	4528	61523

(c) *hill climbing* consists of trying to improve a given solution incrementally. If the new solution is better than the previous one, it is kept. Otherwise, it is discarded and the previous solution is kept. In AFB, when a *walk* move leads to a better solution than the one memorized by a bird, a second walk is systematically performed in the next cycle.

(d) *accepting negative moves* consists of keeping a new solution that is worse than the previous one; it is somehow the opposite of hill climbing. In AFB, when a *walk* move does not lead to a better solution, a second walk is still possible, but not systematic.

(e) *population-based search* consists of considering multiple agents. AFB considers a population of artificial birds, each bird having its own position and memory.

In particular, when using specific parameter values, AFB performs like well-known algorithm. If $p_1 = 1$ and $p_2 = p_3 = p_4 = 0$, AFB performs a random walk. If $p_3 = 1$ and $p_1 = p_2 = p_4 = 0$, AFB performs like a hill climbing algorithm.

AFB presents some similarities with ABC. In both algorithms, agents perform the four following tasks: random exploration of the solution space, local search, reversion to the best solution found so far, and concentration of several agents on the most promising solutions. In AFB, these tasks correspond to the four moves of the birds: move 2 allows random exploration, move 1 local search, move 3 reversion to the best solution, and move 4 allows a bird to "join his force" with another bird and eventually to adopt his best solution. In ABC, these tasks are associated with the three types of bee. Scouts are in charge of the random exploration, workers of local search with reversion to the best position found in case of failure, and onlookers allow concentrating more agents on the best solutions. However, there is an important difference between ABC and AFB: in ABC, local search and reversion to the best position are grouped in the worker behavior, while in AFB, we separated them in two distinct moves (1 and 3). This separation allows accepting negative moves (i.e., "diagonal moves" in nonlinear optimization), and we have seen in Sect. 5.4 that it improved the results for some benchmark functions. While our $walk()$ function for global nonlinear optimization was inspired by ABC, our metaheuristic differs, and hence the difference observed in the results.

The inspiration source of AFB is particular at two levels. First, in Sect. 2, we noticed that most inspiration sources were exceptional or extraordinary animal behaviors, such as light emission. Here, on the contrary, we successfully inspired ourselves from a very trivial behavior: birds searching for food. From an evolutionary point of view, the most efficient behaviors lead to a higher chance of survival and thus, they are expected to be encountered more frequently. Consequently, it should be more interesting to inspire ourselves from very common behaviors, widely spread over many species, rather than exceptional behaviors. However, this hypothesis needs additional verification.

Second, most inspiration sources include communication between animals, using chemical signs (ants), dances (bees), or light (fireflies). On the contrary, we did not observe communication when pigeons are feeding; their behavior seemed to us rather "individualistic". For example, when a pigeon finds some food, he does not seem to call other pigeons. When a pigeon joins another one, it is not following a call, but rather following a simple observation ("another bird is there, he seems to feed, let's get closer!"). In consequence, in AFB, observation replaces communication. In Algorithm 1, each agent accesses only his own information and variables, as well as the position of other agents, which can be obtained through simple observation. On the contrary, an agent never accesses to the best position found by another agent. Surprisingly, when we modified the metaheuristic by adding communication of the best position, i.e., changing move 4 so as it moves the bird on the best position found by another bird ($x_i = X_j$ and $f_i = F_j$) rather than his current position, the results did not improve significantly (and they were even poorer for $Rosenbrock()$).

The first strength of AFB is its extreme simplicity. Nature-inspired algorithms are often surprisingly simple, with regards to their performance [23]. This is especially true for AFB: the metaheuristic (Algorithm 1) is very simple and, in particular, does not need complex computations, contrary to many other algorithms (e.g., ABC has a complex formula for computing the probability of onlooker bees to choose a given food source).

The second strength of AFB is its generic nature. It can be run on any optimization problem that can be defined by a (*cost*, *fly*, *walk*) triplet of functions, and we have shown that it performs well on very different kinds of problems. In the presentation of our algorithms, we clearly separated the AFB metaheuristic from its adaptation to the two problems (global nonlinear optimization and ordering problems). This separation greatly facilitates the adaptation to new problems, since one only has to define the two functions *fly*() and *walk*(). Usually, in other metaheuristics, the separation between the problem-specific and the problem-independent part of the algorithm is not so clear. Furthermore, AFB makes no assumption about the optimization problem and the solution space, and in particular, it does not require to compute distance between solutions. Distance computation is not trivial in some problems, such as TSP, in terms of computation method and computation time. Finally, AFB is rather insensitive to parameter values, since we used the same default values across all our experiments. This allows to use the metaheuristic without having to tune the algorithm for a specific problem.

Perspectives of this work include the adaptation of the AFB metaheuristic to other optimization problems, such as clustering, and its use in real-life applications. AFB could also be improved, for example, by adding additional moves, possibly inspired by other metaheuristics. Finally, the extreme simplicity of AFB could also make it interesting for educational purpose.

References

1. Abraham A (2005) Artificial neural networks. Handbook of measuring system design. Wiley, Chichester
2. Blum C, Li X (2008) Swarm intelligence in optimization. Natural computing series, Swarm intelligence: introduction and applications. Springer, Berlin, pp 43–85
3. Bonabeau E, Dorigo M, Theraulaz G (2000) Inspiration for optimization from social insect behaviour. Nature 406:39–42
4. Brajevic I, Tuba M (2013) Training feed-forward neural networks using firefly algorithm. In Recent advances in knowledge engineering and systems science
5. Darrell W (1994) A genetic algorithm tutorial. Stat Comput 4:65–85
6. Dorigo M, Birattari M, Stutzle T (2006) Ant colony optimization - artificial ants as a computational intelligence technique. IEEE Comput Intell Mag 1:28–39
7. Duan H, Qiao P (2014) Pigeon-inspired optimization: a new swarm intelligence optimizer for air robot path planning. Int J Intell Comput Cybern 7(1):24–37
8. Garnier S, Gautrais J, Theraulaz G (2007) The biological principles of swarm intelligence. Swarm Intell 1(1):3–31
9. Karaboga D (2005) An idea based on honey bee swarm for numerical optimization. Technical report
10. Karaboga D, Basturk B (2007) Artificial Bee Colony (ABC) optimization algorithm for solving constrained optimization problems. Lect Notes Comput Sci 4529:789–798
11. Karaboga D, Ozturk C (2009) Neural networks training by artificial bee colony algorithm on pattern classification

12. Karaboga D, Ozturk C (2011) A novel clustering approach: Artificial Bee Colony (ABC) algorithm. Appl Soft Comput 11(1):652–657

13. Lamy JB, Berthelot H, Capron C, Favre M (2017) Rainbow boxes: a new technique for overlapping set visualization and two applications in the biomedical domain. J Vis Lang Comput 43:71–82

14. Lamy JB, Berthelot H, Favre M (2016) Rainbow boxes: a technique for visualizing overlapping sets and an application to the comparison of drugs properties. In: International conference information visualisation (iV), Lisboa, Portugal, pp 253–260

15. Lamy JB, Berthelot H, Favre M, Ugon A, Duclos C, Venot A (2017) Using visual analytics for presenting comparative information on new drugs. J Biomed Inform 71:58–69

16. Lones MA (2014) Metaheuristics in nature-inspired algorithms. In: Proceedings of the Companion Publication of the 2014 annual conference on genetic and evolutionary computation, Vancouver, BC, Canada, pp 1419–1422

17. Marinakis Y (2009) Heuristic and metaheuristic algorithms for the traveling salesman problem. Encyclopedia of optimization. Springer, Berlin, pp 1498–1506

18. Poli R, Kennedy J, Blackwell T (2007) Particle swarm optimization - an overview

19. Snyder LV, Daskin MS (2015) A random-key genetic algorithm for the generalized traveling salesman problem. Eur J Oper Res 174(1):38–53

20. Stützle T, Hoos HH (2000) MAX-MIN ant system. Future Gener Comput Syst 16(9):889–914

21. Voss F (2009) Metaheuristics. Encyclopedia of optimization. Springer, Berlin, pp 2061–2075

22. Yang XS (2009) Firefly algorithms for multimodal optimization. In: Stochastic algorithms: foundations and applications - lecture notes in computer sciences, vol 5792, pp 169–178

23. Yang XS (2010) Nature-inspired metaheuristic algorithms (second edition). Luniver Press, Frome

24. Yang XS, Deb S (2009) Cuckoo search via Levy flights. In: World congress on nature and biologically inspired, computing, pp 210–214

Application of Nature—Inspired Algorithms in Medical Image Processing

S. Kanimozhi Suguna, R. Ranganathan, J. Sangeetha, Smita Shandilya
and Shishir Kumar Shandilya

Abstract Medical image processing plays an indispensable role in our day-to-day life, as every individual is dependent on it in some of the other aspects. The dependency is quite essential and acts as stepping stone to further advanced applications and scientific endeavors. To achieve better and efficient results the process itself is carried out in multiple phases. For performing segmentation and classification on medical images, in this chapter among the myriad options available in the advanced scientific and technological field, nature-inspired algorithms such as Lion Optimization Algorithm (LOA) and Monkey Search Optimization Algorithm (MSO) is utilized. This chapter concludes with results and discussions of the optimization algorithms, along with a futuristic scope of the algorithm in the medical processing field.

Keywords Lion optimization algorithm · Monkey search optimization algorithm Medical image processing

1 Introduction

The world is unraveling around medical data which has become a sine qua non for human existence. Though the primary means of medical imaging is life-saving and extending the human life expectancy, the medical data processing in an efficient way aids in better understanding of the problem. Since the number of personals utilizing the medical images has increased manifolds; it has prompted the need for effective

S. K. Suguna (✉) · R. Ranganathan · J. Sangeetha
School of Computing, SASTRA University, Thirumalaisamudram, Thanjavur 613401, India
e-mail: kanimozhi.suguna@gmail.com

S. Shandilya
Department of Electrical and Electronics Engineering, Sagar Institute of Research
Technology and Science, Bhopal, Madhya Pradesh, India

S. K. Shandilya
School of Computer Science & Engineering, VIT Bhopal University, Bhopal,
Madhya Pradesh, India

© Springer International Publishing AG, part of Springer Nature 2019
S. K. Shandilya et al. (eds.), *Advances in Nature-Inspired Computing
and Applications*, EAI/Springer Innovations in Communication and Computing,
https://doi.org/10.1007/978-3-319-96451-5_4

implementation of nature-inspired algorithms [6–8, 15–17, 25, 38] for unambiguous definite results.

Multiple domains such as image processing, data mining, networking, navigation, etc., are exploited by nature inspired algorithms. Many optimization techniques are discussed in [1, 9, 14, 19, 21–24, 28, 32, 39–41, 57, 58, 60, 68, 75, 79]. This article focuses on the discussion of two such nature-inspired optimization algorithms namely, Lion Optimization Algorithm (LOA) and Monkey Search Optimization Algorithm (MSO). Most of the nature-inspired algorithms focus on two major behavioral aspects of the specimen (i.e., inspirer): mating and food gathering. Mating refers to the process performed by the specimen to create the next generation, while food gathering refers to acquiring foods for survival and reproduction. The specimen follows 'survival of the fittest,' guarantying of the survival. Hence, this chapter will discuss the process regarding the search for food in chosen algorithms. Extension of this concept in medical image processing can improve accuracy and decrease error rates regarding misclassification of images.

1.1 Overview of the Breast and the Breast Cancer

Breast Cancer is the second leading cancer next to Cervical Cancer [49]. Breast cancer is a type of cancer originating from breast tissue; most commonly form the inner lining of milk ducts or the lobules that supply the ducts with milk. There are two types of breast cancer tumors. The terms cancer and tumor are not synonyms to each other, but they have the close relation to each other. A mass of abnormal tissue is called a tumor. There are two types of breast cancer tumors. They are benign (non-cancerous) and malignant (cancerous) [88].

Benign is a condition in which though the cells are identified [4] with the tumor, it will not spread to the neighboring cells and responds well to the medical treatment. Malignant is a tumor state in which the cells will spread and affect the adjacent cells which further extends to different body parts [89]. These malignant tumors may or may not respond to the medical treatment and often recurs after removal. Breast cancer is a type of cancer originating from breast tissue; most commonly form the inner lining of milk ducts or the lobules that supply the ducts with milk [27]. Figure 1 shows The Breast: cross-section scheme of the mammary gland.

The exocrine gland in the breast associates, the degree of a non-uniform anatomic structure composed of layers of various varieties of tissue; among those predominantly two varieties are fatty tissue and glandular tissue, which affects the lactation functions of the breasts. Figure 2a represents the mammogram image used in this chapter. Features of breast anatomy considered in this research are depicted in the following Fig. 2b.

To make better study and performance in the medical images implement image processing techniques by considering more tissue regions. Among the four views of mammography procedure, MLO view is preferred in this research. Hence, mammogram images are chosen from Mini Mammographic Database from Mammo-

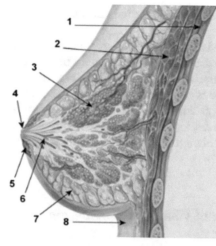

1. Chest wall
2. Pectoralis muscles
3. Lobules
4. Nipple
5. Areola
6. Milk Duct
7. Fatty tissue
8. Skin

Fig. 1 The breast: cross—section scheme of the mammary gland

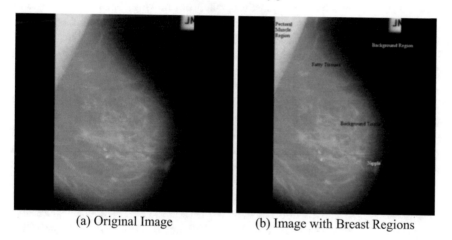

(a) Original Image   (b) Image with Breast Regions

Fig. 2 Mammogram image

graphic Image Analysis Society (MIAS) for this research. MIAS Mini Mammo-graphic Database is the first MIAS (digitized at 50-micron pixel edge) [54] database which is decreased to 200-micron pixel edge and cut/cushioned with the goal that each picture is 1024 pixels × 1024 pixels [91]. Figure 2b represents different regions in the mammogram image. The background region includes labels and other marks having the details of the patient is name or number and some additional information. While performing the scanning process, the pectoral muscle region is also scanned and is highly complicated to avoid it. Remove these two regions before segmenting the region of interest (ROI), as the pixel properties of these regions will be similar to that of the ROI, because of which there might be increased misclassification of the mammogram image.

The background tissue in the breast includes normal breast tissues, abnormal tissues, and nipple region. The normal background of breast tissue can be of three different characteristics such as fatty, fatty-glandular and dense-glandular. The severity of abnormality in this database is benign and malignant, whereas the classes of abnormalities in the database are: calcification, well-defined/circumscribed masses, spiculated masses, ill-defined masses, architectural distortions, asymmetry and normal [3]. Among all the classes of abnormalities, in this chapter they are grouped as normal and abnormal mammogram.

1.2 Different Modalities and Views of Mammogram Images—An Outline

Medical imaging is a technique, method, and art of making visual presentations of the inside of the body for the clinical analysis and medical intervention. Women have the highest possibility of breast cancer. Some of the modalities used in scanning the breast are Digital Mammography, Ultrasonography, Magnetic Resonance Imaging (MRI).

The result of the mammography process is Mammogram, and there are two types: Screening and Diagnostic Mammogram. Screening is performed to identify breast cancer when the patient does not have any symptom of it whereas in the latter method either abnormality is identified in screening or when women face some physiological problem such as nipple discharge, breast pain and so on. This method also helps in examining the nearby tissues which are affected by cancerous cells and hence it is most preferred than others.

The image observed by mammography can be of different views [29], and it is as in the Fig. 3. The direction of scanning is as in the Fig. 3a and the successive images Fig. 3b–e represents the Mediolateral Oblique (MLO), Cranio—Caudal (CC), Medio—Lateral (ML) and Latero—Medial (LM) views of the breast respectively [87]. The reason for choosing MLO is more breast tissue obtained from the upper outer quadrant of the breast and the axilla (armpit). In this MLO view, the pectoral muscle should be pictured obliquely from the top down to the nipple position and furthermore down. Besides these regions, the breast muscle should be curve or bulge outwards, nipple should be depicted and a small stomach fold must be visible [87]. Observing the whole of the breast region aids in the visibility of these regions.

Modalities are implemented for breast cancer screening whereas image file format is the standardized means of organizing and storing digital images. The images of Lossless compression such as JPEG, PNG, etc., are preferred over the Lossy compressed image since it preserves the representation of the original uncompressed image than the latter.

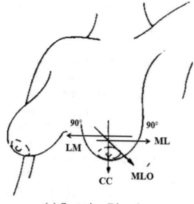

(a) Scanning Direction

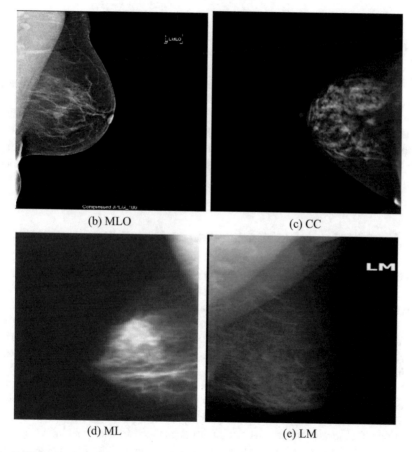

(b) MLO

(c) CC

(d) ML

(e) LM

Fig. 3 Different view of mammogram images

1.3 Medical Images and Processes Involved in Digital Image Processing

In the image received from any of the modalities, image processing techniques cannot be applied directly to it. The image obtained from any of the modality will be in the combination of electrical power and sensor material which is the transformation of incoming energy into voltage. Digitize the output waveform before implementing any of the steps in the medical image. To convert the continuous image into the digital form, sampling the function based on both the coordinates and in amplitude is necessary. Digitizing the coordinate value is called sampling whereas digitizing the amplitude is called as quantization. In an image, it is not mandatory to implement all the processes. The processes involved in digital image processing [52] and its resulting outputs are displayed in the following Table 1.

1.4 Chapter Overview

The chapter arrangement has eight subchapters comprising of this introduction subchapter and seven further subchapters as follows. Section 2 discusses the Machine Learning and Analysis and is supported by Sects. 3 and 4 with the concepts of Monkey Search Optimization (MSO) Algorithm and Lion Optimization Algorithm (LOA) respectively. Forthcoming Sects. 5 and 6 discuss the implementation of an algorithm for Medical Images and Results and Discussion of the implemented concepts and they are followed by Conclusion of Sect. 7.

2 Machine Learning and Analysis

A subfield of software engineering all the more especially delicate figuring, called Machine taking in advanced from the investigation of example acknowledgment and computational learning hypothesis in automated reasoning. Characterization of machine learning as a "Field of concentrate that gives PCs the capacity to learn without being expressly modified" is explained in [35]. Machine learning investigates the examination and development of calculations that can gain from and make forecasts on information.

Machine learning is now and then conflated with information mining, where the last subfield concentrates more on exploratory information examination and is known as unsupervised learning. Machine learning identifies (and regularly covers) computational measurements, which additionally focuses on forecast making using PCs. It has substantial connections to scientific streamlining, which conveys strategies, hypothesis and application spaces to the field.

Table 1 Process involved in digital image processing

S. No	Name of the process	Function of the process	Output for the process
1	Image acquisition	• Pre-processing like scaling	Image
2	Image enhancement	• Highlight certain features of interest in an image • Subjective	Image
3	Image restoration	• Mathematical or probabilistic model for improving the appearance of an image • Objective	Image
4	Colour image processing	• To work on color images	Image
5	Wavelets and multi-resolution processing	• Representation of image over various resolutions • Used for data compression	Image
6	Compression	• Reduces storage for saving the image • Reduces bandwidth for transmitting the image	Image
7	Morphological processing	• Extracting image components for representing and describing the shape	Image based on neighbor's value
8	Segmentation	• Partitioning an image	Image attributes
9	Representation and description	• Raw pixel data • Boundary representation • Regional representation • Feature selection	Image attributes
10	Object recognition	• Label object based on its description	Image attributes

Grouping of machine learning assignments is generally into three general classes, contingent upon the idea of the learning "flag" or "criticism" accessible to a learning framework. Not at all like an arrangement, are the gatherings not known heretofore, making this usually an unsupervised errand. Inside the field of information examination, machine learning is a technique used to devise complex models and calculations that loan themselves to expectation; in business utilize, and it is the prescient investigation. Such calculations work by building a model from a case preparing a set of info perceptions with a specific end goal to settle on information-driven expectations or choices communicated as yields, as opposed to following entirely static program directions. In bunching, information sources are arranged to isolate into gatherings.

Formative learning, expounded for robot learning, creates its particular arrangements (additionally called educational modules) of learning circumstances to in total secure collections of unique abilities through independent self-investigation and social cooperation with human instructors and utilizing direction systems, for example, dynamic learning, development, engine collaborations, and impersonation. Powerful machine learning is troublesome because discovering designs is hard and frequently insufficient preparing information is accessible; accordingly, machine-learning programs regularly neglect to convey. Among the different classes of machine learning issues, figuring out how to learn takes in its particular inductive predisposition given past involvement.

3   Monkey Search Optimization (MSO) Algorithm

3.1   Introduction to Monkey Search Optimization Algorithm

The monkey search optimization proposed [47, 48, 84] in this chapter is derived from the simulation of the monkey's mountain climbing process [31]. The foraging behavior (search of food) of monkeys involves several activities, such as: exploring, climb, watch—jump, cooperation and somersault. The foraging behavior focuses on the search for the local optimal solution in the search space [5, 10]. The activities are performed in sequence by the monkeys. Apply this algorithm to mammogram images for removing the background region, pectoral region removal and for feature extraction, feature selection, and segmentation. After finding the optimal solution, i.e., after the completion of the segmentation process, is completed, the monkey algorithm will be terminated [81].

The critical role of the activities performed by the monkeys is to mark the quantity and quality of the food source. As the algorithm is a population-based algorithm, the monkeys will move in groups. Hence, consider the whole mammogram image as forest and the background region is assumed as the non-edible region by the monkeys. Monkeys are in need to find the boundary between the edible and non-edible regions; hence it will make a natural process to go in search of food.

The behaviors of monkey species such as Howler, Spider, and Squirrel differ with one another based on the food type, movement, searching for food, resting, watching, and doing other activities [31, 84]. Types of food for the monkeys are leaves, fruits, and insects. Howler monkey depends on leaves, Spider monkey focuses on leaves and fruits, whereas Squirrel monkeys have both fruits and insects [35].

3.2 Food Classification for Monkeys

Monkey takes less time for consuming food when compared to the time it has taken for foraging. Food is classified as leaves (a low-quality food) are in abundance, while fruits (a better-quality) is in less abundant, and insects (a high-quality food) are the least abundantly available. In the mammogram image assume the following: leaves found in abundance as the background details, fruits which are less abundant as pectoral muscle region and insects as the region of interest.

Though the leaves and fruits are low-quality and better-quality foods, the chance of having non-edible, low quality and average quality foods are in abundance when compared to that of the high-quality food. Hence, in the mammogram image, consider if an area is having the background details and pectoral muscle regions it has to be removed, whereas if it is the region of interest, i.e., the high-quality food, select the part, extract and segmented for further processes.

3.3 Foraging Behaviour of Monkeys

Monkeys are involved in various activities such as exploration, climb, watch—jump, cooperation, and somersault for finding the local and global optimum [2]. After finding the global optimization, the algorithm gets terminated, showing there are no more processes to be performed. Figure 4 is the representation of activities conducted by the monkeys. In the exploration process, the monkeys explore the new search area, and in the climb, the monkeys make some movement. Climb process can be subdivided into Long-Step and Short-Step climb process, to avoid getting distracted from reaching the target food [31]. Based on the gradient—descent function, choose either of the two climb processes. After finding the individual best, see the local optimum by performing watch—jump process and is followed by the cooperation process to make all the monkeys to gather in a single place. Somersault process is to find the global best. These sequences of step terminate when there is no more best to be seen.

Fig. 4 Activities performed
by monkeys

3.4 Mathematical Explanation of MSO

Segmenting mammogram images can be performed by implementing the Monkey
Search Optimization (MSO) algorithm. In this segmentation process, it involves the
removal of the background region, pectoral muscle region and feature extraction of
the region of interest. All these three processes are carried out with the implementation of MSO algorithm. This section explains the different processes involved in
this MSO algorithm with its application in the segmentation processes, and it is as
follows.

Let us assume the concept of having the decision variable vector $R_i = (r_1, r_2, r_3, \ldots, r_n)^T$ and the objective function for minimization as $f(R)$ [80]. $\Delta R = (\Delta r_1, \Delta r_2, \Delta r_3, \ldots, \Delta r_n)^T$ is a randomly generated vector. The pseudo-gradient of
function $f(R)$ at the point, R can be expressed as $\left(f_1'(R), f_2'(R), \ldots, f_n'(R)\right)^T$ and
is represented by the following Eq. (1).

$$f_j'(R) = \frac{f(R + \Delta R) - f(R - \Delta R)}{2\Delta r_j} \quad j \in \{1, 2, \ldots, n\} \tag{1}$$

The repetitions in the generations of ΔR, the local optimum is found by the
decrease in the objective function $f(R)$ based on the sign caused as a result of slow
replacement of R with $R + \Delta R$ or $R - \Delta R$. Replacement of the value R defines that
the monkey has obtained a better pixel value when compared with the current pixel
value. The value $R + \Delta R$ and $R - \Delta R$ denote the variation in the pixel properties in
forwarding and the backward pass of the transition [35].

Based on the Fourier Transformation, Eq. (2) depicts the energy calculation for
monkeys. Fourier Transformation is performed in the frequency domain to decom-
pose the image into sine and cosine components which aid to calculate the distance of
variation in the properties of the images [90]. This Fourier Transform will be used to
convert the image from the spatial domain to the frequency domain and vice—versa.

As the image processing in the spatial domain focuses on the frequency of change in the pixel properties and also the repetition of the pattern change in the image, image processing in the frequency domain is preferred. Depending on the image processing application and its needs, the domain of implementation can be either spatial or frequency domain.

$$\|E\|_2 = \sqrt{\int_{-\infty}^{\infty} f(r) e^{-ihw} dr} \tag{2}$$

where

$$f(r) = (hw)_{ij} \tag{3}$$

where hw in Eq. (3) is the energy required by each monkey for moving from one place to another [35]. The value e^{-ihw} can be calculated as $e^{-ihw} = (\cos(hw) - i \sin(hw))$. This aids in identifying the pixel with better properties in the curved regions in the mammogram image.

Norm is the mathematical notation for representing the total size or length of all vectors in the vector space or matrices [78]. There are various norms with various forms. In Eq. (2) E is represented in l_2—is the Euclidean Norm which is used to measure the distance in the vector space. As the image is a vector image and the number of monkeys considered is also in vector, it is preferable to use Euclidean Norm for measuring the distance between the vectors. Another advantage of using l_2—norm over other norms is that it extracts the best solution from infinite solutions available in the search space.

Total energy, i.e., the energy for all the monkeys is calculated based on the following Eq. (4) with the boundary conditions as in Eqs. (5)–(7).

$$T(r) = \|E\|_2 (hw)_{ij} = \|E\|_2 \sum_{i=1, j=1}^{M} (hw)_{ij} \tag{4}$$

$$I_m I_v = I_a - I_d \tag{5}$$

$$-I_l^M \leq I_l \leq I_l^M, \quad l \in L \cup L' \tag{6}$$

$$0 \leq r_j \leq r_j^M, \quad j \in \{1, 2, \ldots, n\} \tag{7}$$

where

E is the total energy,
I_m is the image matrix,
I_v is the pixel value,
I_l is the original image or the image where the previous process has occurred,
I_a represents the active pixels in the image,
I_d represents the dense pixels in the image,
I_l^M is the maximum of how many pixels affected in the image,

L   is the set of an existing path for movement,
L'   is the set of a new path for movement.
r_j is the number of monkeys in the correct position, and
r_j^M is the maximum number of monkeys,

The solution representation for the given problem is depicted in Eq. (8).

$$R_i = \left(r_{i,1}, r_{i,2}, r_{i,3}, \ldots, r_{i,n} \right)^T \tag{8}$$

The objective function of MSO is represented as in the Eq. (9), for which the value will be modified for each monkey.

$$\min f(x) = \begin{cases} RE \sum_{i=1, j=1}^{M} (hw)_{ij} + G_0 \sum_{l \in L \cup L'} \max\{0, |RI_l| - RI_l^M\} \\ G_1 \text{ Otherwise} \end{cases} \tag{9}$$

while ($t <$ MaxGeneration) or (segmentation completed).
G_0, gravitational force when monkey jumps from one tree to another,
G_1, the penalty if the monkey is unreachable.
The initial population of M monkeys is generated as in Eq. (10).

$$r_i = (r_1, r_2, r_3, \ldots, r_n)^T \tag{10}$$

(a) Climb Process

Initialize random position for M monkeys as $R_i = \left(r_{i,1}, r_{i,2}, r_{i,3}, \ldots, r_{i,n} \right)^T$. The current position is evaluated to reach the border of the image. For this process, monkeys can select either of large-step or small-step climb process. The following subdivisions describe the generation of the large-step and small-step climb process.

The Large-Step Climb Process
The process of generating the large-step climb process is given in the following points.

$$\Delta R_i = \left(\Delta r_{i,1}, \Delta r_{i,2}, \Delta r_{i,3}, \ldots, \Delta r_{i,n} \right)^T$$

Interval $[-r_L, r_L]$ where r_L is the length of large-step climb process.
Calculate $f(R_i + \Delta R_i)$, $f(R_i - \Delta R_i)$
If $f(R_i + \Delta R_i) < f(R_i - \Delta R_i)$ and $f(R_i + \Delta R_i) < f(R_i)$
Then $R_i = R_i + \Delta R_i$
Else if $f(R_i + \Delta R_i) < f(R_i - \Delta R_i)$ and $f(R_i - \Delta R_i) < f(R_i)$
Then $R_i = R_i - \Delta R_i$
Repeat (i)–(iii) of large-step climb until $N_{C,L}$ has been reached.

The Small-Step Climb Process

The process of generating the small-step climb process is depicted in the following points.

$$\Delta R_i = (0, \ldots, 0, \Delta r_{i,j}, 0, \ldots, 0)^T$$

where $j\{1, 2, \ldots, n\}$ and $\Delta r_{i,j}$ is a non-zero integer with an interval of $[-r_S, r_S]$ where r_S is the length of small-step climb process.

and (iii) steps for the small-step climb process is similar to the (ii) and (iii) steps in large-step climb process.

Repeat (i)–(iii) of small-step climb process until $N_{C,S}$ has been reached [35].

The base for the large-step and small-step climb processes on the images is the pseudo-gradient function which follows the concept of the dynamic sliding window [45]. Usual scanning process invloves 3×3 matrices for identifying the region which has to be segmented. In some circumstances, if there is no variation in the pixel properties, then the monkeys can change its step size to larger and continue to move in the direction. While the monkey is moving with large-step climb process and if it identifies a change in the pixel properties, it chooses the small-step climb process and continues its movement. Every time it checks for a change in the pixel properties. Accordingly, it changes its step from small to large or vice versa.. During these processes, the large-step climb process will have a doubled kernel size of the small-step. This small-step will have the kernel size of 3×3 which is the standard kernel size used in scanning and making modifications in the current position.

(b) **Watch—Jump Process**

Check whether if there is any higher position when compared to current position based on the eyesight b [35]. The role of eyesight in the scanning process is to watch for the better values when compared with its value, i.e., it includes the process of analyzing the values of the personal best with the individual best of neighbor monkeys. If there is available when compared with the personal best of the monkey, the monkey will move on to the individual best of its neighbor. The following points depict this short discussion.

If higher position is available,

Then generate an integer $r'_{i,j}$ for an interval of $[r_{i,j} - b, r_{i,j} + b]$ randomly, where $j \in \{1, 2, \ldots, n\}$ [31].

Let $R'_i = (r'_{i,1}, r'_{i,2}, \ldots, r'_{i,n})^T$ [107].

If $f(R'_i) < f(R_i)$, let $R_i = R'_i$.

Repeat (i)–(ii) until maximum allowable number N_w has been reached.

(c) **Co-Operation Process**

The optimal solution in one iteration is assumed as $R^* = (r_1^*, r_2^*, \ldots, r_n^*)^T$. Let the initial position of the monkeys be $R_i = (r_{i,1}, r_{i,2}, r_{i,3}, \ldots, r_{i,n})^T$ [40]. The main advantage of this process is to maintain the coordination among the monkeys. In this process, each and every monkey's personal best is compared with the individual best of its neighbor [31]. Then the entire monkey's will climb to the tree or position which has better quality food.

Generate real number β in $[0, 1]$ randomly.

Calculate $r_{i,1}''' = \text{round}\left(\beta \, r_j^* + (1 - \beta)r_{i,j}\right)$, $j \in \{1, 2, \ldots, n\}$.

Let $R_i = \left(r_{i,1}, r_{i,2}, \ldots, r_{i,n}\right)^T$.

Set $R_i = R_i''$ and repeat the climb process.

The climb process is continued to reach the better value. Previous points represent the calculation procedure for the cooperation process.

(d) Somersault Process

After all the monkeys have grouped to reach the better quality food as compared with the neighbors, they have to check for still more better quality food in the given search space than the current position. Then the monkeys will do the somersault process for identifying the better quality calcified cells in the search space of the mammogram image. The following points explain the discussed procedures.

Generate the real number α in $[c, d]$.

Calculate $u_j = \frac{1}{M} \sum_{i=1}^{M} r_{i,j}$, $j \in \{1, 2, \ldots, n\}$,

where $\hat{U} = (u_1, u_2, \ldots, u_n)^T$ as the pivot of Somersault process.

For $\forall i \in \{1, 2, \ldots, M\}$, $\forall j \in \{1, 2, \ldots, n\}$,

Calculate $r_{i,j}''' = r_{i,j} + \text{round}\left(\left|u_j - r_{i,j}\right|\right)$

Let $R_i''' = \left(r_{i,1}''', r_{i,2}''', \ldots, r_{i,n}'''\right)^T$.

Set $R_i = R_i'''$ and repeat climb process.

If $r_{i,j}'''$ is new position > upper limit r_j^M,

Then

$$r_{i,j}''' = r_j^M$$

Else if $r_{i,j}''' < 0$

Then

$$r_{i,j}''' = 0$$

After the completion of the somersault process, if there is a calcified cell of higher quality, the monkeys will move to the new position. This somersault process is similar to finding the global optimization value. In the normal money search, the monkeys will climb to the top of the mountain and search for another mountain which is of greater height is compared with the current mountain. If there is the availability of mountain with greater height when compared with the current mountain height, all the monkeys will somersault to the mountain with greater height. As per this process, the money search in image processing for searching the calcified cells, once it reaches the local optimization (Cooperation process), the monkeys will look forward to the global optimization. As the monkeys are scattering around the search space with their cooperated better quality cells, the monkeys will then compare their values with other groups and will repeat the climbing process.

(e) Stochastic Perturbation Mechanism

Once all the monkeys have reached the global optimization, verify the value of each monkey with the equivalent pixel values of other monkeys to confirm that there is no single monkey with a different food source or unwanted food source. This process of cross verification is the stochastic perturbation which and the following points explains it.

When $r_{i,j}$ is same for all monkeys,
Then,

$$u_j = r_{i,j}$$

Hence, $r_{i,j}''' = r_{i,j}$ for monkey $k, k \in \{1, 2, \ldots, M\}$,

Let $r_{k,j} = e$, where, e is a uniformly distributed integer from is $\left[0, r_i^M\right]$

(f) Termination Process

The MSO algorithm will reach its termination when either of the following two criteria is met.

Reaching the stochastic perturbation (or)
Entering the maximum number of iteration.

4 Lion Optimization Algorithm (LOA)

4.1 Introduction to LOA

Lion Optimization Algorithm (LOA) is a Nature—Inspired Search Algorithm by [51] or Nature—Inspired Metaheuristic Algorithm by [44]. This subchapter focuses on the foraging behavior of the lion. In this LOA, consider the population of the Lion have categories among them. They are categorized based on their age and the foraging behavior. This foraging behavior of lion includes different stages and the stages are, initialization, hunting, moving toward the safe place, roaming, mating defense, migration, population equilibrium, convergence [50].

The foraging behavior of most of the metaheuristics algorithm can be either in search of food or mating. This LOA makes a deviation from this approach, i.e., by considering both the behaviors reduces the computational complexity. But this condition need not be mandatory for all the circumstances. Implementation of LOA algorithm for Image Steganalysis and Data Clustering is in [3, 59].

4.2   Foraging Behaviour of LOA

LOA being a variant of GA allows metaheuristic approach to global optimization. Like many other nature-inspired algorithms, this studies the behaviors of the Lion species allowing insight into better optimization. Lions happen to be of particular interest due to their social and robust sexual dimorphic nature. Lions exist instead as nomads or in pride. The switch between lifestyle is possible. Lions hunt together, mainly the females' only hunt in a pride. As for mating, once the young ones sexually mature, they are removed from the pride. Nomads usually roam in the territory.

4.3   Mathematical Explanation of LOA

(a)   Initialization

Lion is represented by a set of values, while computing the cost (fitness value) is the function on the set of values. On initial run, positions are marked with respective best-visited positions.

(b)   Hunting

Hunting forms the main aspect of survival and fitness in the algorithm. From the set of females, selected ones go for hunting of prey. These are then split into three wings namely: left right and center. Center has the highest fitness. There is a regular update on fitness with each iteration and hunt. Opposition Based Learning is applied by the left and right parties to engage the prey. To mimic the hunting and encircling of the location is updated to nearer or required position as given in Eqs. (11) and (12).

$$\text{Hunter}' = \begin{cases} \text{rand}((2 \times \text{PREY} - \text{Hunter}), \text{PREY}), & (2 \times \text{PREY} - \text{Hunter}) < \text{PREY} \\ \text{rand}(\text{PREY}, (2 \times \text{PREY} - \text{Hunter})), & (2 \times \text{PREY} - \text{Hunter}) > \text{PREY} \end{cases} \tag{11}$$

$$\text{Hunter}' = \begin{cases} \text{rand}(\text{Hunter}, \text{PREY}), & \text{Hunter} < \text{PREY} \\ \text{rand}(\text{PREY}, \text{Hunter}), & \text{Hunter} > \text{PREY} \end{cases} \tag{12}$$

Pseudocode as follows
Divide hunters into three subgroups randomly
Generate a prey

Divide hunters into three subgroups randomly
Generate a prey
For i = 1: H(H is number of hunters)
Move ith hunter toward prey according to its relevant group
    If the new place of an ith hunter is better than its last
        Prey escapes from hunter
   End
End

(c) Moving to the safe location

The non-hunting female lions move in search of the better fitness area, i.e., area where their fitness is bigger and secured. The reference to fitness can be done by that of an individual lion or a pride's fitness average. The determination of fitness is an iterative process concerning Eq. (13)

$$S(i, t, P) = \begin{cases} 1 \; \text{Best}_{i,P}^{t} < \text{Best}_{i,P}^{t-1} \\ 0 \; \text{Best}_{i,P}^{t} < \text{Best}_{i,P}^{t-1} \end{cases} \tag{13}$$

where best position is found by a lion I until iteration t. A large number of lions covering at a point identifies the greater number of success. A low value of success shows that the lions are swinging around the optimum solution without significant improvement. This helps in selection of Tournament. Equation (14) determines the Tournament's size.

$$T_j^{\text{Size}} = \max\left(2, \; \text{ceil}\left(\frac{K_j(s)}{2}\right)\right) \quad j = 1, 2, \ldots, P \tag{14}$$

Pseudocode is as follows:
For $i = 1$ to $P(P$ is number of pride)
Calculate tournament size for ith pride
For $j = 1$to $R(R$ is number of remained female in ith pride)
Select a place among pride's territory by the tournament selection
Move jth female toward the selected place
End
End

(d) Roaming

Roaming allows the lion to wander in search of better fitness or better solution. Random selection of lion and its roaming behavior allows updating of the best solution. Equation (15) gives the movement inside the pride.

$$x \sim \text{U}(0, 2 \times d) \tag{15}$$

where d represents distance between male lion's position and target location and x is a random number with uniform distribution. For wider area searching, add the angle θ to this direction.

Pseudocode is as follows,

For i = 1 to RM (RM is the number of resident males)
    Select %R of territory randomly to visit by the male
    For j = 1 to S (S is the number of selected place in the last step)
        Go toward place j^{th}
            If the new place of ith male better than it's best-visited position
                Mark that place as territory (update best-visited position)
        End
    End
    Select the best-visited position by the male as its current position
End
These steps vary for nomad line:
For i = 1 to NN (NN is number of nomad lions)
    Move i^{th} nomad randomly
        If the new place of the ith nomad is better than its best-visited positi
            Update i^{th} nomad's best-visited position)
    End
End

Nomad's adaptive roaming allows prevention of local minima, maintained by using Eqs. (16) and (17).

$$\text{Lion}'_{ij} = \begin{cases} \text{Lion}_{ij} & \text{if rand}_j > \text{pr}_i \\ \text{RAND}_j & \text{otherwise} \end{cases} \tag{16}$$

$$\text{pr}_i = 0.1 + \min\left(0.5, \frac{(\text{Nomad}_i - \text{Best}_{\text{Nomad}})}{\text{Best}_{\text{Nomad}}}\right) \tag{17}$$

$$i = 1, 2, \ldots, \text{number of nomad lions}$$

where pr_i is a probability calculated for each nomad lion independently, and Nomad_i and $\text{Best}_{\text{nomad}}$ Bestnomad represents the cost of the current position.

(e) Mating

Mating allows survival as well as a means to exchange among members. Certain females of the pride mate with randomly selected males to produce offspring. For nomads, the mating operator is a linear combination of parents producing two new offerings as in Eqs. (18) and (19).

$$\text{Offspring}_j 1 = \beta \times \text{Female Lion}_j + \sum \frac{(1 - \beta)}{\sum_{i=1}^{NR} S_i} \times \text{MaleLion}_j^i \times S_i \tag{18}$$

$$\text{Offspring}_j 2 = \beta \times \text{Female Lion}_j + \sum \frac{(\beta)}{\sum_{i=1}^{NR} S_i} \times \text{MaleLion}_j^i \times S_i \tag{19}$$

where j is dimension, S_i decides to select whether male-i is to mating or not; NR represents a number of resident males, β is randomly generated number with some mean value and standard deviation. Of the two offspring, one is randomly assigned male and the other female.

(f) Defense

Beaten males while defending or from combat, leave the pride of becoming a nomad. Vice versa, if the nomad is strong enough it can become resident, making current resident lion nomad. Defense operator involves defending against mature resident males or nomad males.

Pseudocode for Resident Males

Merge new mature males and old males
Sort all males according to their fitness
Weakest males drive out of the pride and become nomad and remained
Males become resident males.

Pseudocode for Nomad Lions
    Merge new mature males and old males
    Sort all males according to their fitness
    Weakest males drive out of the pride and become nomad and remained
    Males become resident males
Pseudocode for Nomad Lions
    For i= 1 to the number of nomad lions
    BT[1, P]= Create a Binary template ($[1 \times P]$) and assign a randomly generated binary ($0 - 1$) to each cell (P is numbers of pride).
    For j = 1 to P
      If j^{th} element of BT==1
        For z = 1 to NR (number of resident males in j^{th} pride)
          If i^{th} nomad male is better than z^{th} resident male in j^{th} pride
                z^{th} resident male in j^{th} pride drive out of the pride, and
                become a nomad and i^{th} nomad becomes resident.
                Go next i
          End
        End
      End
    End
    End

(g) Migration

Migration allows an increase in diversity of the target pride, as well as building bridge for exchange of information. In pride, size to be migrated to nomad is decided as %I of a maximum number of females in the pride. After this step, consider a sorted list of nomad females to satisfy the %S sex ration to be maintained.

(h) Convergence

For stopping condition, control either by the number of iteration in contrast to assumed CPU time.

5 Implementation of Algorithms for Medical Images

The dataset used for comparing the performance of the algorithm is the Mammogram Image Analysis Society (MIAS).

This section explains the experimental setup metrics used in the study of the performance of the algorithms implemented on images. Algorithms involved in the analysis of segmentation and classification processes are Monkey Search Optimization (MSO) Algorithm and Lion Optimization Algorithm (LOA).

5.1 Introduction to Mammogram Image

The performance analysis on segmentation by LOA and MSO algorithms were carried out on the mammogram images of MIAS database. MIAS database has 322 images digitized at 50-micron pixel edge which has been reduced to 200-μm pixel edge [86]. Every image in the database has 1024×1024 pixels [12]. The images in the database have two major categories: (1) 207 images of abnormal and (2) 115 images of normal, making the total of 322.

Analysis of classification by algorithms is trained on three–fourth of the total 322 images in database mammogram and tested on rest of one-fourth count of the image database. The above ratio results to the training of 240 images and testing of 82 images in the database. The database has various types of abnormalities such as calcification, well-defined, circumscribed masses, spiculated masses, ill-defined masses, architectural distortions, asymmetry and normal. The two significant categories of mammogram image are abnormal and normal. In these two categories, normal will have only the normal abnormality, and the remaining other abnormalities are categorized into abnormal. The segmentation and classification analysis of algorithms on the MIAS database which is implemented using the software MATLAB R2009a. The objective of this research is to minimize the energy consumed by the agents of the algorithms MSO, and LOA for better classification.

5.2 Preprocessing on Mammogram Image

The mammographic image database chosen is Mediolateral Oblique (MLO) view as it is the only view which covers the whole breast region while scanning the breast using Mammography modality. For performing functions in the medical images, uncompressed image file format provides better results. Hence, choose the image with lossless compression. The reason for selecting the lossless images is that the images will have better quality without any loss in the pixel. Among various file formats, a medical image with the following file format of lossless compression is preferred: (1) PGM, (2) PPM and (3) PNM (produced by PBM, PGM, and PPM file formats).

5.3 Feature Selection and Extraction of Region of Interest in Mammogram Images

5.3.1 Background Removal

This section discusses the process involved in the removal of the background region. In general methodologies, the process involved in the removal or the suppression of the background region is similar steps as that of the filters used in the process of noise removal [71]. In this background removal, the kernel will take the values of either 0 or 1. When the kernel slides in the image, based on the pixel intensity, replace the pixels with either 0 or 1. In this process, the grayscale image is converted to binary image. After this conversion, the image will have a cluster of pixels. Among those clusters, a cluster with least number of the pixel will represent the non-breast region of labels, tapes, and scanning artifacts. This group of the pixel replaces regarding other value. After this process, select the remaining image from the original image. From the following Fig. 5a–d, explains the pictorial representation of the label removal process. The image obtained from Fig. 5c will be mapped to the original image (Fig. 5a) to obtain the image Fig. 5d.

The process next to the noise removal is the removal of the background region. This background region consists of various features such as labels, machine parts, some marks made by the radiologist (example patient name, numbers) and so on. In some circumstances, the pixel properties of this region might be equivalent to the pixel properties of other areas. These properties may lead to misclassification. To avoid this problem, the background region is suppressed before getting into the next processes.

5.3.2 Pectoral Muscle Removal

This section focuses on the removal of the pectoral muscle region [62, 64, 65, 67] after the removal of the background region in the mammogram images. Pectoral muscles are the regions in mammograms that contain brightest pixels [85]. Before performing the pectoral muscle segmentation, it is necessary to find the orientation of the breast as left or right. Determining the breast orientation is a little difficult task. When the pectoral muscle region is pointed at the top left corner of the mammogram image, it indicates that the breast region orients towards the right and hence it is

(a) Original Image.   (b) Binary Converted Image.

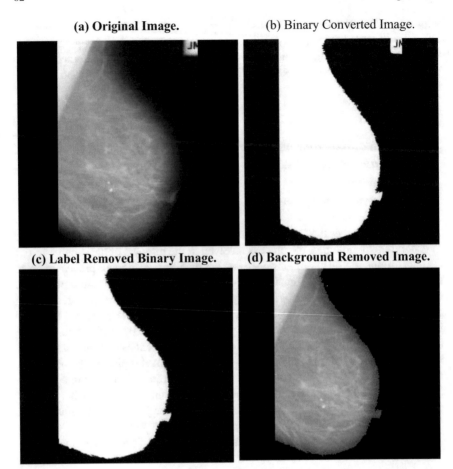

(c) Label Removed Binary Image.   **(d) Background Removed Image.**

Fig. 5 **a** Original image. **b** Binary converted image. **c** Label removed binary image. **d** Background removed image

called as the right orientation breast. When the pectoral muscle region points to the top right corner of the mammogram image, then the breast region will be pointed towards the left and hence the orientation of the breast will be on the left [30]. There are various methods involved in identifying the orientation of the breast.

In the first method, divide the mammogram image into two equal columns. Then count the number of non-zero pixels, and compare the value with each other. The subdivided column which has higher number of non-zero pixels will be determined to have the pectoral muscle region.

In the second method, discussion about the pectoral muscle detection and removal is in [30]. In his dissertation, the sum of the pixel intensities of the first five columns and last five columns are taken. Then the values are compared with simple if-then rules, based on which the group of columns with higher intensity the orientation of the breast is determined. From the Fig. 6, the hypotenuse (x_1, y_1) and (x_2, y_2) has to be determined based on the origin $(0, 0)$. These two points (x_1, y_1) and (x_2, y_2) which is in the format of (row, column) in the digital images are obtained by iterating the rows and columns of the binary image. The process involved in identifying the orientation of the right or the left breast is performed using the following steps.

(a) In the binary mammogram image, the point (x_1, y_1) is obtained by checking the first row of the image for the occurrence of a 1, as the value 0 represents the black pixels or the background region of the image. The point (x_1, y_1) is marked when there is the first occurrence of the value 1 in the first row of the image.

(b) Similar to the previous step, to determine the point (x_2, y_2), the first occurrence of first 1 is searched in the first column of the binary mammogram image, and the point (x_2, y_2) is marked when the first occurrence of 1 is obtained.

(c) After determining these two points (x_1, y_1) and (x_2, y_2), an offset value of r is specified which transforms these two points (x_1, y_1) and (x_2, y_2) into $(x_1 - r, y_1)$ and $(x_2, y_2 + r)$ for right side breast and on the other hand, the value is specified as $(x_1 + r, y_1)$ and $(x_2, y_2 + r)$.

(d) Next to these points are the straight line equation is employed to represents the hypotenuse of the right-angled triangle which acts as the boundary between the breast profile region and the pectoral region which is about to be segmented. The straight line equation can be represented as given in Eq. (18) in which m represents the slope of the line or the steepness of the line and c is the intercept on the y-axis, i.e., the line which crosses the y-axis.

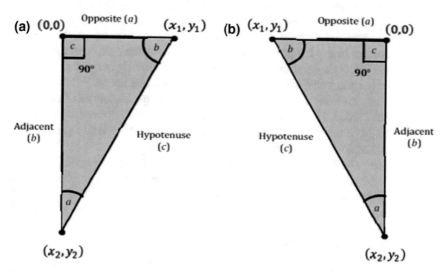

Fig. 6 Right angled triangle for the removal of pectoral muscle region.a Right angle triangle representing the pectoral muscle region of right—oriented breast.b Right angle triangle representing the pectoral muscle region of left—oriented breast

$$y = mx + c \tag{20}$$

(e) In Eq. (20), m represents the slope, and it can be calculated using Eq. (21).

$$m = \frac{y_1 - x_1}{(y_2 - x_2)} \tag{21}$$

where c is an intercept which is calculated using the Eq. (22).

$$c = y_1 - (m * y_2) \tag{22}$$

From Eq. (18) straight line can be obtained with the range of $(x_2 < x < y_2)$ where the value of x_2 is always 1.

(f) The vectors x and y will combine to form the data matrix. This data matrix superimposes on the image for the removal of the pectoral muscle.

The third method is searching for nonzero pixels, which starts at the top left and right corner simultaneously. The width of the image in which the nonzero pixel detected from both the corner is counted and compared. If the left width is smaller than the right width, then it is assumed that the pectoral is on the left side of the image else it is on the right side. From the detected corner pixel, expose the discontinuity in the intensity of the pixel on every column of the same row. If there is an intensity change in the coordinates of the pixel, consider it as the width of the pectoral region.

All the pixels which lie inside the pectoral width and half of the height of the whole image are segmented from the original image. This rectangle shaped image contains the entire pectoral muscles. To extract the pectoral muscles from this image, implement the process of obtaining the binary image by simple thresholding. This binary image contains pectoral muscles and other tissues. To segment, the pectoral muscles from the binary image implement raster scanning the right or left the side of the image to detect the intensity discontinuities. The resulting image contains only the pectoral muscle, and delete this region from the original image.

As the pectoral muscle is of right-angled triangle shaped, analysis of the whole image is not needed. Hence, remove the pixel which is of high intensity nearer to it. There might be three regions after the process of thresholds which represents the pixels of higher, lower, and with intermediate intensities.

The pixels of higher intensity will represent the pectoral muscle region (but it will have region of interest also) and that of the lower intensity will be the background region (this region will be removed before pectoral removal, but this process of pectoral removal is more suitable when the background removal is not performed) whereas the intermediate pixel intensity represents the regions nearer to the pectoral region. The next process is to compute the transitional region between the pectoral and intermediate region, i.e., this intermediate pixel intensity might be a border between the pectoral muscle and the breast.

Apply the gradient operator along with some adjustment calculations (e.g., ignore the pixels which are too far from the sequence of pixels previously recognized as border points) to calculate the hypothetical coordinates of this border. As it was mentioned earlier labels and other artifacts are removed on the occasion of previous operations. Indeed, the image with the region of low-intensity pixels may contain some of the background and skin-air interface. Again if we compute the gradient on this image the approximate breast border region can be estimated. Then the pectoral muscle region is removed from the mammogram image.

5.3.3 Extract the Region of Interest for Classification

The texture is the most important characteristics used in image processing to identify the objects or regions in an image. In [54] the introduction of the textural features which is the only most frequently used second order statistical features is explained. This feature has two steps for the process of feature extraction [67]. At first, compute the co-occurrence matrix and then the calculation of the texture feature based on the co-occurrence matrix as a second process. Some of the segmentation and classification algorithms proposed by various authors are explained in [11, 13, 18, 20, 26, 42, 43, 46, 53, 55, 56, 61, 66, 69, 70, 76, 77, 82, 83].

For the process of feature extraction [37], consider the Gray Level Co-Occurrence Matrix (GLCM) and on the other hand, use Haralick features for the classification of the mammogram image [63, 72]. Both GLCM and Haralick's features have the same working principle of K-Nearest Neighbours (KNN). There are fourteen Haralick's features. Before getting into the process of GLCM, the list of Haralick's features are discussed which are as [73]: Angular Second Momentum, Contrast, Correlation, Sum of Squares—Variance, Inverse Different Moment, Sum Average, Sum Variance, Sum Entropy, Entropy, Difference Variance, Difference Entropy, Information Measure of Correlation 1, Information Measure of Correlation 2, and Maximum Correlation Coefficients.

The process of GLCM will focus on the eight connected graph of the neighborhood in the eight directions such as 0, 45, 90, 135, 180, 225, 270, and 315. Use GLCM in a different order to work with a different function on texture calculations. First order texture will have a statistical calculation on the values obtained from the image, and it will not consider the relationship between the neighbor pixels. Second-order texture calculation will consider the relationship between the neighbor pixels of the original image. Third and higher order pixels will consider the relationship among three or more pixels. But the implementation of the third order will increase the time for the calculation process. In most of the applications, second order texture classification will be performed.

5.4 Hybrid Classification with SVM

The SVM is a popular binary classifier introduced in [74]. These classifier works are based on mapping input points to a high dimensional feature space, which is separated by a hyperplane. It maximizes the distance between the closest patterns and is called a margin. The main advantage of using SVM is simple to use and gives better classification results.

The sample marginal hyper-plane of SVM is shown in Fig. 7. Many possible linear classifiers can separate different data from each other, but these linear classifiers consider the maximization of the margin (maximizes the distance between it and the nearest data point of each class) only. This linear classifier is the maximized margin separating the data as a hyperplane. Intuitively, it is expected that this boundary will generalize better when opposed to other possible boundaries.

In a hyperplane, the distance between the closest point to the origin can be found by maximizing x, as x is on the hyperplane. The process is repeated for other points with a similar scenario. Thus solving and subtracting the two distances will be the summed distance from the separating hyperplane to the nearest point is obtained.

The major advantage of SVM is the training process, which is relatively easy. No local optimal, unlike in neural networks. Scaling is relatively good to high dimensional data and the trade-off between classifier complexity and controlling of error is evident. The main weakness is the need for a good kernel function and sometime SVM often leads to the over-fitting problem from optimizing the parameters to model selection.

There are two major types of classification. One is linear classification whereas the other is non-linear classification. In linear classification, the output for the classification performed in medical images will be categorized only as normal or abnormal. In non-linear classification, the classification is further extended the categorization of abnormal into two more subclasses as benign and malignant. Hence the result of this non-linear classification will have three categories (minimum). They are normal, benign and malignant. Like the types of classification, the methodology of classifying the image has two major types.

Support vector machine (SVM) can be hybridized with other algorithms to perform the classification process. This hybrid classification is varied based on either algorithm or characterization. Algorithm-based hybrid classification [36] focuses on classifying the data based on metrics (properties) of the image. On the other hand,

Fig. 7 Margin
hyper—planes of SVM

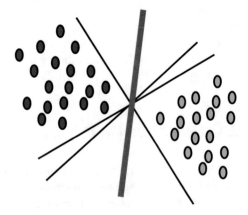

Table 2 Confusion Matrix for Mammogram Image Classification

		Predicted class	
		Abnormal	Normal
Actual class	Abnormal	True positives	False positives
	Normal	False negatives	True negatives

characterization or class-based hybrid classification, for defining the character it is in need of a property. For this purpose, rule generation is performed in class-based hybridization. The rules are framed with a neural network, fuzzy logic or a genetic algorithm. In this research, the hybrid method used for classifying the mammogram is linear classification.

SVM classifier is a simple and a better classifier besides lacking in some problems such as kernel function and overfitting. The hybridization of this classifier is required for efficient classification. Classification accuracy is performed based on two categories. They are confusion matrix and the sensitivity-specificity—accuracy calculation.

Confusion Matrix is also known as a contingency table or an error matrix for visualizing the performance of an algorithm. Each column of the matrix represents the instances in a predicted class, while each row represents the instances in an actual class. The table of confusion or the confusion matrix has two rows and two columns which represent True Negatives (TN), True Positives (TP), False Negatives (FN) and False Positives (FP). Confusion Matrix helps in calculating Sensitivity, Specificity, and Accuracy of the Optimization Techniques. The confusion Matrix for the mammogram image classification can be represented as in the following Table 2.

Positive is the condition in which the samples are identified (as abnormal) and Negative is the condition where the samples are rejected (as normal). True Positive (TP) is the circumstance where the samples are correctly identified as abnormal. False Positive (FP) is the state in which the samples are incorrectly identified as abnormal. If the samples are correctly rejected as normal, the situation is termed as True Negative (TN) else if the samples are incorrectly rejected as normal, the form is False Negative (FN).

Sensitivity relates to the ability to identify a condition correctly by the optimization techniques, hence sensitivity is also termed as the True Positive Rate (TPR). The sensitivity of the test is the proportion of people known to have the disease. The mathematical representation of Sensitivity is as given in Eqs. (23) and (24).

$$\text{Sensitivity} = \frac{\text{Number of True Positives}}{\text{Number of True Positives} + \text{Number of False Negatives}}$$

$$= \frac{\text{Number of True Positives}}{\text{Total Number of Sick Individuals in the given population}} \qquad (23)$$

$$\text{Sensitivity} = \text{probability of a positive test} \qquad (24)$$

Specificity relates to the ability to identify a condition negatively by the optimization techniques, hence specificity is also termed as the True Negative Rate (TNR). The sensitivity of the test is the proportion of people known to have tested with negative results. The mathematical representation of Specificity is as in the following Eqs. (25) and (26).

$$\text{Specificity} = \frac{\text{Number of True Negatives}}{\text{Number of True Negatives} + \text{Number of False Positives}}$$
$$= \frac{\text{Number of True Negatives}}{\text{Total Number of well individuals in the given population}} \quad (25)$$
$$\text{Specificity} = \text{probability of a negative test} \quad (26)$$

Accuracy measures how correct an optimization technique (diagnostic test) identifies and excludes the given condition. Accuracy can be calculated with the mathematical representation as shown in the following Eq. (27).

$$\text{Accuracy} = \frac{\text{True Positives} + \text{True Negatives}}{\begin{array}{c}\text{True Positives} + \text{True egatives} + \\ \text{True Negarives} + \text{False Positives}\end{array}} \quad (27)$$

A Receiver Operator Characteristic (ROC) or ROC curve is a graphical plot illustrating the performance of the binary classifier. ROC curve is plotted with the values of the fraction of True Positive (TP) versus the fraction of False Positive (FP) or True Positive Rate (TPR) versus False Positive Rate (FPR) at different threshold settings. TPR is also known as the Sensitivity, whereas FPR is calculated by one minus the specificity. As ROC makes a comparison of two operating characteristics such as TPR and FPR, ROC is also called as the relative operating characteristic curve

From Fig. 8, x and y axis are plotted as FPR and TPR respectively are noted. As TPR is equivalent to sensitivity and FPR is equal to (1—Specificity), this ROC curve is also called as the Sensitivity—(1—Specificity) curve. ROC curve is plotted based on the predictions obtained from the confusion matrix. The point (0, 1) is called as the Perfect Classification Point. The diagonal line drawn from the left bottom corner to the right top corner based on the random guess is called as the Line of No—Discrimination which is plotted irrespective of the positive and negative rates of classification.

After plotting the classification results, if the curve is above this random guessed line then the algorithm for which the curve is plotted is a good classification algorithm (as A in the Fig. 8) when compared to the random guessed plot. If the plotline lies over the random guessed line it indicates that the algorithm gives 50% accuracy (as B in Fig. 8). In Fig. 7.2, C represents the originally predicted result, where C′ is the mirror of C which is obtained by reversing the methods of C. Though C is in worse state, its mirror method C′, has shifted to the better space which is also better than the original method A. In such a manner, if the method is worse than the random guess, the method's prediction can be reversed to make that method to shift on to the better space.

6 Results and Discussions

6.1 Results of Preprocessing Technique

As this research focuses on the medical image processing, the images obtained from various modalities may have noises. There are various causes for having noises in the images and are discussed in earlier sections. The following Fig. 9a–d depicts the image showing some of the noises implemented on the mammogram image.

This research is based on the medical image processing. Most of the medical image will of digital and the noise obtained in these images might be of Gaussian in nature. The following Fig. 10 depicts the representation of 3×3 averaging mean filter. These figures are followed by Fig. 11a, b showing the mean filter applied to mammogram image. From these two images it can be observed that as the kernel size is getting increased, the smoothing or blurring in the image is also getting increased.

Figure 11c, d gives the result of Median and Gaussian filters applied on the mammogram image.

From the above discussion and results on the noises and filters, it is observed that in medical images the type of noise observed is the Gaussian Noise and also, from

Fig. 8 Sample ROC curve space and plot

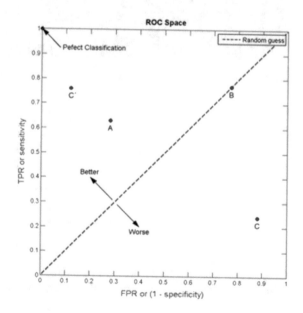

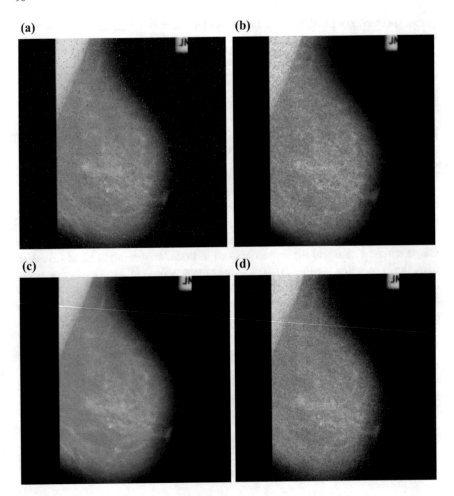

Fig. 9 **a** Salt and pepper noise. **b** Speckle noise. **c** Poisson noise. **d** Gaussian noise

Fig. 10 Averaging mean filter

the resulting image of filters, it is seen that Gaussian Filter provides an image with better quality when compared with other filters. Hence, the images obtained after

(a) **(b)** **(c)** **(d)**

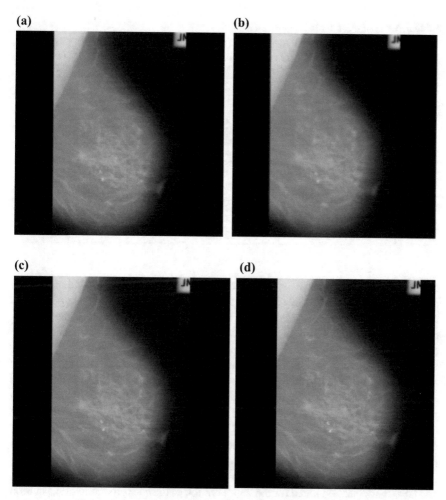

Fig. 11 **a** Mean filter with kernel size of 5 × 5. **b** Mean filter with kernel size 10 × 10. **c** Median filter with kernel size of 3 × 3. **d** Gaussian filter

removing the noises by applying a Gaussian filter is considered for further processes to be performed for segmentation and classification.

6.2 Results on Feature Selection and Extraction of Region of Interest in Mammogram Images

After performing the removal of noise in mammogram images, background region and the pectoral region in mammogram image which have the same properties as that

Table 3 Value of mean for the segmented region

Dataset	Image	Background removal MSO	Background removal LOA	Pectoral removal MSO	Pectoral removal LOA	Region of interest MSO	Region of interest LOA
1	mdb001	43.4	31.9	21.3	40.0	110.5	115.2
2	mdb002	44.2	42.4	45.0	29.0	122.0	122.7
3	mdb006	61.6	58.5	23.0	20.0	0.0	0.0
4	mdb013	67.8	65.1	29.7	20.7	116.3	123.2
5	mdb023	46.7	44.7	42.0	23.7	106.4	118.4
6	mdb027	58.7	57.3	18.3	20.0	0.0	120.5
7	mdb031	48.5	46.6	24.0	46.3	111.4	140.7
8	mdb239	59.4	40.3	44.0	45.7	141.8	143.9
9	mdb240	32.2	41.1	44.7	28.3	152.9	0.0
10	mdb241	24.8	44.9	24.0	43.0	94.1	93.1

of the region of interest have to be removed. In this regard, the background region will be removed and will be followed by the removal of the pectoral region is performed by selecting and extracting the pectoral region. After the extraction of the pectoral region, it is segmented from the image, which helps in the better classification of the cancerous region.

The performance analysis on segmentation by MSO [33, 35] and LOA algorithms are carried out on the mammogram images of MIAS database. MIAS database has 322 images digitized at 50-μm pixel edge which has been reduced to 200-micron pixel edge. Every image in the database has 1024×1024 pixels. The images in the database have two major categories: (1) 207 images of abnormal and (2) 115 images of normal, making the total of 322. Mean, Entropy, CPU Time and Energy were the performance metrics for segmentation.

The following Tables 3, 4, 5, and 6 represents the values of Mean, Entropy, Energy, and CPU Time(sec) obtained for the segmentation of the background region, pectoral muscle region, and region of interest using MSO and LOA respectively. The next sequence of Figs. 12, 13, 14 and 15 are the graphical representation of Tables 3, 4, 5, and 6 respectively.

From Fig. 13 it can be observed that for the image mdb006 there is no line to represent the entropy of ROI with the implementation of LOA. It states that, either the algorithm is unable to find the region of interest or the image itself does not possess the region of interest. Similar, the non-occurrence of the lines for the images mdb027 can be observed for ROI using MSO.

Table 4 Value of entropy of the segmented region

Dataset	Image	Background removal MSO	Background removal LOA	Pectoral removal MSO	Pectoral removal LOA	Region of interest MSO	Region of interest LOA
1	mdb001	2.9	2.8	2.6	3.2	1.0	0.9
2	mdb002	3.0	3.1	0.3	1.2	1.0	0.9
3	mdb006	3.9	4.6	0.7	2.0	0.0	0.0
4	mdb013	3.7	5.7	2.7	2.2	1.0	0.9
5	mdb023	3.1	3.1	2.4	2.3	1.0	1.0
6	mdb027	3.7	4.7	1.0	0.7	0.0	1.0
7	mdb031	3.2	3.9	1.2	2.9	1.0	0.9
8	mdb239	2.9	2.0	1.5	1.5	0.9	0.9
9	mdb240	3.6	5.8	2.9	0.3	0.9	0.9
10	mdb241	3.6	3.0	3.1	2.2	1.0	1.0

Table 5 Value of energy of the segmented region

Dataset	Image	Background removal MSO	Background removal LOA	Pectoral removal MSO	Pectoral removal LOA	Region of interest MSO	Region of interest LOA
1	mdb001	358,235	318,891	305,973	362,818	121,046	121,046
2	mdb002	406,028	330,883	329,453	334,997	101,555	113,807
3	mdb006	261,689	309,753	357,397	324,584	100,470	106,019
4	mdb013	452,222	485,466	326,905	344,719	107,823	106,673
5	mdb023	258,544	271,327	342,787	388,832	101,683	115,697
6	mdb027	407,051	421,382	345,602	334,866	100,839	95,835
7	mdb031	481,935	378,553	396,321	372,020	111,242	121,160
8	mdb239	288,468	296,988	362,776	373,739	106,175	127,302
9	mdb240	468,136	391,520	376,344	379,284	132,627	139,637
10	mdb241	297,979	481,249	399,505	346,912	131,452	144,477

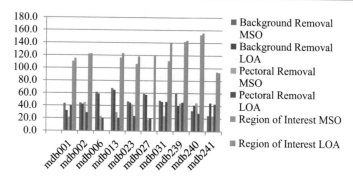

Fig. 12 Graphical representation—mean of segmented region

Table 6 Value of CPU time (sec) for the segmented region

Dataset	Image	Background removal MSO	Background removal LOA	Pectoral removal MSO	Pectoral removal LOA	Region of interest MSO	Region of interest LOA
1	mdb001	13.7	14.7	262.5	250.5	24.0	26.1
2	mdb002	42.2	44.3	280.3	247.5	21.9	25.9
3	mdb006	48.6	50.4	238.4	152.4	20.1	21.6
4	mdb013	49.0	50.7	216.0	257.3	23.4	23.9
5	mdb023	45.4	49.8	235.5	257.3	21.5	22.9
6	mdb027	45.8	46.1	246.9	249.2	20.2	21.6
7	mdb031	42.2	32.6	199.3	249.7	22.4	24.0
8	mdb239	13.0	29.1	220.7	264.1	25.5	28.5
9	mdb240	48.7	55.3	240.4	209.5	25.9	27.9
10	mdb241	31.0	13.8	230.7	259.0	22.8	23.8

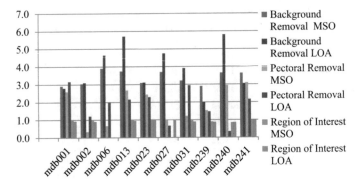

Fig. 13 Graphical representation—entropy of segmented region

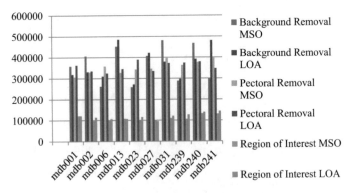

Fig. 14 Graphical representation—energy for the segmented region

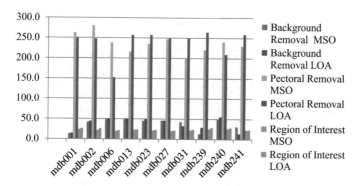

Fig. 15 Graphical representation—CPU (sec) for the segmented region

Table 7 Confusion matrix for MSO—SVM

Actual	Predicted	
	Abnormal	Normal
Abnormal	77	4
Normal	3	76

Table 8 Confusion matrix for LOA—SVM

Actual	Predicted	
	Abnormal	Normal
Abnormal	76	3
Normal	4	77

6.3   Classification Results on Mammogram Images

Classification of the mammogram image is implemented by making a hybrid of the algorithms with SVM, as MSO—SVM and LOA—SVM. Analysis of classification by MSO—SVM [34] and LOA—SVM is trained on three–fourth of the total 322 images in database mammogram and tested on rest of one-fourth count of the image database. The above ratio results to the training of 240 images and testing of 82 images in the database. The different types of mammogram image database will be classified into two major categories as: abnormal and normal. Performance analysis on classification can be performed by using Confusion Matrix and Receiver Operator (ROC) for the variations in the matrix. Table 7 gives the confusion matrix for the MSO—SVM classification results. The classification results of LOA—SVM algorithm is represented in Table 8.

Table 9 gives the value for Sensitivity—Specificity—Accuracy measures using MSO—SVM, and LOA—SVM. From these two tables, it can be observed that the hybrid classifiers result in at most similar accuracy rates. Hence, the classifiers can be the modified in future to obtain increased accuracy.

Table 9 Sensitivity—specificity—accuracy measures

Measure	MSO—SVM	LOA—SVM
Sensitivity	0.9500	0.9500
Specificity	0.9506	0.9625
Precision	0.9620	0.9620
Negative predictive value	0.0500	0.9506
False positive rate	0.0494	0.0375
False discovery rate	0.0375	0.0380
False negative rate	0.9563	0.0500
Accuracy	0.9565	0.9563
F1 score	0.9126	0.9560

7 Conclusion

Breast cancer is the most commonly seen cancer among women which needs at most attention for early treatment. This Chapter focuses on the segmentation and classification of the mammogram images. Among the four types of views, the MLO view is considered to scan as it covers the maximum breast tissue which includes the nipple and pectoral muscle region. The detail of the image has to be stored based on the compression techniques, and hence lossless compression technique is considered for it. Thus the images of PNG and PGM formats are used.

The images obtained from the modality cannot be directly used for performing any processing. Digitize the image before implementing any process on it. The mammogram image obtained from the modalities may have noises and hence remove it. Most of the digital mammography will be affected by Gaussian noise. Among the various types of filters, the Gaussian filter is implemented for noise reduction.

In this chapter, the segmentation process is applied for removing the background or the unwanted region, removing the pectoral muscle region and extracting the selected feature in the mammogram image. For these processes, the MIAS mammographic image database is considered. This database consists of 322 images of different abnormalities. Optimization algorithms MSO and LOA are implemented in all the segmentation processes. Performance analysis of these segmentation processes is performed based on the metrics such as mean, entropy, energy and CPU time. The results obtained after segmenting different regions in mammogram image using MSO and LOA are approximately similar to each other with minor differentiation. Hence, these algorithms can be modified in the future to obtain improved segmented results.

In the segmented images, some of the Haralick features such as contrast, correlation, energy and homogeneity values extracted using the optimization technique and the values are trained with the support vector machine making a hybrid system. The system is compared to making confusion matrix, sensitivity—specificity—accuracy ratio and plotting ROC curve. Confusion matrix and sensitivity—specificity—accuracy ratio are derived in two ways (1) training 80% of the image and testing with

20% of the image and (2) testing by choosing a random number of images from the database. From the classification results, it is observed that the proposed MSO—SVM and LOA—SVM have approximately similar. The algorithms can be modified to get still better classification results in the future. Some iterations considered for this testing are 120. All processes on mammogram images are implemented executed and tested using MATLAB.

References

1. Abu-Srhan Ala'a, Al Daoud Essam (2013) A hybrid algorithm using a genetic algorithm and cuckoo search algorithm to solve the travelling salesman problem and its application to multiple sequence alignment. Int J Adv Sci Technol 61:29–38
2. Amato KR, Onen D, Emel SL, May HC (2007) Comparison of foraging behaviour between howler monkeys, spider monkeys, and squirrel monkeys. Ecol Dartmouth Undergraduate J Sci 28–31
3. Anita CJ, Ramesh R, Vaishali D (2016) Bio-inspired computational algorithms for improved image steganalysis. Indian J Sci Technol 9(10)
4. Arpita D, Mahua B (2008) GA based neuro-fuzzy techniques for breast cancer identification. In: International machine vision and image processing conference
5. Banavar JR, Maritan A, Micheletti C, Trovato A (2002) Geometry and physics of proteins 47:315–322
6. Basturk B, Karaboga D (2006) An artificial bee colony (ABC) algorithm for numeric function optimization. In: IEEE swarm intelligence symposium, Indianapolis, Indiana, USA
7. Birattari M, Di Caro G, Dorigo M (2002) Toward the formal foundation of ant programming. In: Proceedings of third international workshop, ANTS 2002, Belgium, pp 188–201
8. Blum C (2004), Theoretical and practical aspects of ant colony optimization. Dissertations in Artificial Intelligence, Akademische Verlagsgesellschaft Aka GmbH, Berlin, Germany
9. Borg IY, Smith DK (1969) Calculated X-ray powder patterns for silicate minerals. Geological Society of America Memoirs
10. Ceci G, Mucherino A, D'Apuzzo M, di Serafino D, Costantini S, Facchiano A, Colonna G (2007) Computational methods for protein fold prediction: an ab-initio topological approach, data mining in biomedicine. In: Optimization and its applications, vol 7
11. Chandramouli K (2007) Particle swarm optimization and self-organizing maps based image classifier. In: Proceedings of 2nd international workshop semantic media adaptation and personalization, pp 225–228
12. Deepa S, Bharathi VS (2015) Analysing the discriminative capability of texture features in mammogram image classification with adaptive ROI extraction. Int J Appl Eng Res (2015)
13. Desai SD, Megha G, Avinash B, Sudhanva K, Rasiya S, Linganagouda K (2013) Detection of microcalcification in digital mammograms by improved-MMGW segmentation algorithm
14. Dorigo M, Di Caro G (1999) New ideas in optimization. McGraw Hill, London, UK
15. Dorigo M, Di Caro G, Gambardella LM (1999) Ant algorithms for discrete optimization. Artif Life 5(2):137–172
16. Eberhart RC, Shi Y (2001) Particle swarm optimization: developments, applications and resources. In: Proceedings of congress on evolutionary computation
17. Epitropakis MG, Plagianakos VP, Vrahatis MN (2012) Evolving cognitive and social experience in particle swarm optimization through differential evolution: a hybrid approach. Inf Sci Elsevier 216:50–92
18. Fahd M, Mohiy H, Kamel M, Khalid A (2012) A new image segmentation method based on particle swarm optimization. Int J Inf Technol 9(5):487–493

19. Felipe T, Ricardo JB, Paulo G, Marco M (2017) Efficient exploitation of the Xeon Phi architecture for the ant colony optimization (ACO) metaheuristic. J Supercomputing
20. Gopi Raju N, Nageswara Rao PA (2013) Particle swarm optimization methods for image segmentation applied in mammography. Int J Eng Res Appl 3(6):1572–1579
21. Hamada A-ARH (2011) Soft computing in medical diagnostic applications: a short review. 2011 national postgraduate conference, 09/2011
22. Helena RL, Martin OC, Stutzle T (2003) Iterated local search. In: Handbook of metaheuristics, international series in operations research and management Sciences. Kluwer Academic Publishers
23. Helena RL (2010) Iterated local search: framework and applications. International series in operations research and management science
24. Holland JH (1975) Adaptation in natural and artificial systems. University of Michigan Press, Ann Arbor
25. Hui W, Yong L, Sanyou Z, Hui L, Changhe L (2007) Opposition based particle swarm algorithm with cauchy mutation. In: Proceedings of IEEE congress on evolutionary computation, Singapore, pp 4750–4756
26. Humberto O, Osslan V, Vianey C, Efren G (2009) A hybrid system based on a filter bank and a successive approximations threshold for microcalcifications detection. J Comput 4(8):691–696
27. Humberto F, de Valdes LFI (2013) Chapter 8 uncommon clinical manifestations of cysticercosis. InTech
28. HIPR2 Explore with Java (2000) Image processing learning resources. Available from: http://homepages.inf.ed.ac.uk/rbf/HIPR2/mean.htm#guidelines, October 2000
29. Imaginis, How Mammography is performed: Imaging and Positioning 2009, The Women's Health Resource, Available from: http://www.imaginis.com/mammography/how-mammography-is-performed-imaging-and-positioning-2, 23 Nov 2009
30. Jawad N (2011) The application of image processing and machine learning techniques for detection and classification of cancerous tissues in digital mammograms. Master of Computer Science Dissertation, University of Malaya, Kuala Lumpur
31. Jingran W, Yixin Y, Yuan Z, Wenpeng L (2010) Discrete monkey algorithm and its application in transmission network expansion planning. IEEE PES GeneralMeeting
32. Jothi G, Inbarani HH, Azar AT (2013) Hybrid tolerance rough set. Int J Fuzzy Syst Appl 3:15–30
33. Kanimozhi Suguna S, Maheswari SU (2014) Removal of pectoral muscle region in mammogram image by metaheuristic algorithm—monkey search optimization (MSO). Asian J Inf Technol 13(4):252–259
34. Kanimozhi Suguna S, Maheswari SU (2014) Classification of feature extracted, selected and segmented mammogram image using hybrid algorithm—monkey search optimization (MSO) and support vector machine (SVM). Res J Appl Sci 9(2):110–118
35. Kanimozhi Suguna S, Maheswari SU (2014) Performance analysis of feature extraction and selection of region of interest by segmentation in mammogram images between the existing meta-heuristic algorithms and monkey search optimization (MSO). WSEAS Trans Inf Sci Appl 11:72–88
36. Karnan M, Thangavel K, Ezhilarasu P Ant colony optimization and a new particle swarm optimization algorithm for classification of microcalcifications in mammograms
37. Karnan M, Thangavel K (2008) Feature extraction and classification microcalcifications in mammograms. Int J Appl Comput 1(1):17–32
38. Kennedy J, Eberhart RC (1995) Particle swarm optimization. In: Proceeding of IEEE conference on neural networks IV, New Jersey
39. Kirkpatrick S, Gelatt CD Jr, Vecchi MP (1983) Optimization by simulated annealing. Science 220:671–680
40. Kumar RP, Palani S (2012) An optimal energy and power model for dynamic voltage scaled multiprocessor systems. Int J Bus Inf Syst 11:461–477
41. Lavan O, Dargush GF (2009) Multi-objective evolutionary seismic design with passive energy dissipation systems. J Earthq Eng 13:758–790

42. Lin K-C, Hung JC, Huang L-D (2014) A novel feature selection method for support vector machines using a lion's algorithm. Lecture notes in electrical engineering
43. Lv G (2005) Fault diagnosis of power transformer based on multi-layer SVM classifier. Electr Power Syst Res
44. Maziar Y, Fariborz J (2016) Lion optimization algorithm (LOA): a nature-inspired metaheuristic algorithm. J Comput Design Eng Elsevier 6:24–36
45. Mithun B, Ranjan P (2012) Character recognition using dynamic windows. Int J Comput Appl 41(15):47–52
46. Moise G, Vladoiu M, Constantinescu Z (2014) MASECO: a multi-agent system for evaluation and classification of OERs and OCW based on quality criteria. Studies in computational intelligence, 2014
47. Mucherino A, Seref O Monkey search: a novel metaheuristic search for global optimization
48. Mucherino A, Seref O, Pardalos, PM (2013) Simulating protein conformations through global optimization. Optimization and Control (math.OC). arXiv:0811.3094 [math.OC]
49. PINK Indian Statistics 2014, Breast Cancer India. Available from: http://www.breastcancerindia.net/bc/statistics/stat_global.htm. 27 Jan 2014
50. Singh P, Dutta M, Aggarwal N (2017) A review of task scheduling based on meta-heuristics approach in cloud computing. Knowl Inf Syst 52:1–51
51. Rajakumar BR (2012) The lion's algorithm: a new nature—inspired search algorithm. In: 2nd international conference on communication, computing and security, SciVerse ScienceDirect Procedia Technology, vol 6, pp 126–135
52. Rafel CG, Richard EW (2002) Digital image processing. Pearson Education, Singapore
53. Ramirez-Villegas JF, Lam-Espinosa E, Ramirez-Moreno DF microcalcification detection in mammograms using difference of gaussians filters and a hybrid feedforward—Kohonen neural network
54. Robert MH, Shanmugam K, Dinstein IH (1973) Textural features for image classification. IEEE Trans Syst Man Cybern 3(6):610–621
55. Ronaldo DARB (2007) Analysis of mammogram using self-organizing neural networks based on spatial isomorphism
56. Roselin R, Thangavel K, Velayutham C (2011) Fuzzy-rough feature selection for mammogram classification. J Electro Sci Technol 9(2):124–132
57. Sakr NA, ELdesouky AI, Arafat H (2016) An efficient fast-response content-based image retrieval framework for big data. Comput Electr Eng 54:522–538
58. Sara M (2005) Algorithm with learning and behaviour ants specialized for scheduling cars. Master of Computer Science, the University of Quebec at Montreal, Canada
59. Satish C, Vijaya P, Praveen D Fractional lion algorithm—an optimization algorithm for data clustering. J Comput Sci 12(7):323–340
60. Serra P, Stanton AF, Kais S (1997) Pivot method for global optimization. Phys Rev E 55(1)
61. Shayesteh MG, Shabanifard M, Akhaee MA (2013) Forensic detection of image manipulation using the Zernike moments and pixel-pair histogram. IET Image Process 7:817–828
62. Shanmugavadivu P, Sivakumar V (2013) Segmentation of pectoral muscle in mammograms using the fractal method. In: 2013 international conference on computer communication and informatics
63. Simonthomas S, Thulasi N (2013) Automated diagnosis of glaucoma using Haralick texture features. IOSR J Comput Eng 15(1):12–17
64. Sonali B, Yash B, Anikta P (2012) Removal of pectoral muscle in mammograms using statistical parameters. Int J Comput Appl 43(6):1–4
65. Sreedevi S, Sherly E (2015) A novel approach for removal of pectoral muscles in digital mammogram. Procedia Comput Sci 46:1724–1731
66. Srivastava S, Sharma N, Singh SK, Srivastava R (2013) Design, analysis and classifier evaluation for a CAD tool for breast cancer detection from digital mammograms. Int J Biomed Eng Technol 13:270–300
67. Subash CB, Thangavel K, Daniel DAP (2013) Automatic mammogram image breast region extraction and removal of pectoral muscle. Int J Sci Eng Res 4(5):1722–1729

68. Thangavel K, Karnan M, Pethalakshmi A Performance analysis of rough reduct algorithms in mammogram. GVIP
69. Thangavel K, Karnan M, Sivakumar R, Kaja Mohideen A (2005) Ant colony system for segmentation and classification of microcalcification in mammograms. ICGST
70. Thangavel K, Roselin R (2012) Fuzzy-rough feature selection with Π—membership function for mammogram classification. Int J Comput Sci Issues 9(3):361–370
71. Thangavel K, Manvalan R, Aroquiaraj IL (2009) Removal of speckle-noise from ultrasound medical image based on special filters: comparative study. ICGST-GVIP J 9(3):25–32
72. Thukaram P, Saritha SJ (2013) Image edge detection using improved ant colony optimization. Int J Res Comput Commun Technol 2(11):1256–1260
73. Tuan AP (2010) Optimization of texture feature extraction algorithm. M.Sc. Thesis, Delft University of Technology, The Netherlands
74. Vladimir VN (1998) Statistical learning theory. Wiley, New York
75. Veerakumar K, Ravichandran CG (2013) Applying ant colony optimization algorithms and variants for lung nodule detection. Pensee J 75(11):283–301
76. Velayutham C (2011) Unsupervised feature selection in digital mammogram image using rough set based entropy measure. In: 2011 world congress on information and communication technologies, 12/2011
77. Vijaya Kumar S, Lazarus MN, Nagaraju C (2010) A novel method for the detection of microcalcifications based on multi-scale morphological gradient watershed segmentation. Int J Eng Sci Technol 2(7):2616–2622
78. Wang J, Yu Y (2010) Discrete monkey algorithm and its application in transmission network expansion planning. IEEE, 978-1-4244-6551-4
79. Wattanit H (2012) Book of Rorasa. Available from: http://rorasa.wordpress.com/2012/05/13/l 0-norm-l1-norm-l2-norm-l-infinity-norm/. 13 May 2012
80. Weimin B, Wei S, Simin Q (2014) Flow updating in real-time flood forecasting based on runoff correction by a dynamic system response curve. J Hydrol Eng 19:747–756
81. Xin C, Yongquan Z, Qifang L (2014) A hybrid monkey search algorithm for clustering analysis. Hindawi Publishing Corporation
82. Yan T, Wang L, Wang J (2014) Method to enhance degraded image in dust environment. J Softw 9:2672–2677
83. Yeung CW (2009) An improved particle swarm optimization algorithm and its applications. M.Phil., Thesis, The Hong Kong Polytechnic University
84. Zhao R, Tang W (2008) Monkey algorithm for global numerical optimization. J Uncertain Syst 2(3):165–176

Web References

85. www.ijera.com
86. www.ijimai.org
87. www.imaginis.com
88. www.info.trinityhealth.com
89. www.reachinformation.com
90. www.ruc.udc.es
91. www.ukessays.com

Firefly Algorithm Applied to the Estimation of the Parameters of a Photovoltaic Panel Model

Ricardo Augusto Pereira Franco, Gilberto Lopes Filho
and Flávio Henrique Teles Vieira

Abstract The computational simulation of photovoltaic panels is essential to design photovoltaic systems. However, the datasheet of the panels does not provide all information necessary for the computational simulation. Therefore, it becomes necessary to develop methods to estimate the unknown model parameters of the panels. This work addresses the application of the firefly algorithm to estimate model parameters of photovoltaic systems. The objective function of the firefly algorithm corresponds to the minimization of the mean square error between the I–V curve provided by the datasheet and the I–V curve generated by the estimated parameters. The firefly algorithm provides the smallest error compared to some methods present in the literature.

Keywords Firefly algorithm · Parameters estimation · Photovoltaic panel
Optimization techniques

1 Introduction

The Firefly Algorithm (FA) is an optimization algorithm that is based on the firefly behavior. Each firefly has a characteristic brightness that can attract other fireflies. This brightness decreases in relation to distance. Thus, a set of fireflies in a given environment will group according to the brightness of each one and the distance between them. These features can be modeled to solve function minimization or maximization problems such that brightness represents an objective function.

The modeling of photovoltaic panels generally consists of obtaining an equivalent electrical circuit that represents their behavior. However, there are some undetermined parameters in the models. This problem of undetermined parameters is multivariable, nonlinear, non-convex, with several local minimums and can be rep-

R. A. P. Franco (✉) · G. L. Filho · F. H. T. Vieira
School of Electrical, Mechanical and Computer Engineering,
Federal University of Goiás, Goiania, Goiás, Brazil
e-mail: ricardofranco3@gmail.com

© Springer International Publishing AG, part of Springer Nature 2019
S. K. Shandilya et al. (eds.), *Advances in Nature-Inspired Computing
and Applications*, EAI/Springer Innovations in Communication and Computing,
https://doi.org/10.1007/978-3-319-96451-5_5

resented as a minimization problem. Thus, it is necessary to use techniques to perform estimation of the parameters of photovoltaic panels so that they represent the panel in the best possible way.

This chapter is organized as follows: initially, the firefly algorithm is presented; the problem of the estimation of parameters of photovoltaic panels is discussed; and finally, the firefly algorithm developed to solve the problem of estimation of parameters is presented, showing the obtained results and comparing it to other methods proposed in the literature.

2 Firefly Algorithm

The firefly algorithm was developed by Yang [21] and it is based on the observation of light from blinking fireflies. In [14], it is said that the signs of bioluminescence, which is a biochemical process to produce light from fireflies, have the following purposes: to serve as a mating ritual; method to attract prey; or even as a warning sign of nearby predators. Therefore, the light signs of fireflies are very important for their survival. Based on the behavior of the fireflies [21] modeled, an optimization heuristic algorithm is called firefly algorithm.

There are several types of heuristic algorithms, which it is estimated to be more than 40 types [19]. Many of them are inspired by the natural behaviors of some animals, such as swarms of particles, colony of ants, bees, beetles, and birds [16]. These algorithms have as the main idea to use a social behavior observed in some a species and from it, with some simplifications, to elaborate mathematical codes to solve engineering problems [19].

An optimization problem refers to the maximization or minimization of an objective function by assigning appropriate values to the variables that are in this function. This type of problem is not present only in the scientific community, but also in the daily routine. For example, when a person wants to go from some place to another one, there are several possible paths and you have to make a decision on which route to choose. The decision can be based on the time spent on the trip, the distance to be traveled or the fuel that will be consumed. This kind of problem with few possible solutions can be solved by analyzing each solution separately. However, in more complex problems, there are not always a finite or small number of solutions. Therefore, different methods are proposed to find the solution based on the behavior of the problem.

Many conventional algorithms are deterministic and, among them, there are some based on the information of the gradient function. An example, it can be cited the Newton–Raphson method which has good performance for well-behaved functions [6]. However, for highly nonlinear, non-convex, non-differentiable, and non-smooth problems, these deterministic methods based on gradient function present convergence problems and they are often stuck in optimum locations.

To solve this problem, it is necessary to use algorithms that are not based on the gradient function such as heuristic (or stochastic) algorithms [21, 22]. These algorithms make some random searches for the best solution. Usually, this search is guided and it is ensured that the method converges to a good solution. Even if these algorithms do not provide an optimal solution, they can approach the global optimum and provide a good solution. Another positive point is that they are less affected by the behavior of each problem, making them more robust for various applications [7, 21].

As already mentioned, real fireflies are flying insects that shine using biolumines-cence, presumably to attract partners. Each firefly can shine with a different intensity. In firefly algorithm, fireflies that are better, that is, which have a smaller error, emanate a light with greater intensity. The better the representation of its objective function in relation to the problem to be optimized, the closer this firefly is to the global minimum and the more intense its luminosity. In this way, other fireflies will also be attracted to brighter fireflies (close to the global minimum) and will distance themselves from fireflies with less luminous intensity (farther from the global minimum). Figure 1 shows the flowchart of firefly algorithm.

It is known that the light intensity decreases with the square of the distance [22]. Therefore, fireflies, although attracted by light, have a limited view of the biolumi-nescence of other fireflies. Equation 1 represents the luminous intensity $G(r)$ as a function of a distance r.

$$G(r) = \frac{G_s}{r^2} \tag{1}$$

where

G_s is the light intensity of the source;
r is the distance from the source.

Considering a light absorption coefficient γ for a fixed distance r, we can write that the luminous intensity G according to Eq. 2:

$$G = G_0 e^{-\gamma r} \tag{2}$$

where

G_0 is the original light source.

The expression G_s/r^2 has a singularity at $r = 0$. Then, combining this expression that considers the effect of the light scattering as a function of the inverse of the square of the distance, with the expression of the luminosity absorption, can be approximated by a Gaussian form according to the following equation:

$$G(r) = G_0 e^{-\gamma r^2} \tag{3}$$

The attractiveness of a firefly x_1 by other fireflies $v_1, v_2, v_3, \ldots, v_n$ is propor-tional to the luminous intensity of fireflies $v_1, v_2, v_3, \ldots, v_n$ that the firefly x_1 is able to see. We can define the attractiveness β of a firefly by another as

$$\beta = \beta_0 e^{-\gamma r^2} \tag{4}$$

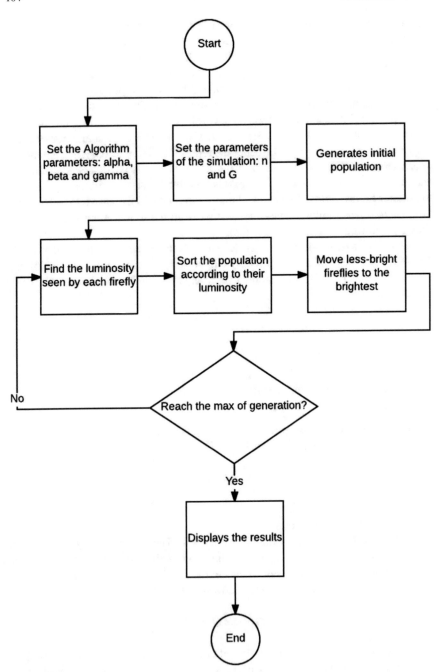

Fig. 1 Flowchart of firefly algorithm

where

β_0 is the attractiveness at $r = 0$.

Depending on the complexity of the problem, we can simplify the calculation of β with the following expression:

$$\beta = \frac{\beta_0}{1 + \gamma r^2} \tag{5}$$

We can still define Γ as being the characteristic distance, according to Eq. 6:

$$\Gamma = \frac{1}{\sqrt{\gamma}} \tag{6}$$

where the expression of attractiveness is $\beta_0 e^{-1}$ or $\beta_0/2$, depending on the considered formulation for the attractiveness.

Generally, we can say that the expression of attractiveness can be any continuous and monotonous function, so that it obeys the following:

$$\beta = \beta_0 e^{-\gamma r^m} \tag{7}$$

where

m is any real number greater than or equal to 1.

To calculate the distance between two fireflies any v_i and v_j, we use the Euclidean distance in a three-dimensional space, given by

$$r_{ij} = \sqrt{\left(x_i - x_j\right)^2 + \left(y_i - y_j\right)^2 + \left(z_i - z_j\right)^2} \tag{8}$$

where $p_i = (x_i, y_i, z_i)$ and $p_j = (x_j, y_j, z_j)$ are the rectangular coordinates of the fireflies v_i and v_j, respectively.

The distance r_{ij}, depending on the problem, may be given by a non-Euclidean equation. Theoretically, we can define r_{ij} in any n-dimensional hyperspace.

The movement of a firefly i to nearest a firefly j (which has greater luminosity) is defined by

$$p_i = p_i + \beta_0 e^{-\gamma r_{ij}^2}\left(p_i - p_j\right) + \alpha \epsilon_i \tag{9}$$

The second term of Eq. 9 is relative to the attractiveness. The third term (Eq. 10) refers to a randomness added to the move, such that α is a parameter of randomness (weight), and ϵ_i is a vector of random numbers obtained from any probability distribution function, as a Gaussian (normal) or uniform [5]. If the parameter α is equal to zero, the motion has no randomness. On the other hand, if β_0 equals zero, all movement is random.

The choice of firefly algorithm parameters depends on the problem to be optimized. However, there are some suggestions in the literature that apply to most cases. It is suggested that $\beta_0 = 1$, $\alpha \in [0, 1]$, and $\gamma \in [0.1, 10]$.

It is interesting to note that in the case $\gamma \to 0$, the attractiveness β tends to become constant with a value equal to β_0. This would be equivalent to saying that there is no absorption of luminosity by light, that is, luminosity does not decrease in all space. Therefore, the light of a firefly can be seen throughout the environment and a great location can be easily found. On the other hand, if $\gamma \to \infty$ then $\beta \to 0$. This is the equivalent of saying that fireflies have no attractiveness to each other, and their movements are only random.

Some improvements can be added to the firefly algorithm so that it converges faster. This is done by decreasing the randomness of the drive over the generations. This means that as we move to the global optimum of the solution, the value α is decremented. An example is given by

$$\alpha = a_\infty + (\alpha_0 - a_\infty)e^{-g} \tag{10}$$

where

g is the generation number;
a_∞ is the parameter of randomness in the last generation;
α_0 is the parameter of randomness in the first generation.

Other formulations for the decreasing α can be made, as in

$$\alpha = \alpha_0 \theta^t \tag{11}$$

where
θ is a randomization reduction constant $(0 < \theta \leq 1)$.

It can be said that the objective function to be optimized is associated with the flashing light of the fireflies. For a simplification of the model, three rules were adopted: (a) it is assumed that all individuals are attracted by all, so that there is no difference as to sexuality; (b) the greater the brightness of a firefly, the greater its attractiveness, in such a way that it decreases with increasing distance, due to the absorption of light through the medium; (c) the brightness of a firefly is affected by its evaluation function (objective function), that is, the better it evaluates, the brighter it will be. The lower bright fireflies move (are attracted) to those of higher brightness.

The firefly algorithm can be synthesized in a pseudocode that describes its steps. The pseudocode is shown in Table 1.

3 Photovoltaic Solar Energy

The growing demand for electric power shows the dependence of society on this type of energy. The electrical sector has been developing new systems to increase reliability in the electrical network, improving its functionalities and its infrastructure. In addition to the fact that crises in the main means of production of the energy matrix bring problems for the whole society. There is also concern about the environmental

Table 1 Pseudocode of the firefly algorithm

Set the algorithm parameters: α, β and γ
Set the simulation parameters: population size (n) and maximum iterations (G)
Generate the initial population randomly
for iteration = 1 : G
       Calculate the brightness of each firefly
       Sort the population according to brightness
       for i = 1 : n -1
           for j = i+1 : n
               if Evaluation (j) > Evaluation (i)
                    Move the firefly i in the direction of the firefly j
               end if
           end for j
       end for i
end for iteration
Show the best solution

impact on power generation, in which the elevation of the Earth temperature is one of the topics. Therefore, it can be noted that the generation of energy through sunlight is a measure that will benefit all parts of the electrical system, that is, the utility, the consumer, and the environment.

Solar energy is a renewable energy source. This generation is carried out through the capture of the solar radiation. It is made the conversion into electric energy through the photovoltaic effect in the photovoltaic cell. In this effect, there is the excitation of electrons, due to the characteristics of some semiconductor materials in the presence of sunlight, taking the photovoltaic effect and generating the electric current of this system. Photovoltaic cells form a photovoltaic panel.

3.1 Equivalent Electric Circuit

A photovoltaic cell or photovoltaic panel can be modeled by an electric circuit that behaves similar to the cell/panel. This modeling is essential, because photovoltaic systems projects are expensive and it is important to perform an accurate and reliable simulation of the system to be installed. Therefore, it is interesting for photovoltaic systems to develop models that simulate the real characteristics of the panels [15]. There are several models in the literature that represent a photovoltaic panel with different complexities. The complexity level of the model will define which mathematical expressions are most suitable for the model [1].

There are four models in the literature that define the equivalent electrical circuit of a photovoltaic panel. The models differ in relation to their complexity, starting from a simpler representation to a more complex representation. The models are Ideal Model (IM), Single Diode Model Simplified (SDMS), Single Diode Model (SDM), and Double Diode Model (DDM).

These models were developed to represent mathematically the photovoltaic panel and to generate computationally the $I–V$ characteristic curve (also called characteristic curve or $I–V$ curve). An example of this curve is shown in Fig. 2.

The equivalent circuit of the ideal model consists of a current source, which generates the photogenerated current, and a diode, which generates the current of the diffusion diode. The current generated by the photovoltaic panel is the difference between the current photogenerated by the current of the diffusion diode. The IM is shown in Fig. 3.

The equation describing the IM is presented by

$$I = I_{irr} - I_d \tag{12}$$

where

I is the current generated by the photovoltaic panel (A);
I_{irr} is the photogenerated current (A);
I_d is the current of the diffusion diode (A).

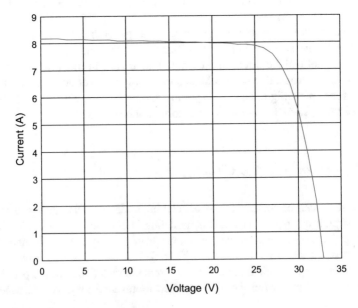

Fig. 2 Example of an IV characteristic curve

Fig. 3 Equivalent circuit of
the ideal model

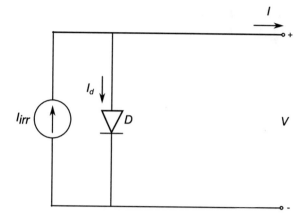

The diffusion diode current is defined by

$$I_\mathrm{d} = I_0\left[e^{\frac{V}{N_s n V_t}} - 1\right] \tag{13}$$

where

V is the output voltage (V);
I_0 is the reverse saturation current of the diffusion diode (A);
N_s is the number of cells in series that make up the photovoltaic module;
n is the diode ideality factor;
V_t is the thermal voltage (V).

The thermal voltage is calculated by Eq. 14:

$$V_t = \frac{kT_c}{q} \tag{14}$$

where

k is the Boltzmann constant (1.380×10^{-23} J/K);
q is the elementary charge of the electron (1.609×10^{-19} C);
T_c is the temperature of the module (K).

Substituting Eq. 13 into Eq. 12, we have that the current generated by the photovoltaic panel is given by

$$I = I_\mathrm{irr} - I_0\left[e^{\left(\frac{V}{N_s n V_t}\right)} - 1\right] \tag{15}$$

Some parameters of Eq. 15 should be estimated because they are not known or provided by the panel manufacturer. The parameters to be estimated in the IM model

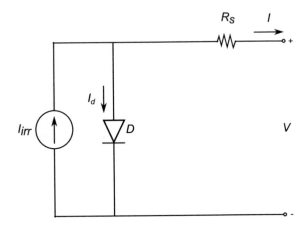

Fig. 4 Equivalent circuit of the single diode model simplified

are photocurrent current (I_{irr}), diode reverse saturation current (I_0), and diode (n) ideality factor.

The second model to be presented is the SDMS. The SDMS has a series resistance, which represents the losses of the metal contacts of the photovoltaic module. Figure 4 shows the SDMS.

The equation of the SDMS is similar to Eq. 15 with the adding the influence of the resistance in series. This resistance is related to the slope of the characteristic curve of the panel at the point of the characteristic curve where the voltage is maximum, that is, in the open-circuit voltage ($I = 0$ and $V = V_{ca}$). The current generated by the photovoltaic module is expressed by Eq. 23.

$$I = I_{irr} - I_0 \left[e^{\left(\frac{V + I R_s}{N_s n V_t} \right)} - 1 \right] \tag{16}$$

where

R_s is the series resistance of the photovoltaic panel (Ω).

The parameters to be estimated in the SDMS model are the same of those of the IM model and the resistance in series, that is, the parameters to be estimated are photocurrent current (I_{irr}), diode reverse saturation current (I_0), factor diode ideality (n), and series resistance (R_s).

The SDM is one of the most used models in the literature [18]. This model presents a shunt resistance, which represents the losses related to the parasitic currents that circulate in the photovoltaic panel. Figure 5 shows the equivalent electric circuit for the SDM.

The equation describing the equivalent circuit of the SDM model is given:

$$I = I_{irr} - I_0 \left[e^{\left(\frac{V + I R_s}{N_s n V_t} \right)} - 1 \right] - \frac{V + I R_s}{R_{sh}} \tag{17}$$

Fig. 5 Equivalent circuit of
the single diode model

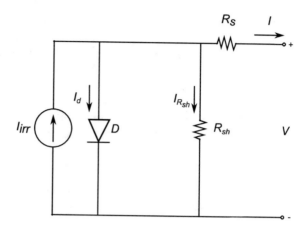

where

R_{sh} is the shunt resistance of the photovoltaic panel (Ω).

The slope of the characteristic curve at the point where the voltage tends to zero ($I = I_{sc}$ and $V = 0$) is related to the value of the shunt resistance.

In this model, there are five parameters to be estimated as they are not provided by the PV panel manufacturer. The parameters are photocurrent current (I_{irr}), diode reverse saturation current (I_0), diode ideality factor (n), series resistance (R_s), and shunt resistance (R_{sh}).

Finally, the last model is the DDM. The DDM has the same components as the SDM model with the addition of another diode. This model has the important characteristic of modeling the carrier recombination losses in the depletion region [18]. The DDM is shown in Fig. 6.

Fig. 6 Equivalent circuit of
the double diode model

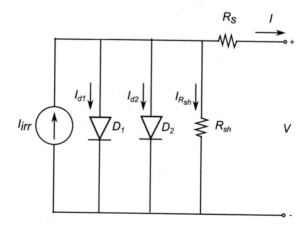

The equivalent circuit of the DDM is expressed by

$$I = I_{irr} - I_{01}\left[e^{\left(\frac{V+IR_s}{N_s n_1 V_t}\right)} - 1\right] - I_{02}\left[e^{\left(\frac{V+IR_s}{N_s n_2 V_t}\right)} - 1\right] - \frac{V + I R_s}{R_{sh}} \tag{18}$$

where

I_{01} is the reverse saturation current of the diffusion diode (A);
n_1 is the ideality factor of the diffusion diode;
I_{02} is the reverse saturation current of the recombination diode (A);
n_2 is the ideality factor of the recombination diode.

This model has six parameters to be estimated, since they are not provided by the PV panel manufacturer. The parameters are photocurrent current (I_{irr}), reverse saturation current of the diffusion diode (I_{01}), the diffusion diode ideality factor (n_1), reverse saturation current of the recombination diode (I_{02}), the recombination diode ideality factor (n_2), the series resistance (R_s), and the shunt resistance (R_{sh}).

The model chosen for parameter estimation was the SDM. Although the DDM is more accurate, the complexity to obtain all parameters increases and the results do not improve as complexity increases. On the other hand, the mathematical equation of the SDM has high complexity while providing adequate representation of the photovoltaic panel.

4   Firefly Algorithm Applied to Parameter Estimation of a PV Model

A photovoltaic system consists of an association of photovoltaic panels connected in series and/or parallel. The behavior of the panel depends on the materials used for its production. The producers provide some information about the photovoltaic panel in its datasheet. However, there are some model parameters that are not provided by the datasheet. Thus, the values of these parameters are unknown and, consequently, must be determined or estimated for the modeling and simulation of the photovoltaic panel.

Several studies address the estimation of the values of these parameters. There are works that use analytical methods, numerical methods, optimization techniques, or combinations between them. Some studies that use the analytical approach are [8, 17]; some studies that use the numerical approach are [2, 3, 11]; in [15] analytical and numerical techniques are combined for parameter estimation; and other works use optimization techniques such as particle swarm [9, 23], genetic algorithms [11], and differential evolution [4, 8, 10]. In [12] and [13], analytical methods are used to reduce the search space and the complexity of this problem, together with optimization techniques to estimate the parameters of the photovoltaic panel. Since the estimation of the parameters is a nonlinear, non-convex, multivariate, multimodal problem with several good locations [12, 15], new techniques may be more satisfactory outcome of the used until now.

To perform the estimation of the parameters of the photovoltaic panels with the firefly algorithm, we considered the standard test condition (STC). The STC is defined for tests of photovoltaic modules with irradiance of 1000 W/m², temperature of 298 K, wind speed equal to 1.5 m/s, and the standard spectral distribution for the air mass of 1.5.

Recalling that the SDM was the model chosen to represent the equivalent circuit of the photovoltaic panel, Fig. 7 and Eq. 19 are shown again below:

$$I = I_{\mathrm{irr}} - I_0 \left[e^{\left(\frac{V + I R_s}{N_s n V_t} \right)} - 1 \right] - \frac{V + I R_s}{R_{\mathrm{sh}}} \tag{19}$$

In this chapter, the firefly algorithm is implemented with the purpose to estimate the parameters n and R_s of the circuit shown in Fig. 7. To this end, the equations proposed by [12] are used to estimate the variables I_0, I_{irr}, and G_{sh}, where $G_{\mathrm{sh}} = (1/R_{\mathrm{sh}})$. These equations are in function of n and R_s. For each iteration of the firefly algorithm, the two parameters (n and R_s) are estimated, and then the other three parameters (I_0, I_{irr}, and G_{sh}) are calculated.

The authors of [12] propose to obtain analytical values to estimate some parameters by manipulating Eq. 19 at the open-circuit point ($I = 0$, $V = V_{\mathrm{ca}}$), at the short-circuit point ($I = I_{\mathrm{sc}}$, $V = 0$), at the maximum power point ($I = I_{\mathrm{mp}}$, $V = V_{\mathrm{mp}}$) and imposing the derivative of the power, with respect to the voltage, equal to zero at the maximum power point $\left(\frac{dP}{dV} \big|_{V = V_{\mathrm{mp}}} = 0 \right)$. Generating equations for three of the five parameters to be estimated. The values of the photogenerated current (I_{irr}), the diode saturation current (I_0) and the parallel conductance (G_{sh}) are, respectively, given by

Fig. 7 Equivalent circuit of the single diode model

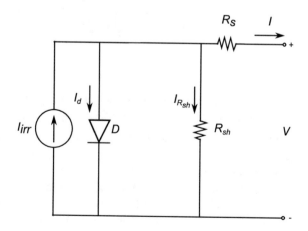

$$I_{\text{irr}} = \frac{I_{sc}V_{oc}\left(e^{\left(\frac{V_{mp}+R_sI_{mp}}{N_snV_t}\right)}-1\right) + I_{sc}V_{mp}\left(1-e^{\left(\frac{V_{oc}}{N_snV_t}\right)}\right) + I_{mp}V_{oc}\left(1-e^{\left(\frac{R_sI_{sc}}{N_snV_t}\right)}\right)}{A_1e^{\left(\frac{R_sI_{sc}}{N_snV_t}\right)} + A_2e^{\left(\frac{V_{mp}+R_sI_{mp}}{N_snV_t}\right)} + A_3e^{\left(\frac{V_{oc}}{N_snV_t}\right)}}$$

$$(20)$$

$$I_0 = \frac{V_{oc}(I_{sc}-I_{mp}) - V_{mp}I_{sc}}{A_1e^{\left(\frac{R_sI_{sc}}{N_snV_t}\right)} + A_2e^{\left(\frac{V_{mp}+R_sI_{mp}}{N_snV_t}\right)} + A_3e^{\left(\frac{V_{oc}}{N_snV_t}\right)}} \tag{21}$$

$$G_{sh} = \frac{(I_{mp}-I_{sc})e^{\left(\frac{V_{oc}}{N_snV_t}\right)} + e^{\left(\frac{V_{mp}+R_sI_{mp}}{N_snV_t}\right)} - (I_{mp})e^{\left(\frac{R_sI_{sc}}{N_snV_t}\right)}}{A_1e^{\left(\frac{R_sI_{sc}}{N_snV_t}\right)} + A_2e^{\left(\frac{V_{mp}+R_sI_{mp}}{N_snV_t}\right)} + A_3e^{\left(\frac{V_{oc}}{N_snV_t}\right)}} \tag{22}$$

where $A_1 = (V_{mp} + R_sI_{mp} - V_{oc})$, $A_2 = (V_{oc} - R_sI_{sc})$, and $A_3 = (R_sI_{sc} - R_sI_{mp} - V_{mp})$.

Equations (20), (21), and (22) depend on two parameters of the SDM model (only n and R_s). It reduces the number of parameters to be estimated and, consequently, the computational complexity of the problem. Therefore, optimization techniques based on heuristics, such as the genetic algorithm and the firefly algorithm, can be used to estimate the remaining parameters.

The fitness function used by the firefly algorithm was the Root Mean Square Error (RMSE), defined in Eq. 23. The RMSE was calculated in relation to the sampled points of current and voltage of the I–V curve (available in the datasheet of the photovoltaic panel) and the I–V curve made with the estimated parameters. That is, for each value of parameter n and parameter R_s an I–V curve is plotted, and the RMSE is calculated between this curve and the I–V curve sampled in datasheet. The parameters will be more adequate when the RMSE was smaller and thus, the obtained curve will be closer to the real curve.

$$\text{RMSE} = \sqrt{\frac{1}{M}\sum_{m=1}^{M}(I_s(V) - I_e(V))^2} \tag{23}$$

where

$I_s(V)$ is the current sampled by the datasheet;

$I_e(V)$ is the estimated current through the firefly algorithm;

M is the number of points sampled in I–V curve of the datasheet.

The stopping criterion of this algorithm was specified as the number of generations. In this implementation, the maximum number of generations was adopted equal to 100. Table 2 presents the pseudocode of the firefly algorithm developed to estimate the parameters of the photovoltaic panel.

The firefly algorithm search process can be illustrated in Fig. 8. This graph represents a fictitious problem in which there are only two input values, X and Y. The overall minimum in this problem is $X = 0$ and $Y = 0$, and a population is three fireflies. Let us assume that the "firefly 0" is in position (2, 1) and therefore is the closest

Table 2 Implemented firefly algorithm pseudocode

Set the parameters of the algorithm: α, β and γ
Set the simulation parameters: population size and maximum iterations
(n and G)
Generate the initial population randomly
for iteration = 1 : G
    for i = 1:n
        Calculate the luminous intensity of source I_0 (for each
        firefly) proportionally to the objective function of the
        problem:
$$I_0 \propto Function\_Error\ (n, RS)$$
    end for i
    for i = 1:n
        for j = 1:n
            Calculate the distance between the firefly *i* and
            the firefly *j*:
$$r_{ij} = \sqrt{(x_i - x_j)^2 + (y_i - y_j)^2 + (z_i - z_j)^2}$$
            Calculate the brightness of the firefly *i* in
            relation to the firefly *j*:
$$I(r) = I_0 e^{-\gamma r_{ij}^2}$$
        end for j
    end for i

to the correct solution. The "firefly 1" is at $(-4, -4)$, and the "firefly 2", which is furthest from the solution, is at $(-6, 6)$.

The best fireflies in the firefly algorithm have a higher luminous intensity, that is, have an objective function associated with it with a smaller error. In Fig. 8, "firefly 0" has the highest luminous intensity, "firefly 1" has intermediate intensity, and "firefly 2" has low intensity. The basic idea of the firefly algorithm is for a firefly to be attracted to other fireflies that have a higher luminous intensity, and that the attractiveness among fireflies is stronger the closer the fireflies are. The greater the attractiveness, the greater the distance the firefly will travel toward the more luminous firefly. In Fig. 8, the "firefly 0" has the highest intensity, so it will not move. The "firefly 1" is closer to the "firefly 0" than the "firefly 2", so "firefly 1" will move a distance greater than the "firefly 2".

The algorithm was implemented using the MATLAB software, and the results were compared to those of four papers in the literature, [3, 12, 15], and [20]. We

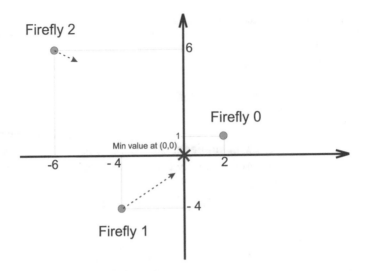

Fig. 8 Illustration of the basic idea of the firefly algorithm

Table 3 Characteristics of Kyocera KC200GT panel

Characteristic	Value
Maximum power (W)	200
Maximum power voltage (V)	26.3
Maximum power current (A)	7.61
Open-circuit voltage (V)	32.9
Short-circuit current (A)	8.21
Number of cells	54

chose the Kyocera KC200GT photovoltaic panel for estimating its parameters. The main characteristics of Kyocera KC200GT panel are in Table 3.

The results were obtained by assigning arbitrarily the values of the variables: alpha (α), beta (β), and gamma (γ). Three test conditions were determined to analyze the impact of these firefly algorithm variables on parameters estimation. In the First Test, which was defined as standard experiment, it is considered a low randomness and average light absorption across the medium. In the Second Test, the randomness of the movement of the fireflies is increased in relation to the First Test, keeping the other values the same as in First Test. In the Third Test, only the absorption of light through the medium is modified in relation to the First Test. The chosen values are presented in their respective sections.

The firefly algorithm is a heuristic algorithm, and it provides a result every time it runs. Therefore, the algorithm was run 10 times in all cases to ensure reliability in the solution obtained with the algorithm. By averaging the output values provided by the algorithm, it is avoided that the obtained parameters are in a local minimum of the *RMSE* surface.

The parameters n and R_s were initialized with values equal to 1.3 and 0.15, respectively. The ideality factor (n) has a well-defined interval ($1 \leq n \leq 2$), that is, it can be initialized with any value within that interval. The series resistance (R_s) has a value different but close to zero ($R_s \neq 0$ and $R_s \approx 0$), so R_s can be initialized with a value close to zero.

4.1 The First Test

The variables of the firefly algorithm for the First Test are presented in Table 4. These values provide low randomness (α) and a medium level of mean light absorption (γ) to the firefly algorithm.

Figure 9 shows the I–V characteristic curves generated with the estimated parameters and that given by the sampled points from the datasheet. The circle points were extracted from the datasheet, and the continuous line is the curve proposed by the First Test. It can be noted that they are very close. The circle points were extracted from the datasheet, and the continuous line is the curve proposed by the First Test.

We also analyzed the statistics of the $RMSE$ to show that it converges to a certain value. Even the highest obtained value, that is, in the worst case, the performance of the firefly algorithm was still better than other algorithms in the literature, as will be shown below. Table 5 shows the obtained parameters and a statistical analysis of the results. The $RMSE$ value shown in Table 5 also justifies the proximity between the curves observed in Fig. 9.

Table 4 Parameter of the First Test

Parameters	Value
Number of fireflies	20
Number of generations	100
α	0.1
β	1
γ	5

Table 5 Statistics for the parameters estimated in the first case

Parameter	Mean	Standard deviation	Variance	Maximum	Minimum
n	1.2254	5.344×10^{-2}	2.857×10^{-3}	1.4079	1.079
$R_s(\Omega)$	0.1484	1.606×10^{-2}	2.580×10^{-4}	0.1919	0.09375
$R_{sh}(\Omega)$	92.59	8.144	66.33	125.48	75.24
$I_0(A)$	3.491×10^{-8}	6.354×10^{-8}	4.038×10^{-15}	38.63×10^{-8}	0.225×10^{-8}
$I_{irr}(A)$	8.223	2.415×10^{-3}	5.834×10^{-6}	8.231	8.216
$RMSE$	9.80×10^{-2}	2.820×10^{-11}	2.255×10^{-6}	10.46×10^{-2}	9.73×10^{-2}

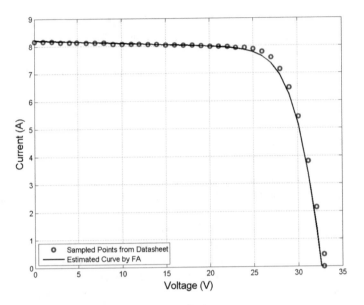

Fig. 9 *I-V* curve of the average cases in the First Test

Table 6 Comparison of the parameter estimation obtained with the proposed method and other methods proposed in the literature—first case

Works	RMSE	n	R_s (Ω)	R_{sh} (Ω)	I_0 (10^{-8}A)	I_{irr} (A)
Proposed firefly	9.80×10^{-2}	1.2254	0.1484	92.59	4.491	8.2230
Laudani et al. [13]	9.76×10^{-2}	1.2645	0.1374	98.08	5.6710	8.2215
Cubas et al. [3]	25.01×10^{-2}	1.3000	0.2309	594.63	9.6930	8.2132
Villalva et al. [20]	21.77×10^{-2}	1.3000	0.2300	566.90	9.7561	8.2135
Majdoul et al. [15]	21.17×10^{-2}	1.3000	0.2310	598.00	9.6896	8.2100

Table 6 presents comparisons of the values of the error (*RMSE*) and the parameters of the photovoltaic panel obtained in this work with other methods of the parameter estimation present in the literature. It can be noted that the *RMSE* obtained by this work was the second minor value achieved until now in the literature. It is also observed that the value of the series resistance (R_s) is closer to the value obtained by [12] showing similarity between the values.

Figures 10, 11, and 12 present the histograms of the ideality diode factor (*n*) and the series resistance (R_s) values. The histograms show that the values obtained for these two parameters are concentrated close to the mean (shown in Table 5). These

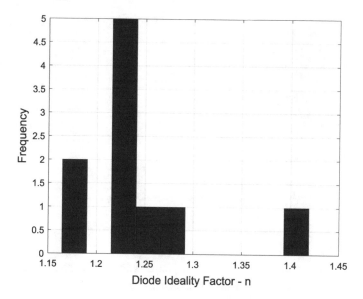

Fig. 10 Histogram of diode ideality factor in the First Test

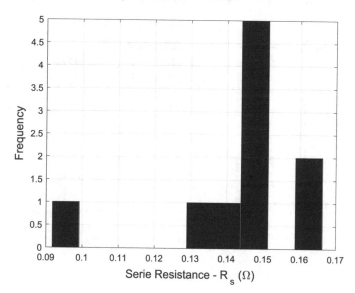

Fig. 11 Histogram of the series resistance in the First Test

findings confirm that the standard deviation is small, in the order of 10^{-3} as shown in Table 5.

In order to show the efficiency of the proposed firefly algorithm, the convergence curve is shown in Fig. 13. These values were compared with the smaller error obtained

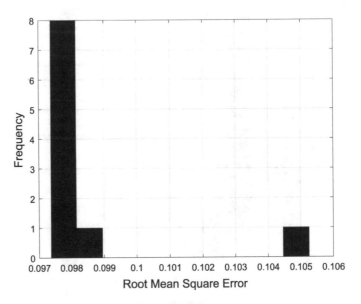

Fig. 12 Histogram of the *RMSE* value in the First Test

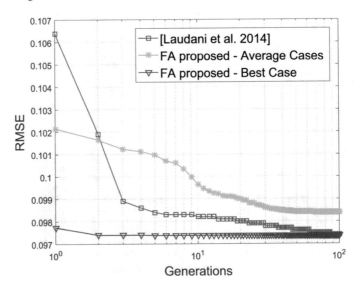

Fig. 13 The convergence curve of *RMSE* values—First Test

in the literature [12]. Analyzing Fig. 13, it is possible to notice that the $RMSE$ values obtained with the best case of firefly algorithm in the First Test tend to be smaller than that presented in [12] in all generations. The logarithmic scale was adopted in y-axis to provide better visualization.

Table 7 Parameters of the Second Test

Parameters	Case 1	Case 2
Number of fireflies	20	20
Number of generations	100	100
α	0.5	0.8
β	1	1
γ	5	5

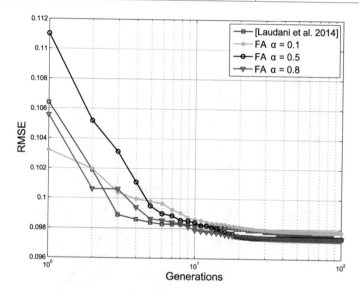

Fig. 14 The convergence curve of *RMSE* values—Second Test

4.2 The Second Test

In the Second Test, two variations were analyzed for the randomness (α) of the firefly algorithm. Thus, the fireflies will have more probability to walk randomly. With the analysis of the results, it will be possible to observe if greater randomness influences positively in the estimation of the parameters of the photovoltaic panels. The values for the implemented are shown in Table 7.

Figure 14 shows the convergence curve of the results obtained for the Second Test compared to the standard case (First Test) and the work [12]. The initial values for the Case 2 ($\gamma = 0.8$) were adequate, providing a small error at the beginning. The greater randomness in the firefly algorithm provides a better search at search space, and it finds a smallest error compared to the errors obtained by others' works considered in this chapter.

Table 8 Statistics of the parameters estimated in Case 1 of the Second Test

Parameter	Mean	Standard deviation	Variance	Maximum	Minimum
n	1.2277	7.455×10^{-6}	5.557×10^{-11}	1.2277	1.2277
$R_s(\Omega)$	0.1477	2.427×10^{-6}	5.892×10^{-12}	0.1477	0.1477
$R_{sh}(\Omega)$	92.31	1.194×10^{-3}	1.426×10^{-6}	92.31	92.31
$I_0(A)$	3.223×10^{-8}	3.795×10^{-12}	1.441×10^{-23}	3.224×10^{-8}	3.224×10^{-8}
$I_{irr}(A)$	8.223	3.448×10^{-7}	1.189×10^{-13}	8.223	8.223
$RMSE$	9.737×10^{-2}	2.820×10^{-11}	7.986×10^{-22}	9.737×10^{-2}	9.737×10^{-2}

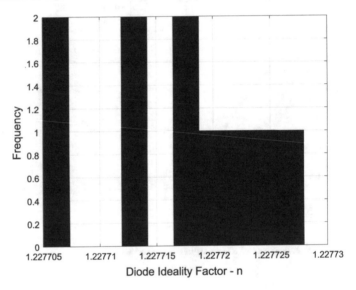

Fig. 15 Histogram of diode ideality factor in the Second Test, Case 1

4.2.1 Case 1

The statistical indicators for case $\alpha = 0.5$ are presented in Table 8. The indicator values show that the parameters converged to the same values in all runs. This fact happens because the parameters have small standard deviation and variance.

Figures 15, 16, and 17 show the histogram of the diode ideality factor (n), series resistance (R_s), and $RMSE$ value for $\alpha = 0.5$. The distribution presented by $RMSE$ was concentrated close to value $RMSE = 0.09737$. The diode ideality factor and series resistance also presented a low variability. The histograms are similar to the results obtained in Table 8.

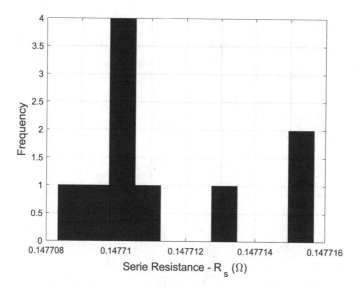

Fig. 16 Histogram of series resistance in the Second Test, Case 1

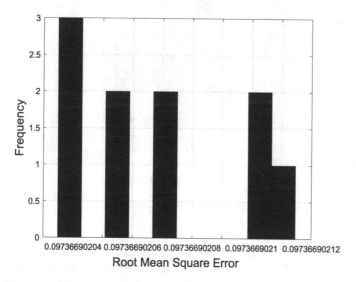

Fig. 17 Histogram of *RMSE* value in the Second Test, Case 1

4.2.2 Case 2

The statistical indicators for the case $\alpha = 0.8$ are presented in Table 9. The presented results are similar to the Case 1 ($\alpha = 0.5$). However, as the randomness is higher than Case 1, the standard deviation and variance increased.

Table 9 Statistics of the parameters estimated in Case 2 of the Second Test

Parameters	Mean	Standard deviation	Variance	Maximum	Minimum
n	1.2277	1.221×10^{-5}	1.4923×10^{-10}	1.2277	1.2277
$R_s(\Omega)$	0.1477	3.643×10^{-6}	1.327×10^{-11}	0.1477	0.1477
$R_{sh}(\Omega)$	92.31	1.194×10^{-3}	3.797×10^{-6}	92.31	92.31
$I_0(A)$	3.223×10^{-8}	6.222×10^{-12}	3.871×10^{-23}	3.224×10^{-8}	3.224×10^{-8}
$I_{irr}(A)$	8.223	5.611×10^{-7}	3.149×10^{-13}	8.223	8.223
$RMSE$	9.737×10^{-2}	7.570×10^{-11}	5.732×10^{-21}	9.737×10^{-2}	9.737×10^{-2}

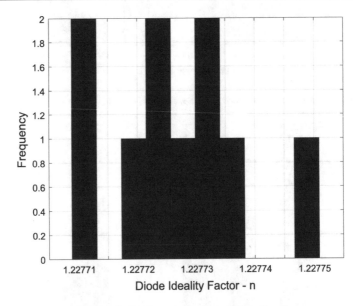

Fig. 18 Histogram of the diode ideality factor in the Second Test, Case 2

Figures 18, 19, and 20 show the histograms of the diode ideality factor (n), series resistance (R_s), and $RMSE$ value for $\alpha = 0.8$. Despite the increase in the randomness of firefly algorithm, the results presented by the histograms are similar in all runs. However, a higher concentration of $RMSE$ value and greater dispersion of R_s is observed in relation to the previous case ($\alpha = 0.5$).

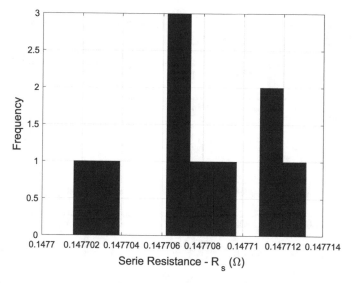

Fig. 19 Histogram of the series resistance in the Second Test, Case 2

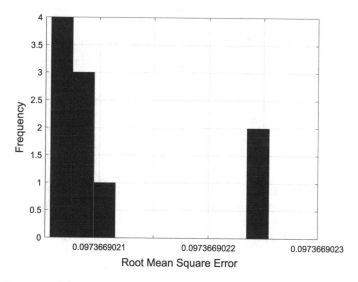

Fig. 20 Histogram of the *RMSE* value in the Second Test, Case 2

4.2.3 Final Considerations

Table 10 shows the results obtained by the randomness variation of the firefly algorithm with the other works. We observed that the lowest *RMSE* values were obtained by the firefly algorithm.

Table 10 Comparison of the parameter estimation obtained with the proposed method and other methods proposed in the literature—Second Test

Works	RMSE	n	R_s (Ω)	R_{sh} (Ω)	$I_0(10^{-8}$ A)	I_{irr}(A)
Proposed firefly	9.80×10^{-2}	1.2254	0.1484	92.59	4.491	8.223
Firefly Second Test Case 1	9.737×10^{-2}	1.2277	0.1477	92.31	3.223	8.223
Firefly Second Test Case 2	9.737×10^{-2}	1.2277	0.1477	92.31	3.223	8.223
Laudani et al. [13]	9.76×10^{-2}	1.2645	0.1374	98.08	5.6710	8.2215
Cubas et al. [3]	25.01×10^{-2}	1.3000	0.2309	594.63	9.6930	8.2132
Villalva et al. [20]	21.77×10^{-2}	1.3000	0.2300	566.90	9.7561	8.2135
Majdoul et al. [15]	21.17×10^{-2}	1.3000	0.2310	598.00	9.6896	8.2100

Table 11 Parameters of the Third Test

Parameters	Value 1	Value 2
Number of fireflies	20	20
Number of generations	100	100
α	0.1	0.1
β	1	1
γ	2	5

The results obtained by the two variations of firefly algorithm randomness obtained the lowest errors among the works in the literature. The values obtained by the firefly algorithm are close to the values obtained by [12] proving that the firefly algorithm result is valid. Figure 21 shows the I–V curve for the Case 2 in the Second Test.

4.3 The Third Test

In the Third Test, the absorption of light by the medium is modified. As defined in the literature, the absorption of light by the medium varies between $0 \leq \gamma \leq 10$. In the First Test, a mean absorption ($\gamma = 5$) was considered. In the Third Test, the value of γ is varied to analyze its influence on the parameter estimation problem. Table 11 shows the values of γ utilized in this section.

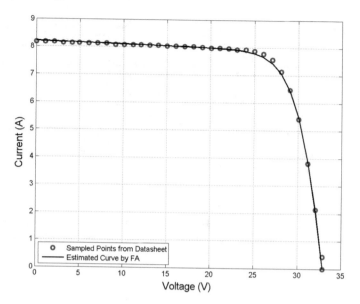

Fig. 21 *I–V* curve of Case 1 in the Second Test

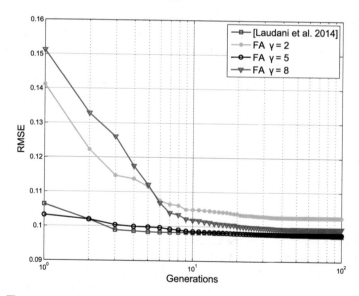

Fig. 22 The convergence curve of *RMSE* values—Third Test

Figure 22 shows the *RMSE* values for the gamma (γ) variation in firefly algorithm, comparing them to those of the standard case (First Test) and the work [12]. Note that with high light absorption through the medium, the firefly algorithm converges more slowly in relation to first case and [12].

Table 12 Statistics of the parameters estimated in Case 1 of the Third Test

Parameters	Mean	Standard deviation	Variance	Maximum	Minimum
n	1.2267	4.573×10^{-2}	2.091×10^{-3}	1.2979	1.1127
$R_s(\Omega)$	0.1479	1.362×10^{-2}	1.854×10^{-4}	0.1815	0.1261
$R_{sh}(\Omega)$	92.47	5.927	35.133	102.8	78.39
$I_0(A)$	3.746×10^{-8}	2.183×10^{-8}	4.767×10^{-16}	9.203×10^{-8}	4.348×10^{-8}
$I_{irr}(A)$	8.223	2.239×10^{-3}	5.015×10^{-6}	8.229	8.220
$RMSE$	9.780×10^{-2}	9.953×10^{-4}	9.907×10^{-7}	10.04×10^{-2}	9.736×10^{-2}

The convergence in the Third Case is justified by the fact that fireflies with low brightness move more slowly to fireflies with high brightness. This provides a delay for the fireflies to reach the global optimum.

4.3.1 Case 1

The statistical indicators for the Case 1 ($\gamma = 2$) are presented in Table 12. The indicators show that there was greater variation of the parameters in relation to the Second Test results. This fact demonstrates that the algorithm converges rapidly, but sometimes without converging to the global optimum.

Figures 23, 24 and 25 show the histograms of the diode ideality factor (n), series resistance (R_s), and $RMSE$ value to $\gamma = 2$. The histograms show that the firefly algorithm converges more times to a same value; however, the firefly algorithm also converges to a high range of different values (for example, $n = 1.2979$ and $n = 1.1127$). Thus, the higher the randomness of the Second Test, the more the difficult to get stuck in local minima.

4.3.2 Case 2

The statistical indicators for the Case 2 ($\gamma = 8$) are presented in Table 13. The indicator values show that the Case 2 provided higher $RMSE$ value compared to Case 1 ($\gamma = 2$), besides converging more slowly than Case 1 (see Fig. 16). Table 13 shows that lower standard deviation and variance were obtained than the previous case ($\gamma = 2$).

Figures 26, 27, and 28 show the histograms of the diode ideality factor (n), series resistance (R_s), and $RMSE$ value for $\gamma = 8$. The histogram values confirm that the variation between the obtained values was smaller than Case 1 ($\gamma = 2$).

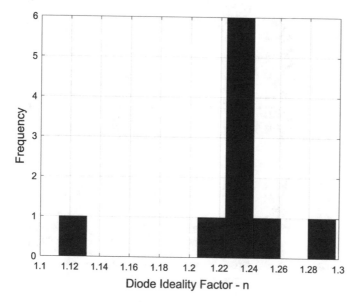

Fig. 23 Histogram of the diode ideality factor in the Third Test, Case 1

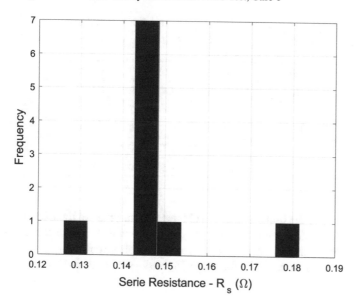

Fig. 24 Histogram of the series resistance in the Third Test, Case 1

4.3.3 Final Considerations

Table 14 presents the results generated by the variation of the luminosity of the medium for the FA, comparing to those of other works.

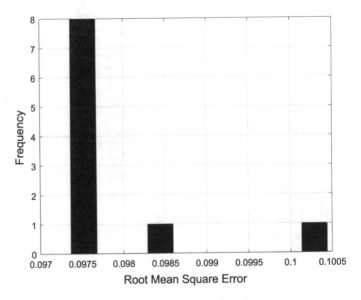

Fig. 25 Histogram of the *RMSE* value in the Third Test, Case 1

Table 13 Statistics of the parameters estimated in Case 2 of the Third Test

Parameters	Mean	Standard deviation	Variance	Maximum	Minimum
n	1.2519	3.184×10^{-2}	1.01×10^{-3}	1.3232	1.2277
$R_s(\Omega)$	0.1418	6.692×10^{-3}	4.478×10^{-5}	0.1477	0.1327
$R_{sh}(\Omega)$	97.03	7.986	63.77	118.21	92.31
$I_0(A)$	5.207×10^{-8}	3.102×10^{-8}	9.624×10^{-16}	13.10×10^{-8}	3.223×10^{-8}
$I_{irr}(A)$	8.222	1.335×10^{-3}	1.782×10^{-6}	8.223	8.219
RMSE	9.790×10^{-2}	1.123×10^{-3}	1.509×10^{-6}	10.13×10^{-2}	9.736×10^{-2}

The results of the firefly algorithm obtained by varying the light absorption coefficient (γ) presented errors close to the error obtained by the algorithm proposed by [12]. The parameters estimated are coherent and, also, they are close to the values obtained by other works in the literature. Figure 29 shows the *I–V* curve for the Case 2 in the Third Test.

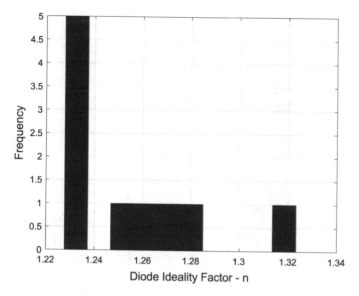

Fig. 26 Histogram of the diode ideality factor in the Third Test, Case 2

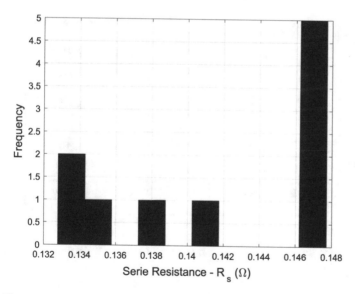

Fig. 27 Histograms of the series resistance in the Third Test, Case 2

5 Conclusion

We proposed in this chapter to estimate the parameters of photovoltaic panels using the firefly algorithm. In the proposed approach, the firefly algorithm performs the

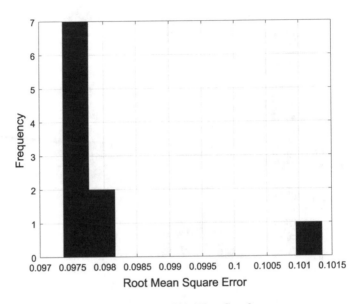

Fig. 28 Histogram of the *RMSE* values in the Third Test, Case 2

Table 14 Comparison of the parameter estimation obtained with the proposed method and other methods proposed in the literature—Third Test

Works	RMSE	n	R_s (Ω)	R_{sh} (Ω)	I_0 (10^{-8} A)	I_{irr}(A)
Proposed firefly	9.80×10^{-2}	1.2254	0.1484	92.59	4.491	8.223
Firefly Third Test Case 1	9.780×10^{-2}	1.2267	0.1479	92.47	3.746	8.223
Firefly Third Test Case 2	9.790×10^{-2}	1.2519	0.1418	97.03	5.207	8.222
Laudani et al. [13]	9.76×10^{-2}	1.2645	0.1374	98.08	5.6710	8.2215
Cubas et al. [3]	25.01×10^{-2}	1.3000	0.2309	594.63	9.6930	8.2132
Villalva et al. [20]	21.77×10^{-2}	1.3000	0.2300	566.90	9.7561	8.2135
Majdoul et al. [15]	21.17×10^{-2}	1.3000	0.2310	598.00	9.6896	8.2100

estimation of two parameters of the SDM model of the photovoltaic panel while the others are estimated by analytical equations. Certainly, the estimation of two unknown parameter values reduces the search space and decreases computational complexity compared with estimation of five unknown parameters.

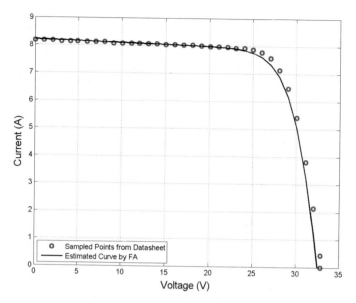

Fig. 29 *I–V* curve of Case 2 in the Third Test

The results of the simulation show that the firefly algorithm provides the smallest error among the considered algorithms. The elevation of randomness impacted positively the estimation of the parameters allowing a better analysis of the search space near to the global minimum, as shown in the Second Test. The decrease in light absorption by the medium allowed the firefly algorithm to converge more quickly to the global minimum. However, the low light absorption can be harmful if the firefly algorithm is stuck to a local minimum, as shown in the Third Test.

The firefly algorithm is a powerful optimization technique that can generate good results. In order to accomplish that, it is necessary an excellent description of the objective function to be solved and the fine adjustments of its parameters. Over 20 new firefly algorithm variants have been developed, and new applications and studies are emerging almost daily. It is no surprise that firefly algorithm has been used in almost every area of sciences, engineering, and industry.

In this work, the firefly algorithm provided good results, that is, the root mean squared error of the *I–V* curve is the smallest registered in the literature. In future works, we intend to apply the firefly algorithm in other equivalent circuit models of the photovoltaic panel, and test other heuristic algorithms.

References

1. Blas MA, Torres JL, Prieto E, García A (2002) Selecting a suitable model for characterizing photovoltaic devices. Renew Energy 25:371–380

2. Chegaar M, Ouennoughi Z, Guechi F (2004) Extracting dc parameters of solar cells under illumination. Vacuum 75:367–372
3. Cubas J, Pindado S, Victoria M (2014) On the analytical approach for modeling photovoltaic systems behavior. J Power Sources 247:467–474
4. da Costa W, Fardin J, Simonetti D, Neto LV (2010) Identification of photovoltaic model parameters by differential evolution. In: Industrial technology (ICIT), 2010 IEEE international conference, pp 931–936
5. Fister I, Fister I Jr, Yang XS, Brest J (2013) A comprehensive review of firefly algorithms. Swarm Evol Comput 6 (in press). http://dx.doi.org/10.1016/j.swevo.2013.06.001
6. Galántai A (2000) The theory of Newton's method. J Comput Appl Math 124(1–2):25–44, ISSN 0377-0427. https://doi.org/10.1016/S0377-0427(00)00435-0
7. Gandomi AH, Yang XS, Alavi AH (2013) Cuckoo search algorithm: a meteheuristic approach to solve structural optimization problems. Eng Comput 29(1):17–35. https://doi.org/10.1007/s00366-011-0241-y
8. Gong W, Cai Z (2013) Parameter extraction of solar cell models using repaired adaptive differential evolution. Sol Energy 94:209–220
9. Huang W, Jiang C, Xue L, Song D (2011) Extracting solar cell model parameters based on chaos particle swarm algorithm. In: 2011 international conference on electric information and control engineering (ICEICE), pp 398–402
10. Ishaque K, Salam Z (2011) An improved modeling method to determine the model parameters of photovoltaic (PV) modules using differential evolution (DE). Sol Energy 85:2349–2359
11. Jervase JA, Bourdoucen H, Al-Lawati A (2001) Solar cell parameter extraction using genetic algorithms. Meas Sci Technol 12:1922–1925
12. Laudani A, Fulginei FR, Salvini A (2014) High performing extraction procedure for the one-diode model of a photovoltaic panel from experimental I–V curves by using reduced forms. Sol Energy 103:316–326
13. Laudani A, Mancilla-David F, Riganti-Fulginei F, Salvini A (2013) Reduced-form of the photovoltaic five-parameter model for efficient computation of parameters. Sol Energy 97:122–127
14. Lukasik S, Zak S (2009) Firefly Algorithm for continuous constrained optimization tasks. In: Ngugen NT, Kowalczyk R, Chen S-M (eds) ICCCI 2009. Lecture notes in artificial intelligence, vol 5796. Springer, Berlin, pp 97–106
15. Majdoul R, Abdelmounim E, Aboulfatah M, Touati AW, Moutabir A, Abouloifa A (2015) Combined analytical and numerical approach to determine the four parameters of the photovoltaic cells models. In: 1st International conference on electrical and information technologies ICEIT'2015, pp 263–268
16. Negnevitsky M (2005) Artificial intelligence: a guide to intelligent systems. Pearson Education Limited, England, New York
17. Ortiz-Conde A, Sanchez FJG, Muci J (2006) New method to extract the model parameters of solar cells from the explicit analytic solutions of their illuminated I–V characteristics. Sol Energy Mater Sol Cells 90:352–361
18. Petrone G, Ramos-Paja CA, Spagnuolo G (2017) Photovoltaic sources modeling. Wiley, London
19. Tilahun SL, Ong HC (2015) Prey-predator algorithm: a new metaheuristic algorithm for optimization problems. Int J Inf Technol Decis Mak 14(6):1331–1352
20. Villalva MG, Gazoli JR, Ruppert Filho E (2009) Comprehensive approach to modeling and simulation of photovoltaic arrays. Trans Power Electron 24(5):1198–1208
21. Yang XS (2008) Nature-inspired metaheuristic algorithms. Luniver Press, Bristol, UK
22. Yang XS (2009) Firefly algorithms for multimodal optimization, In: Stochastic algorithms: foundations and applications, SAGA 2009, lecture notes in computer science, vol 5792, pp 169–178
23. Ye M, Wang X, Xu Y (2009) Parameter extraction of solar cells using particle swarm optimization. J Appl Phys 105:09450

Realization of PSO-Based Adaptive Beamforming Algorithm for Smart Antennas

Rathindra Nath Biswas, Anurup Saha, Swarup Kumar Mitra and Mrinal Kanti Naskar

Abstract A novel beamforming technique based on Particle Swarm Optimization (PSO) algorithm and its subsequent implementation on Xilinx Virtex4 Field-Programmable Gate Arrays (FPGA) board is described. A prescribed limit in Side-Lobes Level (SLL), Beamwidth between the First Nulls (FNBW) and depth of the nulls steered at various interfering directions are considered as beam controlling attributes in this work. All these criteria are included first in two dissimilar reference templates using Dolph–Chebyshev polynomial and Cosine function. Stochastic process is used next to optimize the physical and electrical parameters of a linear antenna array satisfactorily complying with the desired pattern features altogether. System design using Finite State Machine with Datapath (FSMD) modeling and suitable COordinate Rotation DIgital Computer (CORDIC) functional blocks are prepared for its final realization on a dedicated hardware. Its performance is then evaluated with several fixed-point simulations in terms of beamforming accuracy and computational overheads under both Additive White Gaussian Noise (AWGN) and Rayleigh fading channel conditions. These results corroborate its competency comparable to the existing beamforming methods of smart antennas.

Keywords Adaptive beamforming · COordinate rotation DIgital computer (CORDIC) · Field-programmable gate arrays (FPGA) · Finite state machine with datapath (FSMD) · Particle swarm optimization (PSO) · Smart antennas

R. N. Biswas
A.J.C. Bose Polytechnic, North 24 Parganas District, West Bengal, India
e-mail: rathin02@gmail.com

A. Saha · M. K. Naskar (✉)
Jadavpur University, Kolkata, West Bengal, India
e-mail: mrinaletce@gmail.com

A. Saha
e-mail: sahaanurup24@gmail.com

S. K. Mitra
MCKV Institute of Engineering, Howrah, West Bengal, India
e-mail: swarup.subha@gmail.com

© Springer International Publishing AG, part of Springer Nature 2019
S. K. Shandilya et al. (eds.), *Advances in Nature-Inspired Computing and Applications*, EAI/Springer Innovations in Communication and Computing, https://doi.org/10.1007/978-3-319-96451-5_6

1 Introduction

With the advent of Very Large-Scale Integrated Circuits (VLSI) and Monolithic Microwave Integrated Circuits (MMICs) technology in wireless systems design, the communication industry is now evolved enough at its various sectors [1, 2]. Extensive and continuous research efforts are also carried on towards the development of a common platform for them since the last few years. On the other hand, wireless devices and circuits of one generation needs to be further upgraded for using it in the next generation, as per the requirement at most of state-of-the-art communication systems. At the same time, bottlenecks of the wireless transmission such as channel capacity and signal quality, etc., should be improved to adopt the provision of new value-added services in contemporary technologies [3, 4]. However, huge scopes and opportunities are still open to address these issues in wireless systems design at this time. Therefore, it is essentially needed to enhance both algorithmic and architectural attributes developing new methodologies those not only cope up with the recent trends in communication technology but also provide flexibility towards their implementation along with the existing wireless infrastructures. Moreover, the most challenging task remains to maintain security in communications, setting up private point-to-point communication links among the several nodes in multi-hop wireless network architectures [5, 6].

In this context, inclusion of smart antennas with conventional wireless systems might be an elegant and economical strategic approach to enhance the performance of radio propagation characteristics [7, 8]. Basically, smart antennas involve two important features in its operation. Amongst them, beamforming attribute could improve the link capacity remarkably providing wider bandwidth per user channel. In contrast, direction finding capability, also termed as Direction of Arrival (DoA) estimation techniques, could expedite several value-added services in modern communication [9–11]. As a consequence, some effective smart antennas-based intelligent wireless systems are already developed for modern real-world applications. For example, mobile communication systems with smart antennas deployed at either base stations or mobile terminals both in ad hoc networks [12–15] or Wireless Local Area Networks (WLAN) architecture [16–18] and intelligent transportation systems based on Internet of Things (IoT) [19–22], etc., are to be mentioned. Utilizing adaptive beamforming techniques, throughput, and latency of these systems are usually optimized via suppression of interferences from the neighboring nodes and selection of optimal route for relaying data packets from source to sink node [23–27].

However, the conventional beamforming methods are based on either placing deep nulls towards the interfering directions, also known as null steering or concentrating entire radiated power of the beam at the desired node direction, also referred to as beam steering [28–30]. In both schemes, putting on necessary conditions to maintain lower side-lobes level and narrower beamwidth in the beam pattern is overlooked. Thus, these are often prone to show vulnerability at the physical layer attacks in a wireless network environment. In true sense, there always might be a probability of an erroneous DoA estimation and beamforming scheme could stretch coverage to the attackers accordingly [31–33]. Hence, proper balancing between side-lobes level and beamwidth in the optimum pattern is important to provide additional benefit by counteracting the attack scenarios to some extent. Obviously, insertion of these two criteria to the conventional beamforming algorithms is quite impossible. Therefore, new algorithms that must consider prescribed side-lobes level and beamwidth in their beam pattern along with deep nulls placed at appropriate directions of the interferences or attackers should be developed for wireless communications under the harsh environments. On the other hand, addition of this new concept would make the beamforming algorithms multi-objective in nature. It is also a very difficult task to solve them by the traditional approaches. In such cases, considering beamforming functions as optimization problems to any heuristic search method or evolutionary algorithm may be worthwhile to find the optimum solution by computing the array weights vector. PSO technique, in general, possesses higher convergence speed with simple structure than the other methods in literature [34, 35] and hence it is chosen for this work.

However, smart antenna systems usually operate at high frequency (order of few 100 MHz or few GHz) and they require higher sampling rate in their signal processing applications [36, 37]. Processors using traditional von Neumann architecture with several Multiplier-Accumulator (MAC) stages are not suitable for computing such complex beamforming algorithms and there is a need of high-speed parallel processor architectures. Then again, recent growth in microelectronics and digital technology makes the FPGA families as flexible platforms to develop digital signal processors for the smart antennas [38, 39]. Development of an efficient PSO-based adaptive beamforming algorithm and its subsequent realization on scalable hardware architecture of Virtex4 FPGA chip is the main theme of this chapter.

The remaining portions of this chapter are structured as follows. In Sect. 2, the related research works in the areas of adaptive beamforming algorithm using traditional methods or evolutionary algorithms and their scope of implementation on FPGA platforms are briefly discussed. In Sect. 3, a speculative design principle of smart antenna system suitable for advanced wireless communication services is briefly described. The proposed PSO-based adaptive beamforming method and procedural steps for its implementation onto FPGA hardware are demonstrated in Sect. 4 and Sect. 5, respectively. In Sect. 6, simulation results along with proper explanations are demonstrated. In Sect. 7, the chapter is concluded keeping a track towards its future extension.

2 Related Works

A tremendous technological growth of smart antennas is found over the past decades [40, 41]. It has become possible, in fact, through a continuous research and development of various pioneering adaptive algorithms during this time. Some renowned beamforming methods and their hardware implementation on FPGA board are briefly reviewed in this section. Ward et al. [42] described a digital beamforming scheme with linear adaptive combiners in which weight vectors are determined by QR decomposition of Recursive Least Square (RLS) algorithm using Givens rotations. They also proposed an efficient pipelined architecture for its implementation to achieve high performance in digital domain. Using an eight-channel L-band digital phased array receiver, Nuteson et al. [43] explained that the calibration and FPGA implementation methodologies have direct influence on the performance characteristics of digital beamforming process in smart antenna applications. Choi and Shim [44] proposed an alternative weight vectors computation technique based on maximization procedure using Lagrange multiplier. Employing the smart antenna array on the base station of Code Division Multiple Access (CDMA) mobile communication system, its ability is also verified to counteract multipath fading with least error in Signal-to-Interference-plus-Noise Ratio (SINR) and Bit Error Rate (BER). Dikmese et al. [36, 37] demonstrated implementation of some CDMA compatible beamforming algorithms such as Least Mean Square (LMS), Constant Modulus (CM), and Space Code Correlator (SCC), etc., on FPGA platform. It was observed that approximately 500 times faster speed can be reached in computing weight vectors rather than their Digital Signal Processor (DSP) implementation counterpart. Sun et al. [45] presented fast beamforming in an Electronically Steerable Parasitic Array Radiator (ESPAR) by tuning the load reactance at parasitic elements surrounding the active central element. This is based on the criterion of simultaneous perturbation stochastic approximation with a maximum cross correlation coefficient. Based on the Newton downhill method, Li et al. [46] developed a beamforming algorithm that can shape the desired beam pattern in a quick and stable manner. Khodaei et al. [47] proposed an adaptive beamforming algorithm that performs well in tracking the mobile users over a wide angle spread environment. Using Memetic algorithms for perturbations of phases and amplitudes, Hsu [48] described an uplink Multiple-Input Multiple-Output Spatial Division Multiple Access (MIMO-SDMA) optimization technique of smart antennas. Using the Eigenspace method to compute the optimal weight vector, Lee and Choi [49] presented a stable adaptive beamforming technique suitable for wide angle spread circumstances.

However, almost all these beamforming methods described above fail to perform satisfactorily under various adversarial attacks in wireless network environment. Adversarial attack is usually assumed to compromise few nodes either by accessing their control system directly or tampering their radio environment in an indirect way to introduce error in DoA estimations. Thus, the necessity of maintaining lower side-lobes level and narrower beamwidth in beam patterns is essential. Otherwise, there always is a chance of stretching coverage to the attackers or the interferers. This situation expedites the motivation towards the development of a new beamforming scheme for smart antennas to be described in this chapter. Considering all the necessary perspective of establishing a private communication link between the nodes in wireless ad hoc networks, this method involves PSO algorithm to compute the array weights vector for generating the final beam patterns in accordance with two separate templates such as Dolph–Chebyshev and Cosine function. Such templates include all the criteria of beamforming attributes to keep the adjacent nodes, except the desired one, beyond its communication coverage treating them as the interferers.

3 Smart Antennas Preliminaries

In wireless networks, most of the radiated power is wasted while nodes set up communication links with their conventional dipole antenna systems producing omnidirectional patterns. Such unused power may create interferences to the neighboring nodes deteriorating overall Quality of Service (QoS) in the networks. In contrast, smart antennas can enhance the SINR in the data links mitigating multipath fading signals by spatial separation at its signal processing unit. This improves channel capacity in the wireless links making them stable and secure to guarantee higher throughput in data transmission [50, 51].

3.1 Theory of Smart Array Design

Smart antennas comprise with several spatially isolated antenna elements, usually in the form of an array configuration at particular geometry such as linear, circular, or planar, etc., followed by a digital signal processor [52, 53]. Beam patterns in such antenna structures are normally governed by the Array Factor (AF), also known as Beamforming Function (BF). Let us consider a linear array with uniform inter-element spacing (Δ) and nonuniform amplitude excitations (I_n) primarily to derive the expression for AF. Such an array with even number (2 N) of antenna elements along the Y-axis is also shown in Fig. 1. Now, AF is typically expressed [54] as

$$\mathrm{AF}(\theta) = \sum_{n=0}^{2N-1} I_n e^{jn\psi} \tag{1}$$

where $\psi = k\Delta(\sin\theta - \sin\theta_d)$ and $k = \frac{2\pi}{\lambda}$.

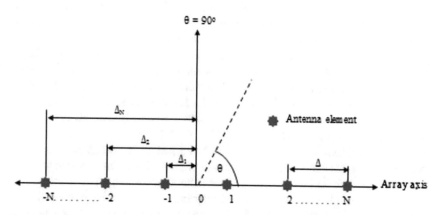

Fig. 1 A symmetrical configuration of linear array with 2N elements

Normally, ψ is called progressive phase shift between the elements. Also, k is termed as the wave numbers and λ is wavelength of the transmitted signal. Magnitude of current for the nth element in array is denoted by I_n. Likewise, scanning angle θ is measured about the array axis in the range $[-90°, 90°]$ and θ_d is a squinted angle at which the main-lobe orientation is to be done. An alternative representation of AF in Eq. (1) can be made in matrix form as

$$AF(\theta) = \phi^T a(\theta) \tag{2}$$

where $\phi^T = \begin{bmatrix} I_0 \ I_1 \ \ldots \ I_{2N-1} \end{bmatrix}$ is called array weights vector and $a(\theta) = \begin{bmatrix} 1 \ e^{j\psi} \ \ldots \ e^{j(2N-1)\psi} \end{bmatrix}^T$ is known as array steering vector.

Assuming symmetry in current distributions of the array elements about the origin, AF in Eq. (1) is now modified and normalized as

$$\mathrm{AF}_p(\theta) = \sum_{n=1}^{N} I_n \cos[k\Delta_n(\sin\theta - \sin\theta_d)] \tag{3}$$

where $\Delta_n = \frac{(2n-1)\Delta}{2}$. Now, Eq. (3) can be easily extended for linear array configuration with nonuniform inter-element spacing as well.

3.2 Principles of Operation

Smart antennas processes intercepted signals iteratively to produce an optimum beam pattern with higher directive gain in a chaotic environment and thus increases coverage area significantly throughout the networks [55]. Its principle of operation can be explained as more likely to a MIMO system that utilizes spatial diversity effect of the antenna arrays, normally termed as SDMA technique [56]. Hence, its channel capacity or data transmission rate over the channel can be improved further by installing large number of array elements and transmitting the space-time block coded waveform [57]. Smart antenna systems are classified into two categories as switch-beam antennas and adaptive array antennas. The latter one is more popular due to its signal processing capability in a fluctuating radio environment [28, 58]. It generally employs smart signal processing algorithms to estimate DoA identifying strengths of the spatial signals incident at various angles. It also involves several efficient beamforming algorithms that utilize DoA information to compute array weights or beamforming vector producing beam pattern in accordance with variable traffic or signal conditions of the environment. Therefore, more stable wireless links can be established with beam patterns focused at the desired direction so that effect of multipath fading and co-channel interference are effectively reduced [59]. In real-time applications, beamforming function of smart antennas, as illustrated in Fig. 2, can be realized with an iterative process of updating the array weights vector

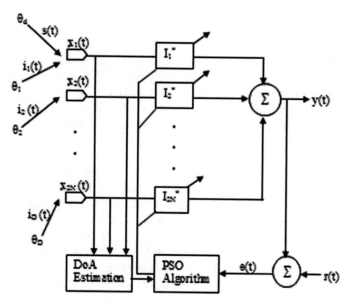

Fig. 2 Schematic diagram of smart antennas

(ϕ^T) to produce an output signal $y(t)$ in accordance with the reference signal $r(t)$. It continues till the error signal $e(t) = r(t) - y(t)$ reaches an allowable limit and can be expressed as

$$y(t) = \phi^T x(t) \tag{4}$$

where the intercepted signals $x(t)$ is a composite input signal vector with desired signal vector $x_s(t)$ arriving at angle θ_d, D number of interfering signal vector $x_i(t)$ arriving from angles θ_1 to θ_D and zero mean AWGN $\eta(t)$ for each channel. This can also be defined as

$$x(t) = a(\theta_d)s(t) + \left[a(\theta_1)\, a(\theta_2) \ldots a(\theta_D) \right] \cdot \begin{bmatrix} i_1(t) \\ i_2(t) \\ \vdots \\ i_D(t) \end{bmatrix} + \eta(t)$$

$$= x_s(t) + x_i(t) + \eta(t)$$

3.3 Beamforming Attributes

Adaptive beamforming means shaping an optimal pattern by steering main-lobe and nulls at appropriate angles as per the DoA information [60]. However, the state-of-the-art in DoA estimation lies on Time series analysis, Spectrum analysis, eigen-structure methods, Parametric methods, or Linear prediction methods, etc. Pattern control algorithms, on the other hand, usually work on the basis of maximizing the Signal-to-Interference Ratio (SIR), minimizing the variance or minimizing the Mean Square Error (MSE), etc. Some popular DoA estimation methods include Bartlett, Capon, MUltiple SIgnal Classification (MUSIC), Root-MUSIC or Estimation of Signal Parameters via Rotational Invariance Techniques (ESPRIT) algorithms, etc. Similarly, LMS, RLS, CM, Sample Matrix Inversion (SMI) or Conjugate Gradient (CG) methods, etc., are examples of few widely used beamforming algorithms.

However, the practical applications of smart antennas are restricted, in many important wireless communication services that require minimum SLL and narrower beamwidth, due to its conventional beamforming attributes as mentioned in the earlier section. Hence, new algorithms should be developed for such applications.

4   PSO-Based Beamforming Methodology

Evolutionary algorithms-based beamforming methods can provide optimum beam patterns successfully as per multiple predefined specifications. However, their versatile uses are restricted, in various services of resource-constrained wireless ad hoc networks, due to huge computational overheads (i.e., time complexity and space complexity). Selection of proper optimization algorithm and formulation of the necessary objective function also have a great impact in the design goal. In optimization process, higher convergence speed and simpler algorithmic structure are always preferred. PSO method involves both the attributes suitably compared to other conventional schemes and hence recommended for energy-efficient computations in practice.

4.1   Overview of PSO Algorithm

PSO is a population-based, stochastic, evolutionary technique introduced by Kennedy and Eberhart in 1995 [35]. It mimics the swarm's behavior in search of collecting honey from the flowers at garden/unknown field and hence the name. Due to its simple structure with higher degree of feasible convergence, it is widely used to solve many multidimensional, discontinuous and complex optimization problems in various fields of application. Moreover, its performance can be improved by adjusting a single parameter, termed as inertia weight (w) or constriction factor, in the velocity update function. Each potential solution is termed as a "particle" in the search space and its acceptability is determined with a fitness value obtained by evaluating the objective function (also termed as cost function or fitness function or error function, etc.). Initially, each particle begins movement with random position (X) and random velocity (V) over the multidimensional problem space. During the search process, it tries to find out the particle position with possibly higher fitness value in an iterative manner. Each time step (t), every particle must update its velocity (V) and position (X) based on the previous knowledge of personal best position (p_{best}) and the global best position (g_{best}) attained so far. The velocity and position update equation is represented as follows:

$$V = w \cdot V + c_1 \cdot \text{rand}_1() \cdot (p_{best} - X) + c_2.\text{rand}_2() \cdot (g_{best} - X) \tag{5}$$

$$X = X + V \tag{6}$$

where t is taken as a unit time step.

The second and third term of Eq. (5) is known as "cognitive" component and "social" component, respectively. Cognitive component encourages each particle to move toward its own best position found so far and social component explores the global optimal solution exploiting the collaborative effect of the particles. Parameters c_1 and c_2 are known as the acceleration constants (typically set to a value of 2.0) that represent the relative weights of stochastic acceleration terms pulling each particle towards p_{best} and g_{best} positions. The $\text{rand}_1()$ and $\text{rand}_2()$ are two random numbers in the range [0, 1] that introduce some randomness to mimic the analogy of real-time scenario. The inertia weight is generally used to balance between the global and local search abilities and set values in the range [0, 1]. Using a linearly variable inertia weight from 0.9 to 0.4, algorithm is found to converge faster. The process of updating velocities and positions will continue till either one of the particles find a location with the possibly highest fitness value or another predefined termination criterion is met. Moreover, boundary conditions are applied to reinforce and stay the particles movement inside the desired domain of interest [26]. The maximum and minimum value of velocity and position is set to the upper and lower limit of the dynamic range of the search space, i.e., (V_{max}, X_{max}) and (V_{min}, X_{min}) respectively. Whenever any particle exceeds this limit, its velocity and position is set forcefully between upper or lower boundary to control the convergence towards the global best solution.

4.2 Proposed Beamforming Scheme

Considering all the aspects (e.g., security, QoS, and resource constraints, etc.) of wireless ad hoc networks, an effective beamforming method is proposed in this chapter. It can successfully produce an optimum pattern with prescribed SLL, FNBW and deep nulls at the direction of interferences (θ_i). Beam patterns can be controlled normally by varying amplitudes and phases of array element excitations as described in the beamforming function of Eq. (3). In this regard, two strategic approaches are carried out: one is to change the phases only in a nonuniformly spaced linear array and another is to adjust the amplitudes in a uniformly spaced linear array. PSO algorithm is used to optimize the element position perturbations keeping the array length unchanged in the former case and amplitude coefficients in the latter case.

4.2.1 Generation of Reference Templates

All the necessary attributes of desired beam patterns are stipulated on the reference templates. Separate templates are considered here in two beamforming mechanisms as mentioned above.

- Dolph–Chebyshev template

  Dolph–Chebyshev patterns normally contain the best possible trade-offs between SLL and FNBW. For specific SLL, amplitude coefficients of a linear and uniformly spaced array are first determined according to the Chebyshev polynomial as

$$I_n = \sum_{q=n}^{N} (-1)^{N-q} (y_0)^{2q-1} \frac{(q+N-2)!\,(2N-1)}{(q-n)!\,(q+n-1)!\,(N-q)!} \tag{7}$$

where $y_0 = \cosh\left[\frac{1}{(2N-1)} \cosh^{-1}(\text{SLL}_d)\right]$.

SLL_d is the desired main-lobe to side-lobe voltage ratio. Then its beam pattern is generated with beamforming function as in Eq. (3). Now, it is further modified to steer nulls of predefined depth (K) and directions (θ_i). Thus, desired template takes the form as

$$\text{AF}_d(\theta) = \begin{cases} K & \text{if } \theta = \theta_i \\ \text{AF}_p(\theta) & \text{elsewhere} \end{cases} \tag{8}$$

- Cosine template

  A simple time-scaled Cosine function, as shown in Fig. 3, is considered for the second case. It can be primarily defined as

$$\text{AF}_d(\theta) = \begin{cases} \cos\left[(\theta - \theta_d)\frac{\pi}{\text{FNBW}_d}\right] & \text{if } |\theta - \theta_d| < \frac{\text{FNBW}_d}{2} \\ K & \text{if } \theta = \theta_i \\ \text{SLL}_d & \text{elsewhere} \end{cases} \tag{9}$$

Here, the desired beamwidth (FNBW_d) and desired side-lobes level (SLL_d) are adjusted by choosing appropriate values for phase angle of the Cosine function and SLL_d, respectively.

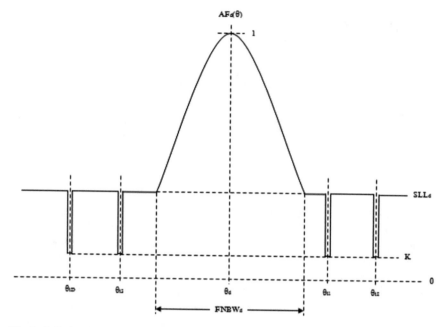

Fig. 3 A Cosine template for desired beamforming

4.2.2 Formulation of Fitness Function

Considering the perspectives of computational complexity in optimization process, a simple formula is adopted here for fitness calculation. Fitness function (F) for the proposed method is defined as

$$F = \min\left\{\sum_{\theta=-90°}^{\theta=90°} D(\theta)\right\} \qquad (10)$$

where $D(\theta) = \begin{cases} 1 & \text{if } AF_d(\theta) < AF_p(\theta) + \text{Tolerance} \\ 0 & \text{otherwise} \end{cases}$.

Typically, $D(\theta)$ represents an weight assigned for each deviation obtained in produced pattern $AF_p(\theta)$ from the desired one $AF_d(\theta)$ at any sample angle(θ) over the range $[-90°, 90°]$. This scheme always tries to minimize the worse solutions count in an iterative manner.

4.2.3 Optimization of Array Weights

Positions and amplitude coefficients of array elements are considered as optimization variable, respectively, in two distinct stages of the proposed scheme. Also,

Table 1 Specifications for the desired patterns

Parameters	Template type	
	Dolph–Chebyshev	Cosine
SLL_d	0.0316 (−30 dB)	0.0316 (−30 dB)
Tolerance	0.0246 (−5 dB)	0 (0 dB)
$FNBW_d$	20°	20°
K	0.00001 (−100 dB)	0.00001 (−100 dB)
θ_d	30°	30°
θ_i	−20°, −5°, 10°, 45°, 60°	−20°, −5°, 10°, 45°, 60°

a symmetrical array configuration with constant length is assumed for both the cases. Now, considering M particles, each having Q dimensions within the solution space, PSO algorithm would produce two $M \times Q$ dimensional matrices for its position (X) vectors and velocity (V) vectors. As the position of N-th array element is fixed, dimensions Q remains equal to ($N - 1$) for the first case. In the second case, dimensions Q are equal to (N). These vectors are randomly initialized for faster convergence of the optimization process. In each iteration, fitness function is evaluated for each particle finding the corresponding p_{best} and g_{best} values as

$$p_{best} = \left[p_{best1} \; p_{best2} \; \cdots \; p_{bestM} \right]^T \text{ and } g_{best} = \min\{p_{best}\}.$$ These values (p_{best} and g_{best}) along with respective positions are recorded and used to update velocity and position of each particle as per the Eqs. (5) and (6). The termination condition for this algorithm is set to a maximum iteration number of 1000. Thus, fitness values are updated as to keep minimum number of deviations obtained for an angle interval of 1° over the range [−90°, 90°] in every iteration. The global best position (g_{best}) achieved at the process termination (through random swarm movements over unknown search field) is defined as the optimum solution. Specifications of the desired parametric settings in an optimum pattern are listed in Table 1. Design and optimization parameters along with their boundary limits are also stipulated in Table 2 for two different cases. Convergence curves of respective optimization method are illustrated in Fig. 4. Also, speed of convergence, denoted by the minimum iteration numbers, are presented with an Empirical Cumulative Distribution Function (ECDF) for 30 runs of the process in Fig. 5. Thus, corresponding patterns obtained for 20 element linear arrays with desired node direction at 30° and five interferences at −20°, −5°, 10°, 45°, and 60° are also shown in Figs. 6 and 7. In both patterns, assuming distance of desired node at the furthest end in beam patterns (d = 1), neighboring nodes are located at 0.9, 0.1, 0.04, 0.5, 0.08, and 0.3, respectively. Corresponding values of SLL_d, $FNBW_d$ and K are also set to −30 dB, 20°, and −100 dB.

Table 2 Design parameters with boundary limits

Parameters	Template type			
	Dolph–Chebyshev		Cosine	
	Lower limit	Upper limit	Lower limit	Upper limit
Frequency (f)	2.4 GHz (Fixed)		2.4 GHz (Fixed)	
Inter-element spacing (Δ_n)	$\Delta_n - 0.25\lambda$	$\Delta_n + 0.25\lambda$	0.5λ (Fixed)	
Element excitation (I_n)	Fixed value as computed in Eq. (3)		0	1
Phase shift (β)	0 (Fixed)		0 (Fixed)	

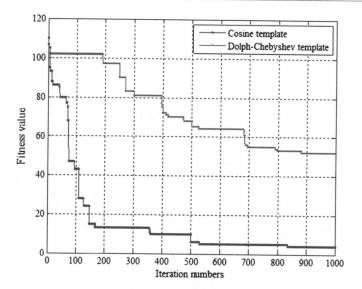

Fig. 4 Convergence curves attained in the optimization process

5 Hardware Realization on FPGA Devices

Two categories of processor architectures, commonly known as general purpose processors and custom processors, are available in the market for implementing the proposed beamforming algorithm on digital platforms [38]. Microprocessor Unit (MPU) and DSP are example of the former type. Although more flexibility in computation of general purpose solutions is achieved with their fixed number of instructions set, but the sequential or serial execution process makes them relatively slow, involving multiple clock cycles to complete any particular task. They also consume more power as these architectures are not optimized for the particular application. In contrast, Application-Specific Integrated Circuits (ASIC) are called dedicated or custom processors. They normally enable computation with a customized architecture and

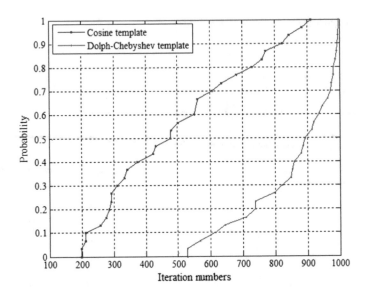

Fig. 5 Convergence speeds achieved in the optimization process

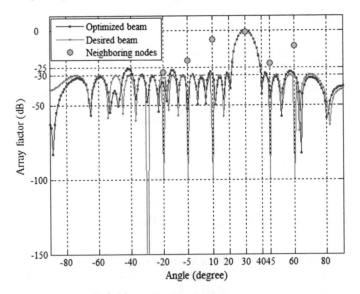

Fig. 6 Optimized pattern with Dolph–Chebyshev template

always offer a specific solution in parallel mode. Hence, processing the entire task in a single clock cycle makes them both fast as well as power efficient. However, they are comparatively more expensive. FPGA family, on the other hand, integrates both the benefits of versatile design solutions and parallel processing capabilities on the same platform. On execution of the algorithms, they often develop reconfigurable

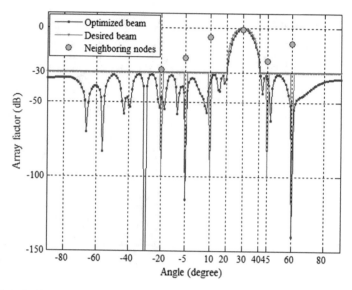

Fig. 7 Optimized pattern with Cosine template

distributed arithmetic structures with their internal logic blocks. This avoids the instruction fetch and data load/store bottlenecks of the conventional von Neumann architecture [39].

This section demonstrates the necessary steps for designing processor architecture of the proposed algorithm on FPGA board. The beamforming function consists of trigonometric Cosine or Sine functions and hence processor architecture, capable of computing fixed-point arithmetic, can be easily developed defining proper CORDIC functional blocks.

5.1 Fundamentals of CORDIC Method

The CORDIC is an efficient computation method, presented first by J. E. Volder in 1959 [61], solving several basic trigonometric and hyperbolic functions. Nowadays, it has become popular due to its various signal processing applications. The key concept of this method is laid on rotating a vector in two-dimensional (2D) coordinate systems for a desired angle (α). Then it is expressed as a sum of several predefined elementary angles (α_i), decomposed by pseudo micro-rotations as depicted in Fig. 8. It turns into simple binary arithmetic with only shift-add operations, setting values to such constituent angles as the power of 2 in an iterative way. This principle is generally expressed in matrix form as

Fig. 8 Rotation vector in 2D
circular coordinates

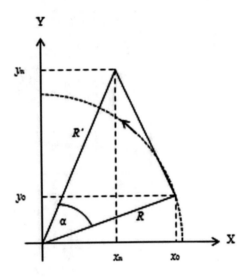

$$\begin{bmatrix} x_n \\ y_n \end{bmatrix} = \begin{bmatrix} \cos\alpha & \sin\alpha \\ -\sin\alpha & \cos\alpha \end{bmatrix} \begin{bmatrix} x_0 \\ y_0 \end{bmatrix} \tag{11}$$

where initial and final positions of the rotation vector (R) in the coordinate systems
are denoted by (x_0, y_0) and (x_n, y_n), respectively.

As rotation angle is decomposed into a set of small angles by pseudo micro-
rotations, hence it can be written as

$$\alpha = \sum_{i=0}^{p-1} \delta_i \alpha_i \tag{12}$$

where δ_i denotes the direction of micro-rotations.

Now, putting $\alpha_i = \tan^{-1}(2^{-i})$, the CORDIC driving equations finally can be
summarized as

$$\begin{cases} x_{i+1} = x_i + \delta_i 2^{-i} y_i \\ y_{i+1} = y_i - \delta_i 2^{-i} x_i \\ z_{i+1} = z_i - \delta_i \alpha_i \end{cases} \tag{13}$$

Here, z is an accumulator to store the effective angle in the iterative process.
Usually, direction of micro-rotations follows the sign of z_i and can be represented as

$$\delta_i = \begin{cases} +1 & \text{if } z_i \geq 0 \\ -1 & \text{otherwise} \end{cases} \tag{14}$$

Through iterations, the rotation vector gets scaled (R') with a factor of about 1.6467605 in circular coordinates. This effect is normally avoided at the final computation of Cosine or Sine function, simply initializing the position of rotation vector at $(x_0 = 0.6072528, y_0 = 0)$ or $(x_0 = 0, y_0 = 0.6072528)$, respectively.

5.2   ASMD Chart

The proposed beamforming method basically operates on a fixed-point PSO processor that computes optimum values of the array weights vector in an iterative way. Towards the development of its system architecture on the Xilinx virtex4 FPGA board, a special schematic representation known as Algorithmic State Machine with Datapath (ASMD) chart is used to translate the sequence of steps in the algorithm, more similar to an FSMD modeling. FSMD architectures normally include both control unit and data unit. Total process should be executed through finite logical operations or states $(S_0–S_{14})$ that imply sequential commands on the controller or Finite State Machine (FSM) to actuate control signals for appropriate operations in the data unit. In the ASMD chart, each building block normally contains several Register Transfer Level (RTL) notations that specify data manipulation and data transfer operations among several registers. Each RTL operation is synchronized with a master clock embedded in the system and thus executes clock-by-clock basis. The ASMD chart for the proposed beamforming algorithm is illustrated in Fig. 9. To synthesize a behavioral model of such FSMD design, Verilog code [62] is used as Hardware Description Languages (HDL), making direct map onto FPGA board.

5.3   Proposed Architecture

The RTL operations described in ASMD chart are normally performed by the data unit. It can be implemented well with registers, multiplexers and other combinational circuits that are required to design several data manipulation functions. To realize RTL operations on hardware, data from the source registers are passed to the combinational circuits for their transformation as per the specific function at the first clock cycle. The updated data is then forwarded to the destination register for its storage at the next clock cycle. Multiplexers are used to route such data between registers and combinational or functional circuits. The state register (that keeps a track to current state of the FSM) is used as the selection signal input for the multiplexers to set appropriate result in any RTL operation. The proposed system architecture on FPGA device is illustrated in Fig. 10. The entire beamforming process can be viewed as to have three basic interrelated functional parts: (i) generation of the beamforming function (AFp) (ii) evaluation of the fitness function (F), and (iii) selection of the global best position (gbest). Beamforming function is characterized by trigonometric Cosine or Sine function and the fitness function is formulated as a sum of the

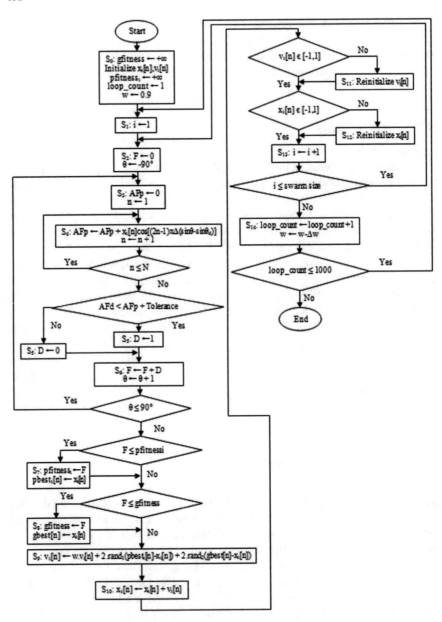

Fig. 9 ASMD chart using Cosine template

deviation weights obtained through difference of two beamforming terms (AF$_d$ and AFp) over finite sample angles. It can be efficiently implemented by formulating suitable CORDIC blocks. The constant design parameters are stored in Read Only

Memory (ROM). Here, Linear Feedback Shift Register (LFSR) is used to generate random numbers for the PSO algorithm.

6 Performance Evaluation

Performance of the beamforming scheme is verified with hardware level fixed-point simulations on its accuracy and computational overheads. In statistical measurement of signals arriving at multipath directions, both of AWGN and Rayleigh fading channel conditions are assumed. Some offline Personal Computer (PC) generated data that closely relates with the real-time wireless radio propagation characteristics are considered here for both noise variance and fading coefficient. The resulting data are obtained with 30 runs of the simulation program on the Xilinx Virtex4 (device: XC4VLX60) FPGA environment. Moreover, analytical discussions on each of these results such as beamforming accuracy, BER, FPGA resource utilization and computation time etc. are also made.

6.1 Simulation Results

Simulation results are presented with degree of accuracy maintained at this beamforming method for both reference templates. Also, an estimation of average bit error rate under harsh channel conditions, assuming Binary Phase-Shift Keying (BPSK) modulation technique in signal transmission over the wireless channel, is described.

6.1.1 Beamforming Accuracy

Beamforming accuracy (A) represents the capability of the scheme, producing beam patterns of higher degree of precision in measurement with the desired templates. Usually, it is a measure of the deviation of optimized pattern from the desired specifications and expressed (in percentage) as

$$A = (1 - \varepsilon) \times 100\% \tag{15}$$

where ε denotes the error associated with the beamforming method. It is here defined as ratio of the aggregated value of deviation weights to the total number of sample points (P) and expressed as

$$\varepsilon = \frac{\sum_{p=1}^{P} D(\theta_p)}{P} \tag{16}$$

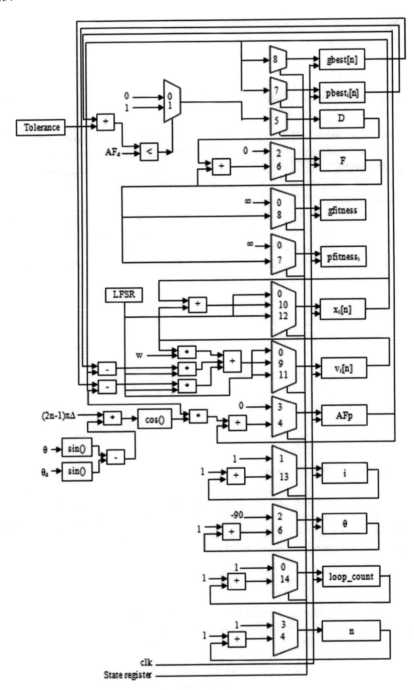

Fig. 10 RTL schematic of the proposed Architecture

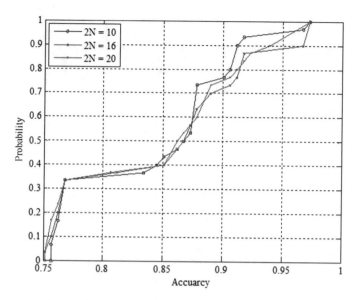

Fig. 11 Accuracy in beamforming using Cosine template (Nulls = 5)

Typically, θ_p is the p-th sample angle over the range $[-90°, 90°]$. The pattern features like beamwidths, SLL, and nulls setting conditions, etc., vary with number of interfering or multipath faded signals and antenna elements in the array. Therefore, accuracy obtained with Cosine template, under various numbers of antenna elements and interferences or nulls are illustrated as their respective ECDF in Figs. 11 and 12. In both the cases, SLL and FNBW are kept constant (SLL = −30 dB and FNBW = 20°). It is obvious that accuracy increases invariably with an increase in the number of array elements or by reducing the interference effects. Another ECDF for accuracy found with Dolph–Chebyshev template, under the similar parametric settings as earlier cases are also presented in Figs. 13 and 14. It is evident that a reasonably lower accuracy is achieved in this case. Magnitudes of optimized element positions and amplitude coefficients in the symmetrical linear array for different number of antenna elements are given in Table 3. Here, other parameters are also considered to have fixed values (SLL = −30 dB, FNBW = 20° and Nulls = 5).

6.1.2 Bit Error Rate

Bit error rate also signifies the ability of system, preserving beamforming performance consistent under various channel noise conditions. It is represented as probability of the error due to various shadowing phenomena in radio propagation. At particular SNR level, it usually depends on the number of array elements in specific channel propagation conditions. Under AWGN channel, statistical measurement of bit error rate is governed by the formula [63] as

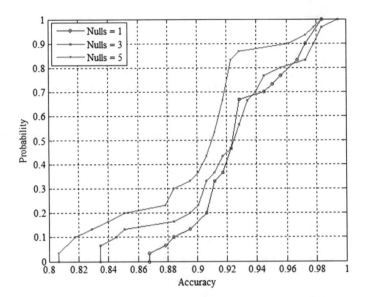

Fig. 12 Accuracy in beamforming using Cosine template (2N = 20)

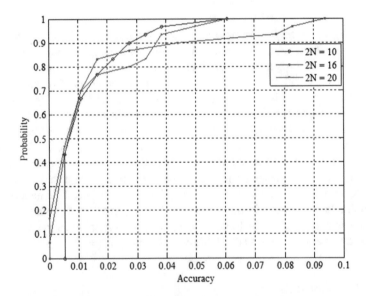

Fig. 13 Accuracy in beamforming using Dolph–Chebyshev template (Nulls = 5)

$$P_e = \frac{1}{2} erfc\left(\sqrt{2N \times \text{SNR}}\right) \approx \frac{1}{2} e^{\frac{-2N \times \text{SNR}}{2}} \tag{17}$$

In contrast, average bit error rate estimation under Rayleigh fading channel due to random varying nature of the fading coefficient, is described in [64] as

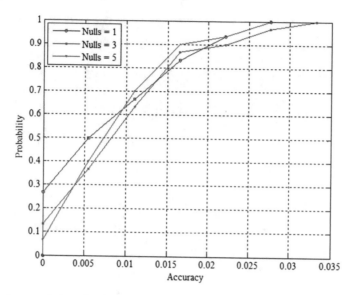

Fig. 14 Accuracy in beamforming using Dolph–Chebyshev template (2N = 20)

Table 3 Values of optimized element positions and amplitude coefficients

Index (n)	Dolph–Chebyshev template			Cosine template		
	Element positions (Δ_n)			Amplitude coefficients (I_n)		
	Number of array elements (2N)			Number of array elements (2N)		
	N = 5	N = 8	N = 10	N = 5	N = 8	N = 10
1	0.2463	0.2450	0.2586	0.9816	0.9212	0.9758
2	0.7535	0.7592	0.7586	0.8124	0.8379	0.8678
3	1.2505	1.2524	1.2590	0.6222	0.7669	0.8480
4	1.7691	1.7605	1.7633	0.3596	0.6678	0.7533
5	–	2.2471	2.2685	0.1530	0.4838	0.6637
6	–	2.7484	2.7575	–	0.3686	0.4693
7	–	3.2453	3.2600	–	0.2500	0.4225
8	–	–	3.7488	–	0.1125	0.3152
9	–	–	4.2543	–	–	0.1870
10	–	–	–	–	–	0.0612

$$P_e = \frac{(1-\gamma)^{2N}}{2} \sum_{m=0}^{2N-1} \binom{2N+m-1}{m} \frac{(1+\gamma)^m}{2} \approx \binom{4N-1}{2N} \frac{1}{(2 \times \text{SNR})^{2N}}$$

(18)

where $\gamma = \sqrt{\frac{\text{SNR}}{2+\text{SNR}}}$

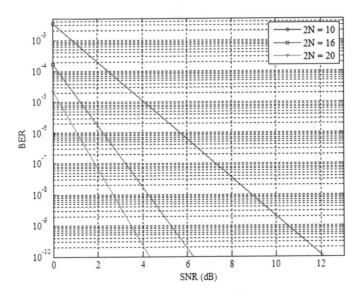

Fig. 15 BER performance under AWGN channel

The average bit error rate with respect to various number of antenna elements under the both channel conditions are plotted in Figs. 15 and 16, respectively. The other parameter values remain same as in the earlier case (SLL $= -30$ dB, FNBW $= 20°$ and Nulls $= 5$). It is apparent that average BER decreases with increase in number of antenna elements for a specific SNR level in the wireless channel.

6.2 Experimental Results

System performance is described here with throughput or latency obtained in the array signal processing. Usually, latency is a composite effect of delays occurring in both beamforming processes as well as mapping techniques involved. Therefore, throughput or latency of the system is validated with its computation time. Hardware complexity in the system architecture is also explained with FPGA resource utilization.

6.2.1 FPGA Resource Utilization

FPGA resource utilization against different number of antenna element settings along the array is given in Table 4. Other parametric values are also kept constant (SLL $= -30$ dB, FNBW $= 20°$ and Nulls or interfering directions $= 5$). It is obvious that less hardware are needed for using Cosine template and only a fixed number of DSPs

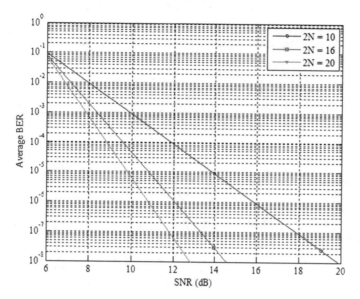

Fig. 16 BER performance under Rayleigh fading channel

or Multipliers are occupied on the system architecture irrespective of the antenna element numbers in this case. Thus, it suits best to the resource-constrained wireless system architectures, not putting extra hardware complexity on them.

6.2.2 Computation Time

For various number of antenna elements along the array, computation time is estimated with a maximum of 1000 iterations at 250 MHz clock and listed in Table 5. Other parametric settings remain the same (SLL = −30 dB, FNBW = 20° and Nulls or interfering directions = 5) as in the earlier case. At FPGA environment, computation time usually depends on variation of the iteration numbers to execute Verilog program under different test conditions. It is noticeable that more time is elapsed to process larger number of array intercepted signals.

Since such system architecture is developed using the serial mode of algorithmic operations, its latency is more prominent. However, latency could be reduced by adopting several parallel and pipelined architectures such as systolic array, Single Instruction, Multiple Data (SIMD) or Multiple Instruction, Multiple Data (MIMD) to its further extension. Behavioral modeling approach in the system design indeed leads to its implementation on FPGA platform with less hardware. It normally optimizes various synthesis tools in the Computer-Aided Design (CAD) flow for making system architecture on the re-configurable FPGA chip.

Table 4 FPGA resource utilization for the proposed beamforming scheme

FPGA resource parameters	Dolph–Chebyshev template			Cosine template		
	No. of array elements (2N)			No. of array elements (2N)		
	$N = 5$	$N = 8$	$N = 10$	$N = 5$	$N = 8$	$N = 10$
No. of occupied slices (26,624)	1278 (4%)	1789 (6%)	2068 (7%)	599 (2%)	831 (3%)	960 (3%)
No. of slice flip flops (53,248)	1212 (2%)	1669 (3%)	2068 (3.88%)	485 (0%)	638 (1%)	733 (1%)
No. of 4 input LUTs (53,248)	2287 (4.29%)	3248 (6.09%)	3865 (7.25%)	1048 (1%)	1497 (2%)	1714 (3%)
No. of bonded IOBs (448)	62 (13%)	70 (15.62%)	70 (15.62%)	13 (2%)	13 (2%)	13 (2%)
No. of DSPs 48 s (64)	18 (28%)	27 (42.19%)	33 (51.56%)	7 (10%)	7 (10%)	7 (10%)

Table 5 Computation time for proposed beamforming scheme

Computation time parameter	Dolph–Chebyshev template			Cosine template		
	No. of array elements (2N)			No. of array elements (2N)		
	$N = 5$	$N = 8$	$N = 10$	$N = 5$	$N = 8$	$N = 10$
No. of clock cycles	26,434,194	41,331,544	51,190,353	23,424,048	35,496,048	43,544,048
Clock time (ns)	4	4	4	4	4	4
Total time (s)	0.1057	0.1653	0.2048	0.0937	0.1420	0.1742

7 Conclusion

In this chapter, design and implementation of a simple PSO-based beamforming method is presented. PSO algorithm is successfully used for this work to determine the optimum array weights producing beam patterns as per the specifications stipulated in the desired templates. Its realization on the Xilinx vitex4 FPGA chip was made with CORDIC functional blocks using FSMD modeling. Beamforming accuracy and BER is verified with several hardware level fixed-point simulations under AWGN and Rayleigh fading channel conditions. Simulation results pertaining higher accuracy and lower computational overheads validate its acceptability in smart antennas, applied with resource-constrained wireless ad hoc networks infrastructures. However, better performance can be assured minimizing the latency problem with parallel and pipelined architectures in its future extension.

References

1. Tsoulos GV, Beach M, McGeehan J (1997) Wireless personal communications for the 21st century: European technological advances in adaptive antennas. IEEE Commun Mag 35(9):102–109
2. Liberti JC Jr, Rappaport TS (1999) Smart antennas for wireless communications: IS-95 and third generation CDMA applications. Prentice Hall, Upper Saddle Rive, New Jersey
3. Sun C, Cheng J, Ohira T (eds) (2009) Handbook on advancements in smart antenna technologies for wireless networks. IGI Global, New York
4. Soni RA, Buehrer RM, Benning RD (2002) Intelligent antenna system for cdma2000. IEEE Signal Process Mag 19(4):54–67
5. Babich F, Comisso M, D'orlando M et al (2006) Interference mitigation on WLANs using smart antennas. Wirel Pers Commun 36(4):387–401
6. Huang X, Wang J, Fang Y (2007) Achieving maximum flow in interference-aware wireless sensor networks with smart antennas. Ad Hoc Netw 5(6):885–896
7. El Zooghby A (2005) Smart antenna engineering. Artech House, London
8. Janaswamy R (2002) Radiowave propagation and smart antennas for wireless communications. Kluwer, New York
9. Bellofiore S, Balanis CA, Foutz J et al (2002) Smart-antenna systems for mobile communication networks. Part 1: overview and antenna design. IEEE Antennas Propag Mag 44(3):145–154
10. Bellofiore S, Foutz J, Balanis CA et al (2002) Smart-antenna systems for mobile communication networks. Part 2: beamforming and network throughput. IEEE Antennas Propag Mag 44(4):106–114
11. Blanz JJ, Papathanassiou A, Haardt M et al (2000) Smart antennas for combined DOA and joint channel estimation in time-slotted CDMA mobile radio systems with joint detection. IEEE Trans Veh Technol 49(2):293–306
12. Winters JH (2006) Smart antenna techniques and their application to wireless ad hoc networks. IEEE Wirel Commun 13(4):77–83
13. Sundaresan K, Sivakumar R (2006) Ad hoc networks with heterogeneous smart antennas: performance analysis and protocols. Wirel Commun Mobile Comput 6(7):893–916
14. Chen K, Jiang F (2007) A range-adaptive directional MAC protocol for wireless ad hoc networks with smart antennas. Int J Electron Commun 61(10):645–656
15. Quintero A, Li DY, Castro H (2007) A location routing protocol based on smart antennas for ad hoc networks. J Netw Comput Appl 30(2):614–636
16. Lim C-H, Wan Y, Ng B-P et al (2007) A real-time indoor Wi-Fi localization system utilizing smart antennas. IEEE Trans Consum Electron 53(2):618–622
17. Roy S, Boudreault J-F, Dupont L (2008) An end-to-end prototyping framework for compliant wireless LAN transceivers with smart antennas. Comput Commun 31(8):1551–1563
18. Perez-Neira A, Mestre X, Fonollosa JR (2001) Smart Antennas in software radio base stations. IEEE Commun Mag 39(2):166–173
19. Falletti E, Presti LL, Sellone F (2006) SAM LOST smart antennas-based movable localization system. IEEE Trans Veh Technol 55(1):25–42
20. Biswas RN, Mitra SK, Naskar MK (2014) A robust mobile anchor based localization system for wireless sensor networks using smart antenna. Int J Ad-hoc Ubiquitous Comput 15(1/2/3):23–37

21. Kang M, Alouini M-S, Yang L (2002) Outage probability and spectrum efficiency of cellular mobile radio systems with smart antennas. IEEE Trans Commun 50(12):1871–1877
22. Amin M, Zhang Y, Mancuso V et al (2002) Applications of smart antennas to rotorcrafts. Digit Signal Proc 12(2–3):159–174
23. Naguib AF, Seshadri N, Calderbank AR (2000) Increasing data rate over wireless channels. IEEE Signal Process Mag 17(3):76–92
24. Bandyopadhyay S, Roy S, Ueda T (2006) Enhancing the performance of ad hoc wireless networks with smart antennas. Taylor & Francis, Boca Raton
25. Huang F, Leung K-C, Li VOK (2010) Transmission radius control in wireless ad hoc networks with smart antennas. IEEE Trans Commun 58(8):2356–2370
26. Babich F, Comisso M (2009) Throughput and delay analysis of 802.11-based wireless networks using smart and directional antennas. IEEE Trans Commun 57(5):1413–1423
27. Khedr AM, Osamy W (2006) A topology discovery algorithm for sensor network using smart antennas. Comput Commun 29(12):2261–2268
28. Godara LC (2004) Smart antennas. CRC Press, Boca Raton
29. Winters JH (1998) Smart antennas for wireless systems. IEEE Pers Commun 5(1):23–27
30. Karmakar NC (ed) (2010) Handbook of smart antennas for RFID systems. Wiley, New Jersey
31. Babich F, Comisso M, D'orlando M et al (2007) Performance evaluation of distributed wireless networks using smart antennas in low-rank channel. IEEE Trans Commun 55(7):1344–1353
32. Kennedy J, Sullivan MC (1995) Direction finding and "smart antennas" using software radio architectures. IEEE Commun Mag 33(5):62–68
33. Kavak A, Torlak M, Vogel WJ et al (2000) Vector channels for smart antennas-measurements, statistical modeling, and directional properties in outdoor environments. IEEE Trans Microw Theory Tech 48(6):930–937
34. Gies D, Rahmat-Samii Y (2003) Particle swarm optimization for reconfigurable phase-differentiated array design. Microw Opt Technol Lett 38(3):168–175
35. Robinson J, Rahmat-Samii Y (2004) Particle swarm optimization in electromagnetic. IEEE Trans Antennas Propag 52(2):397–402
36. Dikmese S, Kavak A, Kucuk K et al (2010) Digital signal processor against field programmable gate array implementations of space-code correlator beamformer for smart antennas. IET J Microw Antennas Propag 4(5):593–599
37. Dikmese S, Kavak A, Kucuk K et al (2011) FPGA based implementation and comparison of beamformers for CDMA 2000. Wirel Pers Commun 57(2):233–253
38. Wood R, McAllister J, Lightbody G et al (2008) FPGA-based implementation of signal processing systems. Wiley, London
39. Grout I (2008) Digital systems design with FPGAs and CPLDs. Elsevier, USA
40. Song YS, Kwon HM, Min BJ (2001) Computationally efficient smart antennas for CDMA wireless communications. IEEE Trans Veh Technol 50(6):1613–1628
41. Haardt M, Spencer Q (2003) Smart antennas for wireless communications beyond the third generation. Comput Commun 26(1):41–45
42. Ward CR, Hargrave PJ, McWhirter JG (1986) A novel algorithm and architecture for adaptive digital beamforming. IEEE Trans Antennas Propag AP 34(3):338–346
43. Nuteson TW, Stocker JE, Clark JS et al (2002) Performance characterization of FPGA techniques for calibration and beamforming in smart antenna applications. IEEE Trans Microw Theory Tech 50(12):3043–3051
44. Choi S, Shim D (2000) A novel adaptive beamforming algorithm for a smart antenna system in a CDMA mobile communication environment. IEEE Trans Veh Technol 49(5):1793–1806
45. Sun C, Hirata A, Ohira T et al (2004) Fast beamforming of electronically steerable parasitic array radiator antennas: theory and experiment. IEEE Trans Antennas Propag 52(7):1819–1832
46. Li J, Jin R, Sheng Y (2003) A fast synthesis algorithm of adaptive beams for smart antennas. Microw Opt Technol Lett 36(6):503–507
47. Khodaei FG, Nourinia J, Ghobadi C (2010) Adaptive beamforming algorithm with increased speed and improved reliability for smart antennas. Comput Electr Eng 36(6):1140–1146

48. Hsu C-H (2007) Uplink MIMO–SDMA optimisation of smart antennas by phase-amplitude perturbations based on memetic algorithms for wireless and mobile communication systems. IET Commun 1(3):520–525
49. Lee W-C, Choi S (2005) Adaptive beamforming algorithm based on eigen-space method for smart antennas. IEEE Commun Lett 9(10):888–890
50. Chang D-C, Hu C-N (2012) Smart antennas for advanced communication systems. Proc IEEE 100(7):2233–2249
51. Hanzo L, Blogh JS, Ni S (2008) 3G, HSPA and FDD versus TDD networking smart antennas and adaptive modulation, 2nd edn. Wiley, London
52. Sarkar TK, Wicks MC, Salazar-Palma M et al (2003) Smart antennas. Wiley, New Jersey
53. Chryssomallis M (2000) Smart antennas. IEEE Antennas Propag Mag 42(3):129–136
54. Balanis CA (2004) Antenna theory analysis and design, 2nd edn. Wiley, New York
55. Zorzi M (2002) On the capture performance of smart antennas in a multicellular environment. IEEE Trans Commun 50(4):536–539
56. Sheikh K, Gesbert D, Gore D et al (1999) Smart antennas for broadband wireless access networks. IEEE Commun Mag 37(11):100–105
57. Cardoso FACM, Fernandes MAC, Arantes DS (2002) Space-time processing for smart antennas in advanced receivers for the user terminal in 3G WCDMA systems. IEEE Trans Consum Electron 48(4):1082–1090
58. Ho M-J, Stuber GL, Austin MD (1998) Performance of switched-beam smart antennas for cellular radio systems. IEEE Trans Veh Technol 47(1):10–19
59. Hartmann C, Nasser N (2009) Modeling and performance analysis of multi-service wireless CDMA cellular networks using smart antennas. Wirel Commun Mobile Comput 9(1):117–129
60. Gross FB (2005) Smart antennas for wireless communications. McGraw-Hill, USA
61. Meher PK, Walls J, Juang T-B et al (2009) 50 years of CORDIC: algorithms, architectures and applications. IEEE Trans Circuits Syst I Regul Pap 56(9):1893–1907
62. Chu PP (2008) FPGA prototyping by verilog examples. Wiley, New Jersey
63. Rao KD (2015) Channel coding techniques for wireless communications. Springer, New Delhi
64. Hong L, Armada AG (2011) Bit error rate performance of MIMO MMSE receivers in correlated Rayleigh flat-fading channels. IEEE Trans Veh Technol 60(1):313–317

A Comparison of Bio-Inspired Approaches for the Cluster-Head Selection Problem in WSN

Karen Miranda, Saúl Zapotecas-Martínez, Antonio López-Jaimes
and Abel García-Nájera

Abstract A Wireless Sensor Network (WSN) is composed of a set of energy and processing-constrained devices that gather data about a set of phenomena. An efficient way to enlarge the lifetime of a wireless sensor network is clustering organization, which structures hierarchically the sensors in groups and assigns one of them as a cluster head. Such cluster head is responsible of specific tasks like gathering data from other cluster sensors and resending it to the base station through the network. Using a cluster-head organization, data gathering process is improved and by extension, the network lifetime is enlarged. However, due to the additional tasks that every cluster head has to perform, their own energy is spent faster than that of the other sensors in the cluster. Each time that a cluster head is out of battery, it is necessary to select a new cluster head from the survival sensors to continue with head duties. In this chapter, we present a performance comparison of three state-of-the-art MOEAs, namely NSGA-II, SMS-EMOA, and MOEA/D for the cluster-head selection problem in Wireless Sensor Networks.

Keywords Cluster head selection · Multi-objective optimization · Wireless sensor networks · Bio-inspired algorithms · Combinatorial optimization

K. Miranda (✉)
Universidad Autónoma Metropolitana Unidad Lerma, State of Mexico, Mexico
e-mail: k.miranda@correo.ler.uam.mx

S. Zapotecas-Martínez · A. López-Jaimes · A. García-Nájera
Universidad Autónoma Metropolitana Unidad Cuajimalpa, Mexico City, Mexico
e-mail: szapotecas@correo.cua.uam.mx

A. López-Jaimes
e-mail: alopez@correo.cua.uam.mx

A. García-Nájera
e-mail: agarcian@correo.cua.uam.mx

© Springer International Publishing AG, part of Springer Nature 2019
S. K. Shandilya et al. (eds.), *Advances in Nature-Inspired Computing and Applications*, EAI/Springer Innovations in Communication and Computing,
https://doi.org/10.1007/978-3-319-96451-5_7

1   Introduction

Wireless Sensor Networks (WSNs) cover a wide range of applications intended to operate in diverse environments, for example, disaster relief operations, smart cities, precision agriculture, and health care. A WSN is composed of a large set of energy- and processing-constrained devices that gather data about a set of phenomena [3]. Therefore, it is desirable that the network be active as long as possible. Typically, sensors are deployed randomly in a prespecified area and they broadcast the collected data over the network toward a base station (BS) located in another position. An efficient way to enlarge the lifetime of a wireless sensor network is to organize the sensors in clusters, which structures the sensors hierarchically in groups and assigns one of them as a cluster head (CH) according to some given rules. Such a cluster head is in charge of specific tasks, like data gathering from other sensors in the cluster and resending that data over the network to a base station.

Hence, using cluster heads, the data gathering process is improved and, by extension, the network lifetime is enlarged. However, due to the additional tasks that every cluster head has to perform, their own energy is spent faster than the neighbor sensors. Each time that a cluster head is out of battery, it is necessary to select a new cluster head from the rest of the sensors in the cluster in order to continue with the head duties. The cluster head selection may be at random or based on well-defined criteria, like residual energy, node distance, signal strength, or connectivity. Therefore, such criteria depend on the objectives to be optimized, e.g., to maximize the network lifetime or to minimize the energy consumption. Moreover, due to every sensor in the network is a candidate to be a cluster head, the selection is a combinatorial problem. The problem of clustering the WSN and the cluster-head selection is known to be NP-hard. In general, the cluster-head selection problem may be seen as a generic multi-objective optimization problem, for example, as a resources allocation with inputs, required outputs, optimization of objective functions, and satisfaction of constraints [22].

Thus, some objectives such as maximizing the lifetime of the network, maximizing coverage, minimizing cost, minimizing energy consumption, or maximizing spectrum utilization are easy to identify. Nevertheless, depending on the particular purpose of the sensor network, we may find a wide variety of objectives that are desired to be optimized. Hence, it is crucial to determine the objectives that are relevant to the problem and those that are optimized by optimizing other objectives, since it is well known that not all of the objectives are necessary [7].

In this chapter, we present a comparison of three state-of-the-art Multi-objective Evolutionary Algorithms, also known as MOEAs, to solve the cluster-head selection problem. On the one hand, we present an analysis of three objectives commonly found in the literature [2]: (*i*) to minimize the distance between sensor members, (*ii*) to minimize the distance of cluster heads and the base station, and (*iii*) to maximize the residual energy of the cluster heads. To this end, we perform an experimental study using the following approaches: (1) Non-dominated Sorting Genetic Algorithm II (NSGA-II) [13], (2) S Metric Selection Evolutionary Multi-objective Optimization

Algorithm (SMS-EMOA) [5], and (3) Multi-objective Evolutionary Algorithm Based on Decomposition (MOEA/D) [46].

These approaches are particularly interesting because they implement different kinds of survival selection mechanisms, namely Pareto-, indicator-, and decomposition-based selection. The Pareto-based selection achieves selection using an approximation of the Pareto frontier, the indicator-based considers the scalar performance indicator, called dominated hypervolume, to approximate the Pareto frontier, and the decomposition-based selection separates a multi-objective optimization problem into several single-objective optimization subproblems. The MOEAs employed in our comparative study handles binary representation, which is a suitable encoding to represent the sensors that are members of the cluster and the cluster head of each cluster. Finally, we believe that the election of an appropriate combination of objectives leading to a suitable multi-objective approach will be of great importance to achieve a good level of energy conservation.

The rest of this chapter is structured as follows. Chapter 2 introduces WSN, as well as the system and energy models. Chapter 3 defines multi-objective optimization problems and the three approaches used in this study. The cluster-head selection problem is explained in Chap. 4. The experimental setup and results are described in Chap. 5. Finally, Chap. 6 presents the conclusions of this investigation.

2 Wireless Sensor Networks in a Nutshell

A typical Wireless Sensor Network (WSN) is composed of a large set of small devices that captures data about a given phenomenon. The sensors works together in order to send over the network the gathered information to a base station located in another position, as depicted in Fig. 1. This task is nontrivial, since the sensors are energy- and processing-constrained. WSNs may be classified by their main goal or by their physical characteristics. In this chapter, we consider a WSN where, once the sensors are deployed, they remain in the same position throughout the network lifetime, i.e., a static sensor network.

Since the main goal of a WSN is to retrieve information about environmental phenomena and in some cases, the sensors are deployed in difficult access areas, it is desirable that the network would keep working as long as possible. To this end, it is important to reduce wireless data communication. Therefore, there exists several data reduction techniques to reduce energy consumption such as in-network processing techniques, data compression techniques, data prediction techniques, and topology control. Specifically, *clustering* is a very effective technique to reduce the data transmitted between the sensors and the base station (BS). Therefore, an efficient way to enlarge the lifetime of a WSN is clustering organization, which structures hierarchically the sensors in groups and assigns one of them as a cluster head (CH) [41].

Each cluster head is in charge of gathering the data from its neighbor sensors and resend it to the BS. Particularly, the cluster-head selection is a crucial decision, since

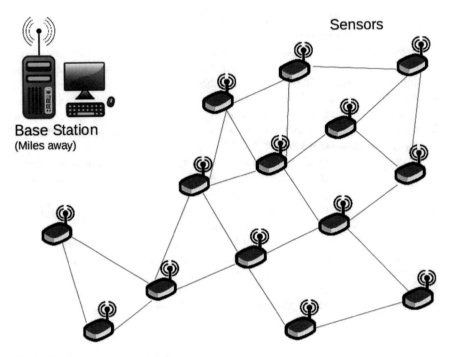

Fig. 1 Wireless sensor network infrastructure

the sensor selected as CH will consume its battery faster than a normal sensor. The cluster-head selection may be simply as to select the sensor with a full battery or with the higher residual energy. Although a simple rule does not consume a significant amount of processing energy, this kind of rules could not satisfy the network specific requirements. Moreover, the cluster-head selection depends on the nature of the network, for example, scalability, fault tolerance, load balancing, enlarging the network lifetime, or enlarging coverage [2].

In the literature, we may find some approaches based on heuristics to solve the cluster-head selection problem. In [38], the authors present a Particle Swarm Optimization approach to find the optimal cluster-heads' positions in order to reduce the distance between the CH and the sensors in the cluster. A Pareto-based method is presented in [4], where it is used to estimate the optimal number of clusters by means of the multi-attribute decision-making process called TOPSIS. In this work, the cluster head is selected according to the energy, the intra-cluster density, and the distance to the base station. In [37], an Ant Colony Optimization (ACO) approach is presented to minimize the number of cluster heads and to maximize the number of members per cluster, and, thus, improving the network coverage. Similarly, the authors in [19] modified the ACO metaheuristic so that each sensor computes the probability to become a cluster head based on the pheromone.

2.1 System Model

A WSN may be represented as a graph $G=\{S, C\}$, where S is the finite set of n sensors s_i and a base station s_B. C represents the finite set of m connections between sensors in the network. The sensors must transmit the gathered information to the base station directly (one-hop communication) or by means of other sensors (multi-hop communication). So, we assume that:

  i. the sensors are randomly deployed in a two-dimensional Cartesian space;
 ii. the sensors remain static once they are deployed;
iii. the sensors' battery is not rechargeable;
 iv. all the sensors have the same initial energy level;
  v. all the sensors have the same processing and communication capabilities;
 vi. the wireless communication links are bidirectional; and
vii. the sensors do not know their own position or the base station position.

2.2 Energy Model

Since in this Chapter, we focus on analyzing the sensor energy consumption when a sensor is selected as cluster head and how this selection may enlarge the network lifetime, we require an energy model that characterizes the main features of a WSN based, mainly, on the distance between the sensors and their cluster head, and the distance between cluster heads and the base station. The network parameters, and their corresponding values, are presented in Table 1.

For convenience, we use the energy consumption model presented in [39] as follows:

$$E_{tx} = k \times E_{elec}$$
$$E_{tx}(k, d) = k \times E_{elec} + \varepsilon_{amp} \times k \times d^2$$
$$E_{CH} = \left(100\mu J + \left(0.2\mu J \times D_{BS}^2\right)\right) + (10\mu J \times N_M)$$
$$M = 100\mu J + \left(0.2\mu J \times D_{CH}^2\right)$$

3 Multi-objective Optimization

In this section, some multi-objective optimization concepts are defined.

Definition 1 (*Multi-objective optimization problem*) A multi-objective optimization problem (MOP) can be defined, without loss of generality, as the minimization problem

Table 1 Network simulation parameters

Parameter	Symbol	Value
Initial sensor energy	E_I	9 J
Number of sensors	n	20–500
Number of sensors per cluster	N_M	
Member	M	100 μJ
Cluster Head	CH	
Position of Base Station	BS	X, Y(0,0)
Packet size	k	2000 bits
Transmission power	E_{tx}	
Electronics energy	E_{elec}	50nJ/bit
Energy used by a cluster head for a report	E_{CH}	
Distance to CH	D_{CH}	
Distance to BS	D_{BS}	
Multiple transmission power	ε_{amp}	
Sensing area	m × m	400×400

$$\min_{x \in \mathfrak{X}} \mathbf{f}(\mathbf{x}) = (f_1(\mathbf{x}), \ldots, f_m(\mathbf{x}))$$
$$\text{subject to } g_i(\mathbf{x}) \leq 0, \ \forall i \in \{1, \ldots, p\}, \tag{1}$$
$$h_j(\mathbf{x}) = 0, \ \forall j \in \{1, \ldots, q\},$$

where $\mathbf{x}$ is a potential solution to the problem, $\mathfrak{X}$ is the domain of solutions, and f_i: $\mathfrak{X} \to \mathbb{R}$, $\forall i \in \{1, \ldots, m\}$, are the m objective functions. The constraint functions $g_i, h_j : \mathfrak{X} \to \mathbb{R}$ delimit the feasible search space.

Definition 2 (*Pareto dominance*) A solution $\mathbf{x} \in \mathfrak{X}$ weakly dominates (or covers) solution $\mathbf{y} \in \mathfrak{X}$, written as $\mathbf{x} \preccurlyeq \mathbf{y}$, if $\mathbf{x}$ is at least as good as $\mathbf{y}$. That is, $\mathbf{x} \preccurlyeq \mathbf{y}$ if and only if $f_i(\mathbf{x}) \leq f_i(\mathbf{y})$, $\forall i \in \{1, \ldots, m\}$.

Solution $\mathbf{x}$ *dominates* solution $\mathbf{y}$, written as $\mathbf{x} \prec \mathbf{y}$, if $\mathbf{x}$ is at least as good as $\mathbf{y}$ and strictly better in at least one objective. That is, $\mathbf{x} \prec \mathbf{y}$ if and only if $\mathbf{x} \preccurlyeq \mathbf{y}$, and $\exists j : f_j(\mathbf{x}) < f_j(\mathbf{y})$. Consequently, solution $\mathbf{x} \in \mathfrak{S} \subseteq \mathfrak{X}$ is non-dominated with respect to subset $\mathfrak{S}$ if $\nexists \mathbf{y} \in \mathfrak{S} : \mathbf{y} \prec \mathbf{x}$.

Definition 3 (*Pareto optimality*) A solution $\mathbf{x} \in \mathfrak{X}$ is Pareto optimal if it is non-dominated with respect to the entire domain of solutions $\mathfrak{X}$.

Definition 4 (*Pareto optimal set*) The Pareto optimal set $\mathfrak{P}_s$ is the set of all Pareto optimal solutions, that is, $\mathfrak{P}_s = \{\mathbf{x} \in \mathfrak{X} | \mathbf{x} \text{ is Pareto optimal}\}$.

Definition 5 (*Pareto front*) The Pareto front P f is the set of the evaluations of the function vector $\mathbf{f}$ at all solutions in the Pareto optimal set, that is, $\mathfrak{P}_f = \{f(\mathbf{x}) \in \mathbb{R}^m | \mathbf{x} \in \mathfrak{P}_s\}$. A Pareto approximation set $\mathfrak{A}$(or simply, Pareto approximation or approximation set) is the result of a heuristic multi-objective optimization process. One important condition of a multi-objective problem is the conflict between its objectives. If the objectives have no conflict between them, then we could solve the problem optimizing each objective function independently.

One important condition of a multi-objective problem is the conflict between its objectives. If the objectives have no conflict between them, then we could solve the problem optimizing each objective function independently. Nonetheless, it has been found that, in some problems, although a conflict exists elsewhere, some objectives behave in a nonconflicting manner. Even though different authors have proposed definitions for conflict between objectives (see e.g., [7, 9, 36]), in this document we used the following definition.

Definition 6 (*Conflict relation*) Let $\mathfrak{S}$ be a subset of $\mathfrak{X}$. According to Carlsson and Fullér [9], two objectives can be related in the following ways (assuming minimization):

1. f_i is in conflict with f_j on $\mathfrak{S}$ if $f_i(\mathbf{x}) \le f_i(\mathbf{y})$ implies $f_j(\mathbf{x}) \ge f_j(\mathbf{y})$ $\forall \mathbf{x}, \mathbf{y} \in \mathfrak{S}$.
2. f_i supports f_j on $\mathfrak{S}$ if $f_i(\mathbf{x}) \ge f_i(\mathbf{y})$ implies $f_j(\mathbf{x}) \ge f_j(\mathbf{y})$ $\forall \mathbf{x}, \mathbf{y} \in \mathfrak{S}$.
3. f_i and f_j are independent on $\mathfrak{S}$ otherwise.

In the cases 2 and 3, those objectives are also called nonconflicting objectives. When $\mathfrak{S} = \mathfrak{X}$, it is said that f_i is in conflict with (or supports) f_j globally. However, in many MOPs the relation among the objectives changes when comparing different subsets of $\mathfrak{X}$.

Although supportive objectives can be optimized independently, in practice it might be useful to include them during the search. For instance, at the early stages of the design of a real-world problem to learn about the relationship among objectives and parameters, or because in some problems the landscape of the search space with additional objectives might make the problem easier.

3.1 Evolutionary Approaches for Multi-objective Optimization

In many cases, the development of a new optimization technique is the result of the need to solve some kind of real-life MOPs. Therefore, the design of those new techniques is oriented to take advantage of the particular characteristics of the given problem. For instance, there are many techniques specialized for solving linear multi-objective optimization problems (see e.g., [46]), and techniques devoted to solve convex MOPs (see e.g., [16]). Because of their flexibility and ease of use, MOEAs have become an alternative to solve a MOP in its most general case. Through the years, several MOEAs from different nature have been proposed, see the comprehensive review reported in [10, 34, 49]. However, these approaches follow a basic principle to obtain a proper approximation to the Pareto set while maintaining a good representation of the Pareto front. In this sense, MOEAs can be classified into three main groups which are briefly described below.

Pareto-based MOEAs

The most common preference relation to compare solutions in $\mathbb{R}^m$ (where m is the number of objectives) is the Pareto dominance. Therefore, it is natural that early proposals for solving MOPs tried to integrate this relation in a straightforward manner. Although there are different approaches to use Pareto dominance, the common goal of most of them is to rank the population to assess closeness to the optimal Pareto front. This way, the rank of a solution can be used as a criterion for mating or survival selection. One of the best-known approaches that use Pareto dominance to rank solutions is the dominance rank [18]. Other approaches are, for example, the dominance count [51] and the dominance depth [41].

As mentioned earlier, a good approximation of the Pareto front has to fulfill two goals simultaneously: convergence and diversity. This means that, in order to distribute the solutions along the entire trade-off curve, Pareto dominance has to be used in cooperation with a second criterion. Thus, some methods that have been proposed to distribute solution along the Pareto optimal front include fitness sharing and niching [12], clustering [51], crowding distance [13], among many others. Although these methods were very popular in the first decade of the 2000s, their use has decreased because of the difficulty of measuring diversity in a set of non-dominated solutions, as pointed out by several researchers [15, 17, 20], particularly in high-dimensional objective spaces [29, 43].

On the other hand, in recent years, several studies [7, 30] have found that Pareto dominance loses their discriminant property as more objectives are aggregated to the MOP. The most accepted explanation for this behavior is the increase of the proportion of non-dominated individuals as the number of objectives increases [7, 30].

Indicator-based MOEAs

Other strategies employed by MOEAs to achieve an adequate representation of the Pareto front are related to performance indicators. With its emergence, the indicator-based evolutionary algorithm (IBEA) [50] posed the possibility to optimize a performance indicator in the evolutionary process of MOEAs. As it is well known, there exist a large number of indicators to assess the performance of MOEAs, see for example [24, 35, 53]. Such indicators are able to assess, in different ways: convergence and diversity, or both of them at the same time. In particular, a good representation of the real Pareto front of a MOP can be achieved using performance indicators such as hypervolume [52], *R2* [21], *IGD* [11], among others.

As it was mentioned earlier, reference indicator-based MOEAs depend on a reference set which, in most of the cases, is difficult to construct. As alternative, IBEAs based on hypervolume have showed flexibility, since they do not need a reference set, instead, one reference vector is only required to compute the hypervolume indicator. However, as pointed out before, the use of these approaches is limited by the high computational cost of the hypervolume indicator, which increases as the number of objectives augments. Nonetheless, an advantage of using

IBEAs based on hypervolume is that they can deal properly with different Pareto front geometries, including convex, concave, mixed, disconnected, and degenerated shapes.

Decomposition-based MOEAs

In the last decade, scalarization functions have been employed by several evolutionary approaches, giving rise to the well-known MOEAs based on decomposition. Decomposition approaches rely on solving a number of scalarization functions, which are formulated by an even number of weight vectors. Such scalarization functions are solved through the search of approximate solutions toward the real Pareto front. Decomposition-based MOEAs have been found to be very efficient in solving complicated test problems, see for example the studies presented in [27, 47, 48]. In addition, having a well-distributed set of weight vectors, a proper representation of the entire Pareto front can be reached in some multi-objective problems. Since in real-world problems, the geometry of the Pareto front is not known beforehand, the distribution of the weight vectors needs to be carefully defined. For instance, a uniform distribution can achieve a poor approximation if the Pareto front is disconnected, degenerated, or a combination of convex and concave shapes. Furthermore, in a many-objective scenario, the number of weight vectors increases exponentially. This weakness of decomposition-based MOEAs becomes inconvenient when dealing with MOPs with difficult Pareto front geometries [28, 45] and high-dimensional objective spaces [23].

4   Multi-objective Cluster-Head Selection

Clustering organization structures hierarchically the sensors in virtual groups and selects one of them as cluster head for each group. Such a cluster head is in charge to gather data from the in-cluster sensors, compress data, and resend it to a base station located in another position, thus, reducing the wireless communication transmissions to the base station and, consequently, saving sensors' energy [32]. Nevertheless, by performing these extra duties, the cluster heads spend their own energy faster than the regular sensors (Fig. 2). Each time that a cluster head is out of battery, it is necessary to select a new cluster head from the remaining sensors to continue with the cluster-head duties.

It is well known that clustering in WSN is an NP-hard problem since, for n sensor in the network, there exist $2^n - 1$ different combinations of solutions and, therefore, the optimal solution implies a search throughout a large set of possible solutions [14]. The cluster-head selection problem may be considered as a generic optimization problem for resource allocation with well-defined input and output, objectives, and constraints [22] such as maximizing the network lifetime, maximizing the network coverage, minimizing the cost, minimizing the energy consumption, and maximizing the spectrum utilization.

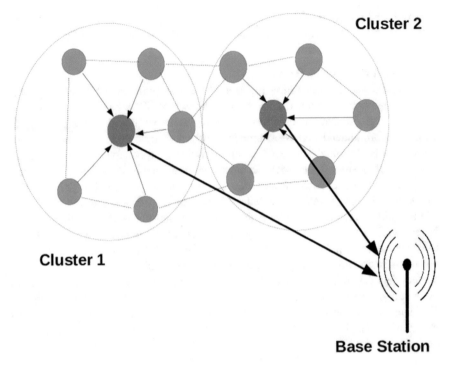

Fig. 2 Clustering communication in WSN

In a WSN, the main goal is to improve the data gathering process and, consequently, to enlarge the network lifetime. However, this particular objective cannot be directly assessed due to that actual lifetime is know when the network stops working. It is often possible to use a network simulator to estimate this time, nonetheless, the computer resources (time and storage) are limited and it is not possible to wait the simulation results each time that a cluster head is replaced. Therefore, we may use proxy functions that contribute to maximize the network lifetime [1, 25]. For example,

1. Minimizing the distance between cluster heads and the base station.
2. Minimizing the distance between sensors and their cluster head.
3. Maximizing the cluster-heads' residual energy.
4. Maximizing the load balance.

These objectives may be optimized individually [22] or simultaneously [2]. In [7, 8], the authors present an analysis of the objectives relevance and they determine that only some objectives are necessary to solve a given problem. In other words, when we have no essential objectives, the Pareto optimal set is the same whether these objectives are included or not.

We represent a solution as a binary string $\mathbf{x} \in \{0, 1\}^n$, where n number of sensors. Therefore, we have $x_i = 1$ if the sensor s_i is a cluster head, and $x_i = 0$ if the sensor s_i is part of a cluster.

We consider that cluster-heads' indices conform the set $H = \{i|x_i = 1\}$ and the position of sensor s_i is $s_i \in \mathbb{R}^2$. Likewise, the base station s_B position is $s_B \in \mathbb{R}^2$.

Thus, in order to determine to which cluster a sensor belongs to, we take into account the closest cluster head. That is, the cluster $C(s_i)$ to which the sensor $s_i = (i = 1, \ldots, n)$ belongs to is given by

$$C(s_i) = \arg \min_{j \in H} \|s_i - s_j\|,$$

where $\| \cdot \|$ is the Euclidean distance between two sensors. The set M_j of sensors indices which are members of cluster j is given by

$$M_j = \{i|C(s_i) = j, i = 1, \ldots, n\}.$$

It is worth noticing that, with this definition, every cluster will have at least one member. Thus, we define the three objectives that we aim at optimizing in this study are as follows:

1. To minimize the average distance between all the cluster heads $j \in H$ and the base station s_B:

$$\text{Minimize } f_1(\mathbf{x}) = \frac{1}{|H|} \sum_{j \in H} \|s_j - s_B\|.$$

2. To minimize the average distance between all members $i \in M_j$ and its corresponding cluster head $j \in H$ overall clusters:

$$\text{Minimize } f_2(\mathbf{x}) = \frac{1}{|H|} \sum_{j \in H} \left(\frac{1}{M_j} \sum_{i \in M_j} \|s_i - s_j\| \right).$$

3. To maximize the average cluster-heads' residual energy is

$$\text{Maximize } f_3(\mathbf{x}) = \frac{1}{|H|} \sum_{j \in H} E_I - E_{CH_j}$$

where E_I is the initial energy of the sensor and E_{CH_j} is the energy spent by the j-th CH as presented in Table 1.

5 Experimental Study

In order to identify the difficulties of solving the multi-objective cluster-head selection (MOCHS) problem, we conduct an experimental study by comparing three state-of-the-art MOEAs over four different instances of the MOCHS problem. Particularly, we are interested in identifying the main obstacles when MOCHS problem is solved by MOEAs based on three different principles. Thus, we focus our experimental study in the comparison of the performance of three evolutionary approaches based on the principles of Pareto dominance, hypervolume, and decomposition for multi-objective optimization. In the following section, we present the MOEAs based on the above principles which, throughout the years, have shown a relative good performance when solving multi-objective optimization problems.

5.1 Adopted MOEAs for Performance Comparison

NSGA-II: Non-dominated Sorting Genetic Algorithm II
The Non-dominated Sorting Genetic Algorithm II (NSGA-II) was proposed by Deb et al. [13]. This evolutionary approach builds a population of competing individuals, ranks and sorts each individual according to its non-domination level, applies evolutionary operators to create a new offspring pool, and then combines the parent and offspring populations before partitioning the new combined pool into fronts. For each ranking level, a crowding distance is estimated by calculating the sum of the Euclidean distances between the two neighboring solutions, from either side of the solution, along with each of the objectives. Once the non-domination rank and the crowding distance are calculated, the surviving individuals are determined using the crowded-comparison operator ($\prec_n$). The crowded-comparison operator guides the selection process at the various stages of the algorithm toward a uniformly spread out Pareto optimal front. Assuming that every individual in the population has two attributes, (1) non-domination rank (i_{rank}) and (2) crowding distance ($i_{distance}$), the partial order $\prec n$ is defined as

$$i \prec_n j : \text{if}(i_{rank} < j_{rank}) \text{ or } ((i_{rank} = j_{rank}) \text{ and } (i_{distance} < j_{distance})) \qquad (2)$$

That is, between two solutions with different non-domination ranks, we prefer the solution with the lower (better) rank. Otherwise, if both solutions belong to the same front, then the solution that is located in a less crowded region is preferred. Algorithm 1 presents the outline of the NSGA-II, which (in the last decade) has been the most popular MOEA, and is frequently adopted to compare the performance of newly introduced MOEAs.

Algorithm 1: General framework of NSGA-II

Input:
N: the population size;
stopping_condition: A stopping criterion.
Output:
P_t: the final approximation to the Pareto set;

```
1   begin
2   |   t = 0;
3   |   Generate a random population P_t of size N;
4   |   Evaluate the population P_t;
5   |   while stopping_condition is not satisfied do
6   |   |   Generate the offspring population Q_t;
7   |   |   Evaluate the offspring population Q_t;
8   |   |   R_t = P_t ∪ Q_t;
9   |   |   Rank R_t by using non-dominated sorting to define ℱ; // ℱ = (ℱ_1, ℱ_2, ...), all
    |   |       non-dominated fronts of R_t
10  |   |   P_{t+1} = ∅ and i = 1;
11  |   |   while (|P_{t+1}| + |ℱ_i| ≤ N) do
12  |   |   |   Assign crowding distance to each front ℱ_i;
13  |   |   |   P_{t+1} = P_{t+1} ∪ ℱ_i;
14  |   |   |   i = i + 1;
15  |   |   Sort ℱ_i by using the crowded-comparison operator;
16  |   |   P_{t+1} = P_{t+1} ∪ ℱ_i[1 : (N − |P_{t+1}|)];
17  |   |   t = t + 1;
18  |   return P_t;
```

SMS-EMOA: S Metric Selection Evolutionary Multi-objective Optimization Algorithm

The S Metric Selection Evolutionary Multi-objective Optimization Algorithm (SMS-EMOA) [5] features a selection operator based on the hypervolume (S metric) difference combined with dominance-depth sorting. The standard SMS-EMOA starts with an initial population of N randomly generated solutions. Then, at each iteration, a new individual is generated by means of stochastic variation operators. This individual is included in the population whose size, therefore, grows to $N + 1$. In order to reduce the population size back to N, and to obtain the next population, one individual has to be removed. The choice of the individual to remove is done in two steps. First, the dominance-depth sorting is applied to assign a non-domination rank to each individual, exactly as it is done in NSGA-II. Then, the individuals having the worst rank are considered (the set of these individuals is denoted by $\mathcal{F}_w$). If there is a single individual in $\mathcal{F}_w$, this one is removed. Otherwise, the contribution of each individual $i \in \mathcal{F}_w$ to the S metric is computed according to:

$$\Delta_S(i, \mathcal{F}_w) = S(\mathcal{F}_w) - S(\mathcal{F}_w\{i\}), \quad (3)$$

where $S(\mathcal{A})$ denotes the hypervolume of the non-dominated solutions in $\mathcal{A}$. Therefore, the individual with a smaller contribution to the hypervolume (i.e., the one with smallest Δ_S value) is removed.

A comprehensive study of several selection variations within the SMS-EMOA framework was presented in [33]. In the present study, we use an improved selection mechanism combining dominance depth, the number of dominating points, and hypervolume. Indeed, it has been shown that this leads to better performance than the standard SMS-EMOA. The difference with the standard SMS-EMOA is that, in the case where there is more than a single non-domination rank, the individual i with the highest number of dominating points among the solutions of the worst ranked front is discarded. The number of dominating points $d(i, P)$ is the number of solutions in P that dominate point i. Otherwise (if there is a single non-domination rank), the hypervolume difference indicator (Eq. 3) is used to select the solution to remove. Algorithm 2 presents the outline of the SMS-EMOA variant used in our study.

Algorithm 2: General framework of SMS-EMOA

Input:
N: the population size;
stopping_condition: A stopping criterion.
Output:
P_t: the final approximation to the Pareto set;

1 begin
2 $\quad$ $t = 0$;
3 $\quad$ Generate a random population P_t of size N;
4 $\quad$ Evaluate the population P_t;
5 $\quad$ while *stopping_condition* is not satisfied do
6 $\quad\quad$ Generate the offspring individual Q_t;
7 $\quad\quad$ Evaluate the offspring Q_t;
8 $\quad\quad$ $P_{t+1} = P_t \cup \{Q_t\}$;
9 $\quad\quad$ Rank P_{t+1} by using non-dominated sorting to define $\mathscr{F}$; //
   $\quad\quad\quad$ $\mathscr{F} = (\mathscr{F}_1, \mathscr{F}_2, \ldots, \mathscr{F}_w)$, all non-dominated fronts of P_{t+1}
10 $\quad\quad$ if $w > 1$ then
11 $\quad\quad\quad$ $i_w = \arg\max_{i \in \mathscr{F}_w} d(i, P_t)$;
12 $\quad\quad$ else
13 $\quad\quad\quad$ $i_w = \arg\min_{i \in \mathscr{F}_1} \Delta\mathscr{S}(i, \mathscr{F}_1)$;
14 $\quad\quad$ $P_{t+1} = P_t \setminus \{i_w\}$;
15 $\quad\quad$ $t = t + 1$;
16 $\quad$ return P_t;

MOEA/D: Multi-Objective Evolutionary Algorithm Based on Decomposition
The Multi-Objective Evolutionary Algorithm Based on Decomposition (MOEA/D) [47], transforms a MOP into N scalar subproblems. Therefore, an approximation of the Pareto set is obtained by solving the N scalar subproblems into which the MOP is decomposed. Considering $W = \{\mathbf{w}^1, \ldots, \mathbf{w}^N\}$ as a well-distributed set of weighting coefficient vectors, MOEA/D finds the best solution to each subproblem defined by each weight vector using a scalarizing function. In our study we adopt the Tchebycheff function in order to formulate such subproblems. Note, however, that other scalarizing approaches could also be adopted, see [31]. More formally, the Tchebycheff approach transforms the vector of function values $\mathbf{F}$ into a scalar optimization problem which is of the form:

$$\text{Maximize } g^{te}(\mathbf{x}|\mathbf{w}, \mathbf{z}) = \max_{1 \leq j \leq k} \{ w_j | f_j(\mathbf{x}) - z_j \}$$

$$\text{s.t} \qquad \mathbf{x} \in \mathbf{\Omega} \tag{4}$$

where $\mathbf{\Omega}$ is the feasible region, $\mathbf{z} = (z_1, \ldots, z_k)^T$, such that $z_j = \min f_j(\mathbf{x}|\mathbf{x} \in \mathbf{\Omega})$ and $\mathbf{w} = (w_1, \ldots, w_k)^T$ is a weight vector, i.e., $\sum_{j=1}^{k} w_j = 1$ and $w_j \geq 0$ for each $j = 1, \ldots, k$. Since $\mathbf{z} = (z_1, \ldots, z_k)^T$ is unknown, MOEA/D states each component z_j by the minimum value for each objective f_j found during the search, $j = 1, \ldots, k$.

In MOEA/D, a neighborhood of a weight vector $\mathbf{w}^i$ is defined as a set of its closest weight vectors in W. Therefore, the neighborhood of the weight vector $\mathbf{w}^i$ contains all the indices of the T closest weight vectors to $\mathbf{w}^i$. Algorithm 3 presents the general framework of MOEA/D.

Algorithm 3: General framework of MOEA/D

Input:

N: the number of subproblems in which the problem is to be decomposed;

W: a well-distributed set of weight vectors $\{\mathbf{w}^1, \ldots, \mathbf{w}^N\}$;

T: the neighborhood size;

stopping_condition: A stopping criterion.

Output:

P: the final approximation to the Pareto set.

1   $\mathbf{z} = (+\infty, \ldots, +\infty)^T$;

2   Generate a random set of solutions $P = \{\mathbf{x}^1, \ldots, \mathbf{x}^N\}$ in $\mathbf{\Omega}$;

3   for $i = 1, \ldots, N$ do

4      $B_i \leftarrow \{i_1, \ldots, i_T\}$, such that: $\mathbf{w}^{i_1}, \ldots, \mathbf{w}^{i_T}$ are the T closest weight vectors to $\mathbf{w}^i$;

5      $z_j \leftarrow \min(z_j, f_j(\mathbf{x}^i))$ // $j \in \{1, \ldots, k\}$

6   while *stopping_condition* is not satisfied do

7   for $\mathbf{x}^i \in P$ do

8   Reproduction: Randomly select two indices k, l from B_i, and then generate a new solution $\mathbf{y}'$ from $\mathbf{x}^k$ and $\mathbf{x}^l$ by using genetic operators.

9   Update of $\mathbf{z}$: $z_j \leftarrow \min(z_j, f_j(\mathbf{y}'))$ // $j \in \{1, \ldots, k\}$

10   Update of Neighboring Solutions: For each index $j \in B_i$, if $g(\mathbf{y}'|\mathbf{w}^j, \mathbf{z}) < g(\mathbf{x}^i|\mathbf{w}^j, \mathbf{z})$, then set $\mathbf{x}^j = \mathbf{y}'$;

11   return P;

5.2 Performance Assessment

The comparison among the algorithms considered in this study was carried out by following the performance assessment experimental setup recommended in [26]. Particularly, the MOEAs were evaluated adopting two performance indicators taken from the specialized literature, which is described below.

Normalized Hypervolume

The hypervolume performance indicator ($\mathcal{H}$) was introduced in [52] to assess the performance of MOEAs. This performance indicator is Pareto compliant [53], and quantifies both convergence and spread of non-dominated solutions along the Pareto approximation. The hypervolume corresponds to the non-overlapped space of all the hypercubes formed by a reference point $\mathbf{r}$ (given by the user) and each solution $\mathbf{a}$ in the Pareto approximation (A). Hypervolume indicator is mathematically stated as

$$\mathcal{H}(A) = \mathcal{L}\left(\bigcup_{a \in A} \{\mathbf{x} | \mathbf{a} \prec \mathbf{x} \prec \mathbf{r}\}\right) \tag{5}$$

where $\mathcal{L}$ denotes the Lebesgue measure and $\mathbf{r} \in \mathbb{R}^m$ denotes a reference vector being dominated by all solutions in A.

Therefore, the normalized $\mathcal{H}$ indicator (denoted here as $\mathcal{H}_n$) measure is defined by

$$\mathcal{H}(A) = \frac{\mathcal{H}(A)}{\prod_{i=1}^{m} |r_i - u_i|} \tag{6}$$

where $\mathbf{u} = (u_1, \ldots, u_m)^T$ is the known ideal vector and M denotes the number of objectives. Thus, $\mathcal{H}_n$ value is given into the range $[0, 1]$. A high value of this performance indicator means that the set A has a good approximation and distribution along the Pareto front.

IGD

The inverted generational distance (IGD) [11] quantifies how far a given Pareto front approximation is from the real Pareto front. Let R be a discretization of the true Pareto front, the IGD for a set of approximated solutions A is calculated as:

$$IGD(A, R) = \left(\frac{1}{R} \sum_{r \in R} \min_{a \in A} d(\mathbf{r}, \mathbf{a})\right)^{1/p} \tag{7}$$

where $p = 2$ and d is defined by

$$d(\mathbf{r}, \mathbf{a}) = \sqrt{\sum_{k=1}^{m} (a_k - r_k)^2} \tag{8}$$

where m is the number of objective functions. A value of zero in this performance measure, indicates that all the solutions obtained by the algorithm are on the Pareto front.

5.3 Experimental Setup

We validate the three different MOEAs (i.e., NSGA-II, SMS-EMOA, and MOEA/D) by comparing their performance over four instances of the MOCHS problem. Particularly, we study the difficulties that the MOCHS problem adheres when it is solved by the abovementioned MOEAs in four different configurations. Thus, we configured the MOCHS using 100, 200, 300, and 400 nodes and they are referred to as MOCHS-100, MOCHS-200, MOCHS-300, and MOCHS-400, respectively.

The parameter settings for all the algorithms is summarized in Table 2, where N represents the number of initial solutions which is implicitly defined by the number of subproblems formulated in MOEA/D. Such subproblems were generated using the simplex-lattice design [37] and the penalty boundary intersection approach (PBI) with a penalty value $\theta = 5$, such as it was suggested by [47]. Therefore, the number of weight vectors is given by $N = C_{H+m-1}^{m-1}$, where m is the number of objective functions. Consequently, the settings of N are controlled by the parameter H. Here, we use $H = 19$ (for three-objective problems), i.e., 210 weight vectors for the three-objective formulation of the MOCHS problem. In Table 2, G denotes the maximum number of generations which was set to $G = 500$. Thus, the search was restricted to perform 105,000 fitness function evaluations for all the adopted MOEAs. P_c and P_m are the ratios for the crossover and mutation operators, respectively. The genetic operators we adopted for all three MOEAs were the well-known two-points crossover and the bit-wise mutation. For MOEA/D, T denotes the neighborhood size. It is worth noting that the parameters for NSGA-II, SMS-EMOA, and MOEA/D were set as suggested by their respective authors [5, 13, 47].

For each instance of the MOCHS, 30 independent runs were performed with each MOEA. The algorithms were evaluated using the $\mathcal{H}_n$ and IGD performance indicators. A statistical analysis was carried out overall the runs in the test problem and the performance indicator under consideration.

Since the features of the MOCHS problem are unknown, the reference Pareto front for computing the IGD performance indicator had to be constructed. In our experimental study, the reference Pareto front for each instance of the MOCHS problem (i.e., MOCHS-100, MOCHS-200, MOCHS-300, and MOCHS-400 problems) was constructed in two steps. (1) The non-dominated solutions found by all the MOEAs over the 30 independent runs were captured; (2) From these solutions, we employed a clustering algorithm to select 6000 non-dominated solu-

Table 2 Parameter settings for NSGA-II, SMS-EMOA, and MOEA/D

Parameter	NSGA-II	SMS-EMOA	MOEA/D
N	210	210	210
G	500	500	500
P_c	0.9	0.9	0.9
P_m	$1/n$	$1/n$	$1/n$
T	–	–	20

tions which define the reference set used by the *IGD* performance measure. In the case of the $\mathcal{H}_n$ performance indicator, the reference vector **r** was obtained by finding the maximum value for each objective in the constructed reference Pareto front. On the other hand, the ideal point was stated by finding the minimum value for each objective in the reference Pareto front for the problem under consideration. In this way, the $\mathcal{H}_n$ performance indicator shall consider, in a better measure, the extreme portions of the Pareto front approximation found by each MOEA.

5.4 Analysis of Results

As indicated before, the results obtained by the adopted MOEAs were compared over the four configurations of the MOCHS. Tables 3 and 4 show the results obtained by the algorithms in the $\mathcal{H}_n$ and *IGD* performance measures, respectively. In each cell, the number on the left is the average indicator value, and the number on the right (in small font size) is the standard deviation. For an easier interpretation, the best values for each performance indicator and test problem are reported in boldface. In order to identify significant differences among the results obtained by the algorithms (NSGA-II, SMS-EMOA, and MOEA/D), we adopt the Mann–Whitney–Wilcoxon [44] non-parametric statistical test with a *p*-value of 0.05 and Bonferroni correction [6]. This way, an algorithm being statistically better than all others can be considered as the best algorithm in the concerned test problem in terms of the performance indicator under consideration, in such case, this value is underlined. On the other hand, italic values correspond to algorithms that do not statistically outperform any other algorithm for a test instance in the concerned performance indicator.

Regarding the normalized hypervolume performance metric, the first thing we can observe from Table 3 is that SMS-EMOA do not statistically outperform any of the other two algorithms. On the other hand, MOEA/D is statistically better than NSGA-II and SMS-EMOA for two instances, MOCHS-300 and MOCHS-400.

With respect to the IGD indicator, we can observe that MOEA/D had the worst performance since results do not statistically outperform any of the other two approaches. On the other hand, NSGA-II had the best performance, since results are statistically better than the other two algorithms for all four instances.

Table 3 Numerical results for the $\mathcal{H}_n$ performance in the problems MOCHS-100, MOCHS-200, MOCHS-300, and MOCHS, respectively

MOP	NSGA-II	SMS-EMOA	MOEA/D
MOCHS-100	0.7085 ± 0.037	*0.4077 ± 0.038*	0.7133 ± 0.080
MOCHS-200	0.5292 ± 0.036	*0.3098 ± 0.031*	0.5282 ± 0.082
MOCHS-300	0.4413 ± 0.037	*0.2878 ± 0.030*	0.4691 ± 0.046
MOCHS-400	0.3823 ± 0.031	*0.2743 ± 0.034*	0.4482 ± 0.043

Table 4 Numerical results for the IGD performance in the problems MOCHS-100, MOCHS-200, MOCHS-300, and MOCHS-400

MOP	NSGA-II	SMS-EMOA	MOEA/D
MOCHS-100	5331.5375 ± 1489.091	64,933.5300 ± 16,785.155	605,072.0120 ± 102,677.021
MOCHS-200	26,722.9184 ± 8307.455	136,692.2040 ± 24,006.472	314,015.4755 ± 62,098.615
MOCHS-300	43,589.1449 ± 13,399.890	149,282.4088 ± 24,531.217	325,778.0605 ± 43,559.412
MOCHS-400	71,852.9357 ± 19,179.930	174,102.9865 ± 25,646.427	272,262.5005 ± 20,653.260

Anytime Behavior

To complement our study, we also analyzed the anytime behavior of the adopted MOEAs considered in our comparative study. For this task, we extracted the convergence plots of the *IGD* performance indicator for each MOCHS along the generations of each MOEA. Figure 3 shows the convergence plots of the *IGD* performance indicator for each MOCHS instance test problem. Each plot shows the convergence of the averaged *IGD* value with a confidence interval of 0.95 which is captured through the generations of each MOEA. From this figure, it is possible to see that, in all the cases, NSGA-II reached a lower *IGD* value in comparison to the other MOEAs. This means that NSGA-II approximated, in a better way, the Pareto front of each problem. From the same figure, it is also possible to appreciate the performance of the other two MOEAs, i.e., SMS-EMOA and MOEA/D. In fact, the behavior for these MOEAs became better or worse depending on the MOCHS problem under consideration. However, it is worth noticing that the convergence of NSGA-II, throughout the generations, became more stable than the other two MOEAs in most of the MOCHS problems considered in our comparative study.

6 Conclusion and Future Work

In this chapter, we have addressed the cluster-head selection problem in Wireless Sensor Networks (WSNs) . We have compared the performance of three state-of-the-art Multi-Objective Evolutionary Algorithms (MOEAs) by means of two multi-objective performance indicators.

Since sensors in a WSN are capturing information from the environment, it is desirable that the network to be active as long as possible. One way to enlarge the network lifetime is to reduce the communication between the devices. This can be accomplished by organizing the sensors in clusters and selecting one sensor in each cluster as the cluster head. The cluster head would be then in charge of collecting information from all the sensors in its cluster and forward it to the base station.

The problem of clustering the sensors and selecting the cluster head is NP-hard. Thus, in order to solve the problem, we have used three MOEAs, namely NSGA-II, SMS-EMOA, and MOEA/D, and compared their performance regarding two

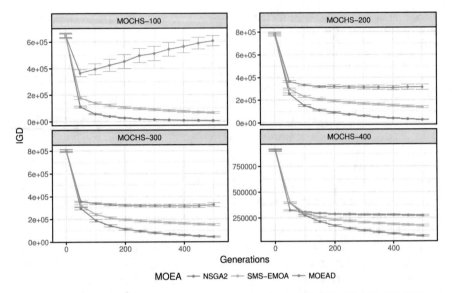

Fig. 3 Convergence plots for the IGD performance value for the MOCHS with 100, 200, 300, and 400 nodes, respectively

proper multi-objective indicators, which are the normalized hypervolume $\mathcal{H}_n$ and the inverted generational distance *IGD*.

Results for the normalized hypervolume indicate that MOEA/D was the statistically best algorithm for two instances out of four, while SMS-EMOA was the statistically worst. On the other hand, the results for the IGD indicator show that MOEA/D exhibits statistically higher *IGD* values than NSGA-II and SMS-EMOA, while NSGA-II was the statistically best for all four benchmark instances. Considering both metrics we can say that, overall, NSGA-II was the algorithms that obtained the best results for the MOCHS instances.

Finally, in this chapter, we focused on three MOEAs, however, we believe that it could be useful to evaluate other kinds of bio-inspired algorithms for the cluster-head selection problem, such as swarm intelligence. Moreover, we are interested into implementing these algorithms in a network simulator like WSNet, NS-2, or NS-3.

References

1. Abo-Zahhad M, Ahmed SM, Sabor N, Sasaki S (2014) A new energy-efficient adaptive clustering protocol based on genetic algorithm for improving the lifetime and the stable period of wireless sensor networks. Int J Energy Inf Commun 5(3):47–72
2. Afsar MM, Tayarani-N MH (2014) Clustering in sensor networks: a literature survey. J Netw Comput Appl 46:198–226
3. Anastasi G, Conti M, Di Francesco M, Passarella A (2009) Energy conservation in wireless sensor networks: a survey. Ad Hoc Netw 7(3):537–568

4. Azad P, Sharma V (2014) Pareto-optimal clustering scheme using data aggregation for wireless sensor networks. Int J Electron 102(7):1165–1176
5. Beume N, Naujoks B, Emmerich M (2007) SMS-EMOA: multiobjective selection based on dominated hypervolume. Eur J Oper Res 181(3):1653–1669
6. Bonferroni CE (1936) Teoria statistica delle classi e calcolo delle probabilita. Pubblicazioni del R Istituto Superiore di Scienze Economiche e Commerciali di Firenze 8:3–62
7. Brockhoff D, Zitzler E (2006) Are All objectives necessary? On dimensionality reduction in evolutionary multiobjective optimization. In: PPSN IX, pp 533–542
8. Carlsson C, Fullér R (1994) Interdependence in fuzzy multiple objective programming. Fuzzy Sets Syst 65(1):19–30
9. Carlsson C, Fullér R (1995) Multiple criteria decision making: the case for interdependence. Comput Oper Res 22(3):251–260
10. Coello Coello CA (1999) A comprehensive survey of evolutionary-based multiobjective optimization techniques. Knowl Inf Syst Int J 1(3):269–308
11. Coello Coello CA, Reyes Sierra M (2004) A study of the parallelization of a coevolutionary multi-objective evolutionary algorithm. In: Monroy R, Arroyo-Figueroa G, Sucar LE, Sossa H (eds) Proceedings of the third mexican international conference on artificial intelligence (MICAI'2004), Lecture notes in artificial intelligence, vol 2972, Springer, Berlin, pp 688–697
12. Deb K, Goldberg DE (1989) An investigation of niche and species formation in genetic function optimization. In: Schaffer JD (ed) Proceedings of the third international conference on genetic algorithms, George Mason University, Morgan Kaufmann Publishers, San Mateo, California, pp 42–50
13. Deb K, Pratap A, Agarwal S, Meyarivan T (2002) A fast and elitist multiobjective genetic algorithm: NSGA–II. IEEE Trans Evol Comput 6(2):182–197
14. Elhabyan R, Yagoub MC (2014) PSO-HC: particle swarm optimization protocol for hierarchical clustering in wireless sensor networks. In: Proceedings of CollaborateCom, Miami, FL, USA, pp 417–424
15. Farhang-Mehr A, Azarm S (2002) Diversity assessment of pareto optimal solution sets: an entropy approach. In: Congress on evolutionary computation (CEC'2002), vol 1, IEEE Service Center, Piscataway, New Jersey, pp 723–728
16. Fliege J (2006) An efficient interior-point method for convex multicriteria optimization problems. Math Oper Res 31:825–845
17. Gee SB, Tan KC, Shim VA, Pal NR (2015) Online diversity assessment in evolutionary multiobjective optimization: a geometrical perspective. IEEE Trans Evol Comput 19(4):542–559
18. Goldberg DE (1989) Genetic algorithms in search, optimization and machine learning. Addison-Wesley Longman Publishing Co., Inc., Boston, MA, USA
19. Gupta V, Sharma SK (2015) Cluster head selection using modified ACO. Springer India, New Delhi, pp 11–20
20. Hallam N, Blanchfield P, Kendall G (2005) Handling diversity in evolutionary multiobjective optimisation. In: 2005 IEEE congress on evolutionary computation (CEC'2005), vol 3, IEEE Service Center, Edinburgh, Scotland, pp 2233–2240
21. Hansen MP, Jaszkiewicz A (1998) Evaluating the quality of approximations to the nondominated set. Technical Report IMM-REP-1998-7, Technical University of Denmark
22. Iqbal M, Naeem M, Anpalagan A, Qadri N, Imran M (2016) Multi-objective optimization in sensor networks: optimization classification, applications and solution approaches. Comput Netw 99:134–161
23. Ishibuchi H, Setoguchi Y, Masuda H, Nojima Y (2017) Performance of decomposition based many-objective algorithms strongly depends on pareto front shapes. IEEE Trans Evolut Comput 21(2):169–190
24. Jiang S, Ong YS, Zhang J, Feng L (2014) Consistencies and contradictions of performance metrics in multiobjective optimization. IEEE Trans Cybern 44(12):2391–2404
25. Kheireddine M, Abdellatif R, Ferrari G (2015) Genetic centralized dynamic clustering in wireless sensor networks. In: Proceedings of CIIA, pp 503–511

26. Knowles J, Thiele L, Zitzler E (2006) A tutorial on the performance assessment of stochastic multiobjective optimizers. 214, Computer Engineering and Networks Laboratory (TIK), ETH Zurich, Switzerland (2006). Revised version
27. Li H, Zhang Q (2009) Multiobjective optimization problems with complicated pareto sets, MOEA/D and NSGA-II. IEEE Trans Evol Comput 3(2):284–302
28. Li, H., Zhang, Q., Deng, J.: Multiobjective Test Problems with Complicated Pareto Fronts: Difficulties in Degeneracy. In: 2014 IEEE congress on evolutionary computation (CEC'2014), IEEE Press, Beijing, China, pp 2156–2163, ISBN 978-1-4799-1488-3
29. Li M, Yang S, Liu X (2014) Diversity comparison of pareto front approximations in many-objective optimization. IEEE Trans Cybern 44(12):2568–2584
30. López Jaimes A, Coello Coello CA (2015) Many-objective problems: challenges and methods. In: Kacprzyk J, Pedrycz W (eds) Springer handbook of computational intelligence, Chap. 51. Springer, Berlin Heidelberg, pp 1033–1046
31. Miettinen K (1999) Nonlinear multiobjective optimization. Kluwer Academic Publishers, Boston, Massachuisetts
32. Miranda K, Ramos V (2016) Improving data aggregation in wireless sensor networks with time series estimation. IEEE Lat Am Trans 14(5):2425–2432
33. Naujoks B, Beume N, Emmerich M (2005) Multi-objective optimisation using S-metric selection: application to three-dimensional solution spaces. In: Evolutionary computation, 2005. The 2005 IEEE congress on, vol 2, IEEE, pp 1282–1289
34. Nedjah N, de Macedo Mourelle L (2015) Evolutionary multi-objective optimisation: a survey. Int J Bio Inspired Comput 7(1):1–25
35. Okabe T, Jin Y, Sendhoff B (2003) A critical survey of performance indices for multiobjective optimization. In: Proceedings of the 2003 congress on evolutionary computation (CEC'2003), vol 2, IEEE Press, Canberra, Australia, pp 878–885
36. Purshouse RC, Fleming PJ (2007) On the evolutionary optimization of many conflicting objectives. IEEE Trans Evol Comput 11(6):770–784
37. Scheffé H (1958) Experiments with mixtures. J Roy Stat Soc Ser B Methodol 20(2):344–360
38. Sett S, Guha Thakurta PK (2015) Multi objective optimization on clustered mobile networks: an ACO based approach. In: Proceedings of India, Kalyani, India, pp 123–133
39. Singh B, Lobiyal DK (2012) A novel energy-aware cluster head selection based on particle swarm optimization for wireless sensor networks. Hum Centric Comput Inf Sci 2(1):13
40. Slavik M, Mahgoub I, Badi A, Ilyas M (2010) Analytical model of energy consumption in hierarchical wireless sensor networks. In: Proceedings of HONET, pp 84–90
41. Srinivas N, Deb K (1994) Multiobjective optimization using nondominated sorting in genetic algorithms. Evol Comput 2(3):221–248
42. Tubaishat M, Madria S (2003) Sensor networks: an overview. IEEE Potentials 22(2):20–23
43. Wang H, Jin Y, Yao X (2017) Diversity assessment in many-objective optimization. IEEE Trans Cybern 47(6):1510–1522. https://doi.org/10.1109/TCYB.2016.2550502
44. Wilcoxon F (1945) Individual comparisons by ranking methods. Biom Bull 1(6):80–83
45. Zapotecas-Martínez S, Coello Coello CA, Aguirre HE, Tanaka K (2017) A new set of scalable multi-objective test problems. Part I: features and limitations of existing benchmarks. Technical Report, Universidad Autónoma Metropolitana (UAM), Cuajimalpa, CDMX, MEXICO
46. Zeleny M (1974) Linear multiobjective programming, vol 95. Lecture notes in economics and mathematical systems. Springer Verlag, Berlin
47. Zhang Q, Li H (2007) MOEA/D: a multiobjective evolutionary algorithm based on decomposition. IEEE Trans Evol Comput 11(6):712–731
48. Zhang Q, Liu W, Li H (2009) The performance of a new version of MOEA/D on CEC09 unconstrained MOP test instances. In: 2009 IEEE congress on evolutionary computation, pp 203–208
49. Zhou A, Qu BY, Li H, Zhao SZ, Suganthan PN, Zhang Q (2011) Multiobjective evolutionary algorithms: a survey of the state of the art. Swarm Evolut Comput 1(1):32–49
50. Zitzler E, Künzli S (2004) Indicator-based selection in multiobjective search. In: Yao X et al (ed) Parallel problem solving from nature—PPSN VIII, lecture notes in computer science, vol 3242. Springer, Birmingham, UK, pp 832–842

51. Zitzler E, Laumanns M, Thiele L (2002) SPEA2: improving the strength pareto evolutionary algorithm. In: Giannakoglou K, Tsahalis D, Periaux J, Papailou P, Fogarty T (eds) EUROGEN 2001. Evolutionary methods for design, optimization and control with applications to industrial problems, Athens, Greece, pp 95–100
52. Zitzler E, Thiele L (1998) Multiobjective optimization using evolutionary algorithms—a comparative study. In: Eiben AE (ed) Parallel problem solving from nature V. Springer, Amsterdam, pp 292–301
53. Zitzler E, Thiele L, Laumanns M, Fonseca CM, da Fonseca VG (2003) Performance assessment of multiobjective optimizers: an analysis and review. IEEE Trans Evol Comput 7(2):117–132

An Energy-Efficient Cluster Head Selection Using Artificial Bees Colony Optimization for Wireless Sensor Networks

Tauseef Ahmad, Misbahul Haque and Asad Mohammad Khan

Abstract In a cluster-based Wireless Sensor Networks (WSN), dividing the network into clusters and choosing an efficient Cluster Head (CH) is a big issue. The selection of CH is a very challenging task, and it affects the energy consumption of the network and also the lifetime of sensors and ultimately network lifetime. This chapter presents a new approach for CH selection based on Artificial Bee Colony (ABC) optimization. This ABC optimization is based upon the remaining energy, intra-cluster distance, and distance from the sink station. The fitness function for ABC is calculated based on three parameters, i.e., residual energy; distance from the sink station; and intra-cluster distance. We optimized the fitness function using ABC optimization. The objective of optimizing the fitness function is to select an optimal CH for each cluster which reduces the energy consumption of the WSN. The proposed model is analyzed through extensive experiments and the outcomes are compared with some famous existing approaches.

Keywords Wireless sensor networks · Cluster head · Optimization · Fitness function · Residual energy · Energy consumption

T. Ahmad (✉)
Department of Information Technology, Rajkiya Engineering College,
Azamgarh, Uttar Pradesh, India
e-mail: tauseefahmad@zhcet.ac.in

M. Haque · A. M. Khan
Department of Computer Engineering, ZHCET,
Aligarh Muslim University, Aligarh, Uttar Pradesh, India
e-mail: misbahul.haque@zhcet.ac.in

A. M. Khan
e-mail: masadiitr@gmail.com

© Springer International Publishing AG, part of Springer Nature 2019
S. K. Shandilya et al. (eds.), *Advances in Nature-Inspired Computing and Applications*, EAI/Springer Innovations in Communication and Computing,
https://doi.org/10.1007/978-3-319-96451-5_8

1 Introduction

Wireless Sensor Networks (WSNs) is a collection of large numbers of tiny Microelectromechanical Systems (MEMS) devices called sensor nodes which are spread over a remote geographical area. These sensor nodes are capable of limited communication, processing, and storage capabilities. These devices are capable of communicating over small distances using radio channels. WSN has wide application in both military and civilian fields. It can be extensively used in environmental monitoring, healthcare (e.g., monitoring patients with Alzheimer's disease), kindergartens, generating alarms for a forest, monitoring of wild animals and detection of infiltration across the line of control (LoC), and tracking an object.

Sensor nodes are battery operated which are non-rechargeable and having limited energy, and these sensor nodes are most often used in the environment where its battery cannot be replaced. This is one of the major concerns while proposing any protocol for WSNs. The proposed protocol should be power efficient, so that the life of the sensor node and ultimately, the WSN can be prolonged.

The basic architecture of a WSN consists of wireless sensor nodes along with sink node. Typically, the sensor nodes collect data from the environment and send them to the sink node [1, 2]. While in cluster-based WSN architecture shown in Fig. 1, a group of sensor nodes along with Cluster Head (CH) forms a cluster. The CH is selected from the group of sensor nodes based on some fitness function. The function of this CH is to collect information from their neighboring sensor nodes and aggregate their information to get some useful information from them. After aggregation, the CH sends the useful information to the base station. Due to this aggregation, the network overhead and overall energy consumption are decreased significantly.

In cluster-based WSN, the whole WSN is divided into the number of clusters and each cluster contains one CH at a time [3]. All the communications outside the cluster is done through CH. Energy consumption increases with an increase in the distance between the communicating nodes. So, with cluster-based WSN, flexibility in long-distance communication is achieved through information aggregation [4]. The main advantages of clustering are: (i) information aggregation is performed at CH level to avoid redundant transmission of information thus energy is saved, (ii) WSN can be easily scalable, and (iii) increase the utilization of the radio link (i.e., bandwidth) [5].

The performance of WSN highly depends upon the selection of the CH. Various cluster-based protocols have been proposed [1, 6–11]. The approach for CH selection varies for different protocols. In LEACH [12], CH selection is randomized and it rotates from one sensor node to another based on some probability to manage power consumption among the sensors of the cluster. Various improvements over LEACH have been proposed in the literature [13]. A comparison table for famous clustering protocols is shown in Table 1.

Various nature-inspired approaches for CH selection have been proposed in the literature. These approaches include Genetic Algorithm (GA) , Particle Swarm

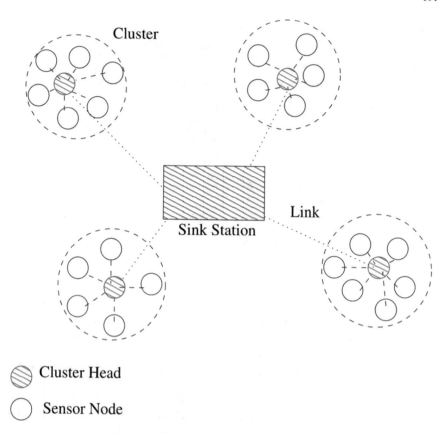

Fig. 1 A cluster-based WSN architecture

Optimization (PSO), and Ant Colony Optimization (ACO). In [14], the author proposed an approach for optimal CH selection based on GA with fitness function based on the latencies. The author did not consider either of intra-cluster distance; sink distance; and residual energy factor. In [15], the author proposed an ACO-based CH selection scheme. It uses residual energy and distance between nodes and sink station as the main parameter for fitness. However, the author ignored the intra-cluster distance for their fitness function. In [16], the author proposed a combination of GA and ACO. The fitness function considers the intra-cluster distance, distance to the base station, and residual energy. The author proposed CH selection based on PSO using fitness function having CH energy consumption and latency in information forwarding in [17]. In this work, the dimension of each particle is shown equivalent to the total number of the sensors in the WSN. In [18], the author uses PSO for optimal selection of the location of the CH. Their focus was on the intra-cluster distance rather than distance to the base station. The author considered both intra-cluster distance as well as sink distance for their fitness function in [19]. But, they do not focus

Table 1 Comparison table for clustering protocols

Protocol	CH selection	Energy efficiency	Scalability
LEACH	Randomly	Poor	Very low
HEED	Residual energy and node degree	Better than LEACH	Moderate
TL-LEACH	Distance based	Better than LEACH	Moderate
M-LEACH	Residual energy, distance	Better than TL-LEACH	Moderate
E-LEACH	Residual energy	Better than M-LEACH	High
LEACH-C	Residual energy, distance	Better than E-LEACH	High
PSO-C	Ratio of total initial energy to total current energy, Intra-cluster distance	Better than LEACH-C	High

on the remaining energy of the node. In [20], the author discussed PSO-C approach for optimal CH selection. They consider intra-cluster distance and residual energy of the nodes for fitness function. However, they do not focus on sink distance, which is vital part for energy consideration.

This chapter discusses a new approach for CH selection based on Artificial Bee Colony (ABC) optimization. The discussed approach is very efficient in the selection of CHs form the sensor nodes of the cluster. The fitness function for ABC is selected based on three parameters, i.e., residual energy, distance from the sink station, and intra-cluster distance. We optimized the fitness function using ABC optimization. The objective of optimizing the fitness function is to select an optimal CH for clusters which reduces the energy consumption of the WSN.

The rest of the chapter is organized as follows. Section 2 provides the overview of the ABC algorithm. Section 3 discusses the proposed optimization model. Section 4 contains results and discussion. The last section contains the conclusion of our work.

2 Overview of ABC Algorithm

Artificial Bee Colony algorithm is a kind of problem optimization approach which mimics the intelligent behavior of honey bee for optimization of numerical problems. These bees form a colony which consists of three kinds of bees: employed bees, onlookers, and scouts [21, 22]. The duty of employed bees is to search where the food exists. After searching the location of the food (destination), they take that food and come back to their home (origin). After coming back to their home, they start waggle dance. The second group of bees, i.e., onlookers, watches the employed bee dancing and starts to follow the employed bees based on the availability of food.

More and more onlooker bees watch the employed bees dancing and started following them. When the employed bee food finishes, it starts behaving like a scout bee and starts searching for food randomly. This behavior of honey bee can be used for many optimization problems. Here, we can divide the total population into two parts (i) employed bees and (ii) onlooker bees. The number of solutions is equal to the total number of onlooker bees or employed bees. Algorithm 1 shows the necessary steps for ABC optimization.

Algorithm 1 Artificial Bee Colony Algorithm

1: **Initialization Phase**
2:      Randomly generate the initial food source for the population.
3: **repeat**
4:      **Employed Bee Phase**
5:          Updates its position if it's better than the previous position, uses greedy approach.
6:      **Onlooker Bees Phase**
7:          Chooses its position depending upon probability value associated with it.
8:      **Scout Bees Phase**
9:          Start searching new position randomly.
10:     Store the best solution achievedso far.
11: **until** Conditions are not met.

Algorithm 1 states that ABC algorithm can broadly be divided into three phases: (a) Initialization phase, (b) Employed bees phase, (c) onlooker bees phase, and (d) scout bees phase. We will now discuss the mathematical background of these phases.

2.1 Initialization Phase

It is the first phase of ABC optimization. In this phase, the whole population B randomly assigned a distributed initial position (food sources). Each food source represents a potential solution to the problem.

$$x_{k,D} = \text{lo} + \text{rnd}(0, 1) * (\text{up} - \text{lo}) \tag{1}$$

where D is the dimension, $k \in (1, 2, \ldots, B)$ and $D \in (1, 2, \ldots, d)$.

where $x_{k,D}$ represents the position of kth bee in dth dimension, $\text{rnd}(0, 1)$ is a number within $[0, 1]$ generated randomly, lo, and up are the minimum and maximum limits of the $x_{k,D}$, respectively.

Here,

$$\vec{x_k} = [(x_{k1}(t), y_{k1}(t)), (x_{k2}(t), y_{k2}(t)) \ldots, (x_{kd}(t), y_{kd}(t))]. \tag{2}$$

2.2 Employed Bees Phase

It is assumed that the number of food sources and employed bees are equal. The employed bees starts to search for a new possible food source $\left(v_{k,D}^{(t+1)}\right)$. The new food can be represented by the expression given follows.

$$v_{k,D}^{(t+1)} = x_{k,D}^{(t)} + \Phi_{k,D}\left(x_{k,D}^{(t)} - x_{l,D}^{(t)}\right) \tag{3}$$

where $\Phi_{k,D}$ is a randomly generated number between $[-1, 1]$ and $k \neq l$. Once the employed bees find a new food source, then they start computing their fitness to the objective function. The fitness function computed is checked whether it is better than the previous food source, and then the previous food source is replaced with the newly chosen food source. The selection of food source is done using greedy approach.

The fitness function used for evaluation of the food sources can be given as follows:

$$f_k t_k\left(\overrightarrow{x_k}\right) = \begin{cases} \frac{1}{1+f_k\left(\overrightarrow{x_k}\right)}, & \text{if } f_k\left(\overrightarrow{x_k}\right) \geq 0 \\ 1 + |f_k\left(\overrightarrow{x_k}\right)|, & \text{otherwise.} \end{cases} \tag{4}$$

where $f_k t_k\left(\overrightarrow{x_k}\right)$ and $f_k\left(\overrightarrow{x_k}\right)$ are fitness function and objective function, respectively.

2.3 Onlooker Bees Phase

The onlooker bees gets the new food information from the employed bees. After getting the food source, they select the food source by computing the probability associated with each of the food sources. The probability Prob_k with food source $\overrightarrow{x_k}$ can be computed as follows:

$$\text{Prob}_k = \frac{f_k t_k\left(\overrightarrow{x_k}\right)}{\sum_{k=1}^{B} f_k t_k\left(\overrightarrow{x_k}\right)} \tag{5}$$

2.4 Scout Bees Phase

When the selected food source does not improve the solutions for a number of iterations, then the selected vector is discarded and scout bees generates a new random vector of food source.

3   Proposed Optimization Model

Our proposed algorithm focuses mainly on CH selection in a cluster. We are assuming that clusters formations are already done. Our objective is to select CHs for each cluster in energy-efficient manner. The CHs selection is performed using ABC optimization. This ABC optimization is based upon the remaining energy, intra-cluster distance, and distance from the base station. These three parameters are used for developing the fitness function for CH selection.

3.1   Deriving the Fitness Function

As we have discussed, our proposed fitness function has the following parameters:

Residual Energy Residual energy of the sensor node is taken as the first parameter for the fitness function. The focus here is maximizing the total residual energy of all CHs of the WSN. Let RE_{CH_D} be the residual energy of CH_D, where $1 \leq D \leq d$, of the selected nodes as CHs. Therefore, we need to maximize the following function:

$$f_1\left(\overrightarrow{x_k}\right) = \sum_{D=1}^{d} RE_{CH_D} \tag{6}$$

Intra-Cluster Distance The second parameter considered for objective function is intra-cluster distance, i.e., distance from all sensor nodes to the CH in a cluster. When nodes communicate they consume energy, so, we need to minimize intra-cluster distance to minimize the energy consumption. This implies that the CH could be chosen such that its distance is least from all the nodes of the cluster. Therefore, we need to minimize the individual objective function f_2 given as follows:

$$f_2\left(\overrightarrow{x_k}\right) = \frac{1}{d}\sum_{D=1}^{d}\left(\frac{1}{T_s}\sum_{i=1}^{T_s} d(S_i, CH_D)\right) \tag{7}$$

where T_s is the number of sensors in the cluster and $d(S_i, CH_D)$ is the distance between sensor S_i and cluster head CH_D.

Distance to Sink Station is defined as the distance between CH_D and the sink station. For information processing, the aggregated information at CH_D from the sensor nodes needs to be transmitted at the sink station (SS). Therefore, we need to minimize the distance between the CHs and the SS. Therefore, we need to minimize the individual objective function f_3 given as follows:

$$f_3\left(\overrightarrow{x_k}\right) = \frac{1}{d} \sum_{D=1}^{d} (d(\text{CH}_D, \text{SS})) \tag{8}$$

The fitness function for our proposed ABC algorithm is to minimize the above three functions, i.e., f_1, f_2, and f_3. Therefore, the fitness function for our algorithm is given as follows:

$$f_k\left(\overrightarrow{x_k}\right) = \beta_1 * \frac{1}{f_1\left(\overrightarrow{x_k}\right)} + \beta_2 * f_2\left(\overrightarrow{x_k}\right) + \beta_3 * f_3\left(\overrightarrow{x_k}\right) \tag{9}$$

where β_1, β_2 and β_3 are constants which provides weights to the objective functions. The sum of these constants should be equal to 1. Our aim is to minimize the cumulative objective function.

A formal algorithm for the proposed Energy-Efficient Cluster Head Selection using Artificial Bees Optimization (EECHS-ABC) for WSN is provided in Algorithm 2.

Algorithm 2 EECHS-ABC Algorithm

1: **Input** Number of employed bees = Number of food sources = Sensors in the network= B. Number of CHs= D. Number of iterations= n.
2: **Output** Optimal position of CHs.
3: Initialize the food sources $x_{(k,D)}, \forall_{k,D}$, k ranges from 1 to B, D ranges from 1 to d.
4: **for** t =1 to n **do**
5:     **for** every employed Bee **do**
6:             Compute new food source $v_{(k,D)}^{(t+1)}$.
7:             Compute $f_k\left(\overrightarrow{x_k}\right)$ for $v_{(k,D)}^{(t+1)}$.
8:             Apply Greedy selection.
9:     **end for**
10:    **for** every onlooker bees **do**
11:            Select $v_{(k,D)}^{(t)}$ based on $Prob_k$ using Roulette Wheel selection.
12:    **end for**
13:    **if** $v_{(k,D)}^{(t)}$ does not improve **then**
14:            Generate new $x_{(k,D)}$ replace with $v_{(k,D)}^{(t)}$.
15:    **end if**
16:    Retain best $v_{(k,D)}^{(t)}$.
17: **end for**

3.2 Sensor Network Model

A wireless sensor network consists of a large number of sensors and a sink station. The sensor network can be represented using graph data structure, G, having sensors as a set of vertices as sensor nodes N and channel as edges C, i.e., $G = (N, C)$.

The number of sensors is greater than the number of the CHs. There exists one sink node and all the information flows to the sink node through the CHs. The location of sensors, CHs, and base station are fixed. The CH collects data from the sensors within the cluster and forwards it to the base station. All the sensors are assumed to be homogeneous and having the same initial energy. Each sensor can be identified by its unique identity.

3.3 Energy Consumption Model

The energy is consumed when communication occurs between the sensor-sensor and sensor-CH in a cluster. The lifetime of the sensor network depends on the overall energy consumption of the sensors. As the energy of a sensors decreases below some threshold value, it is assumed to be dead. The proposed scheme should be energy efficient, so that the sensors lifetime can be prolonged and ultimately, the sensors network lifetime is optimized. The energy consumption model is taken from [23].

If the distance between sensor and CH or CH and base station are assumed to be r, then the total energy consumed for transmitting packets containing b bits is provided as follows:

$$\text{EN}_{\text{TR}}(b, r) = E_e * b + E_a * b * r^\alpha \begin{cases} \alpha = 2, & \text{if } r < r_0 \\ \alpha = 4, & \text{if } r \geq r_0 \end{cases} \tag{10}$$

where r_0 is the threshold distance.

The energy consumed for receiving a packet having b bits of data can be given as follows:

$$\text{EN}_{\text{Re}}(b) = E_e * b \tag{11}$$

where E_e and E_a are energy required by transceiver electronics for its operation and transmission of data, respectively. The residual energy E_{Res} of each sensor is refreshed after each transmission and receiving operation. The sensor is considered to be dead if $E_{\text{Res}} \leq 0$.

Table 2 Network and artificial bees colony parameters with their default values

Parameter	Default value
Network area	$400 \times 400 \text{ m}^2$
Base station	(200, 200), (400, 400), (500, 500)
Initial energy	3 J
E_e	50 nJ/bit
E_a	10 pJ/bit/m^2
r_0	50 m
β_1	0.5
β_2	0.3
β_3	0.2
Number of rounds	2000, 5000

4   Result and Discussion

This section focuses on the analysis of the results obtained through experiments. The proposed scheme is simulated using MATLAB. The performance matrices used are total energy consumption and total number of alive sensors. The various simulation variables with their default values are shown in Table 2. We analyze and compare our results with LEACH, E-LEACH, and PSO-C.

4.1   Effect on Total Energy Consumption

The simulation is performed by taking the number of CHs 15 and 30. The number of sensors is taken to be 500. A total of 5000 rounds of operations are performed. The sink station is assumed to be at the center of the network, i.e., (200, 200).

The result obtained shows that the proposed EECHS-ABC model performs better than the LEACH, E-LEACH, and PSO-C. The total energy consumption achieved is minimum by the proposed model, it is because our fitness function for CH selection also focuses on minimizing the distance between the sensors and CHs, which causes lower consumption of energy and ultimately overall energy consumption decreases. Figure 2 and Fig. 3 show the results for 15 CHs and 30 CHs, respectively.

Effect of Position of Sink Station This section describes the effect of the position of the sink station on the energy consumption of the sensors network. The location of the sink station is taken to be at the center, at the boundary, and outside of the sensor network.

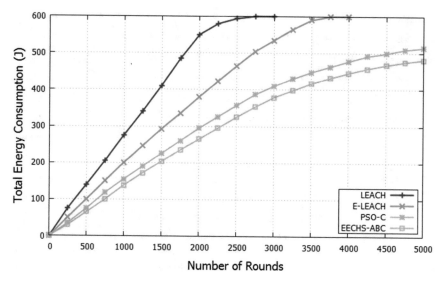

Fig. 2 Total energy consumption with 15 CHs

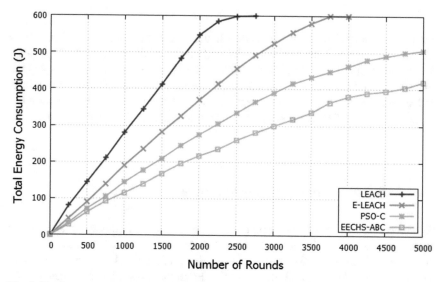

Fig. 3 Total energy consumption with 30 CHs

The result shown in the Fig. 4 shows that when the sink station is at the center, as the number of sensors increases from 100 to 500, the total energy consumption is achieved to be least by the proposed model EECHS-ABC. This is because the fitness function is chosen such that the energy efficiency is improved and residual energy of the sensors is increased. Figure 5 shows the energy consumption at different sink positions while increasing the number of the sensors from 100 to 500. The minimum

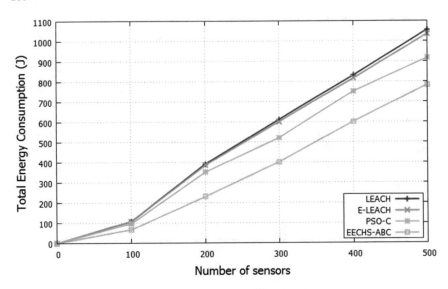

Fig. 4 Energy consumption while sink is at the center of the network

energy is consumed when the sink node is at the center of the network and maximum amount of energy is consumed when sink station is outside the network. The reason behind is that the fitness function focuses on the distance between the CHs and the sink station, and when the sink station is at the center, the CHs needs to transmit their data at the smaller distance and energy is saved. While more energy is required to transmit data from CHs to the sink station at the boundary and outside of the network.

4.2 Effect on Number of Sensors Alive

Figures 6 and 7 show the number the sensors alive for 2000 rounds for 50 and 100 sensors, respectively. The position of the sink station is assumed to at the center of the network, i.e., (200, 200). The value of β_1, β_2, and β_3 are set to be 0.5, 0.3, and 0.2, respectively, for the proposed EECHS-ABC. From the results, it is observed that the proposed model outperforms the other models and lifetime of the network increases. This is because, in our fitness function, we have focused on the distance between the CH and the sensors of the cluster. The CHs are selected such that the distance between the CH and the sensors of the clusters are minimal. It causes less energy consumption while transmitting data from the sensors to the CHs and lifetime of the sensors are increased.

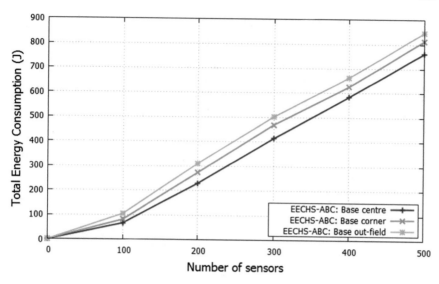

Fig. 5 Energy consumption with different sink positions

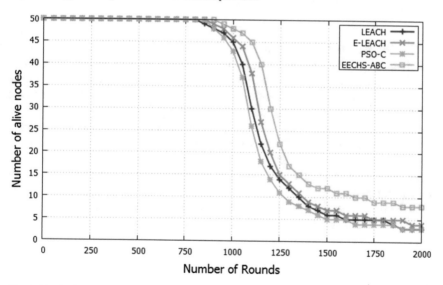

Fig. 6 Total alive nodes (number of sensors = 50)

5 Conclusion

One of the main limitations of the sensor network is the limited power of sensor nodes. This limitation affords that saving energy and increasing network lifetime becomes two main issues in WSNs applications and algorithms.

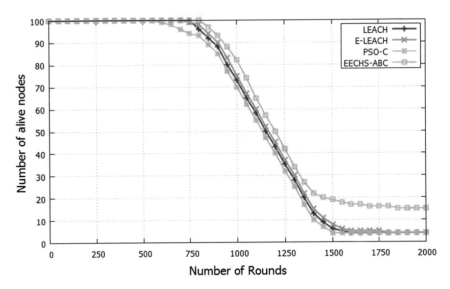

Fig. 7 Total alive nodes (number of sensors = 100)

This chapter discusses a new approach for energy-efficient cluster head selection based on artificial bee colony optimization of the derived fitness function in WSN. The derived fitness function is based on the remaining energy of the sensors, distance between the cluster elements, and distance from the sink station. We performed extensive simulations to verify the model and compared the obtained result with some famous existing models. The result shows that the proposed model performs better than the existing models in terms of energy efficiency and sensor node lifetime. We also examined the effect of the sink station on the total energy consumption and observed that best performance is achieved when the sink station is at the center of the sensor network.

In the future, one may use this model and work on developing some new cluster-based routing algorithm which will be more energy efficient.

References

1. Liu X (2012) A survey on clustering routing protocols in wireless sensor networks. Sensors 12:11113–11153
2. Haque M, Ahmad T, Imran M (2017) Hierarchical routing protocols for wireless sensor networks: a comparative survey. In: Proceedings of the IEEE international conference on wireless communications, signal processing and networking (WiSPNET), pp 2115–2119
3. Abbasi AA, Younis M (2007) A survey on clustering algorithms for wireless sensor networks. Comput Commun 30:2826–2841
4. Mishra H, Kumar V, Shibu S (2015) Cluster based energy efficient routing protocol for wireless sensor network. Eng Univ Sci Res Manag 7(1):1–5

5. Jiang C, Yuan D, Zhao Y (2009) Towards clustering algorithms in wireless sensor networks: a survey. In: IEEE wireless communications and networking conference, Budapest, Hungary pp 1–6

6. Boyinbode O, Le H, Mbogho A, Takizawa M, Poliah R (2010) A survey on clustering algorithms for wireless sensor networks. In: 13th IEEE international conference on network-based information system, pp 358–364

7. Wei C, Yang J, Gao J, Zhang Z (2011) Cluster-based routing protocols in wireless sensor networks: a survey. In: 2011 international conference on computer science and network technology, Harbin, China, pp 1659–1663

8. Marhoon HA, Mahmuddin M, Nor SA (2015) Chain-based routing protocols in wireless sensor networks: a survey. ARPN J Eng Appl Sci 10(3):1389–1398

9. Naeimi S, Ghafghazi H, Chow CO, Ishi H (2012) A survey on the taxonomy of cluster-based routing protocols for homogeneous wireless sensor networks. Sensors 12(6):7350–7409

10. Xu X, Ansari R, Khokhar A, Vasilakos AV (2015) Hierarchical data aggregation using compressive sensing (HDACS) in WSNs. ACM Trans Sensor Netw (TOSN) 11(3)

11. Haque M, Ahmad T, Imran M (2017) Review of hierarchical routing protocols for wireless sensor networks. In: Hu Y-C et al (eds) Intelligent communication and computational technologies, Lecture notes in networks and systems, vol 19. Springer, Berlin, pp 237–246

12. Ran G, Zhang H, Gong S (2010) Improving on LEACH protocol of wireless sensor networks using fuzzy logic. J Inf Comput Sci 7(3):767–775

13. Hani RMB, Ijjeh A (2013) A survey on LEACH-based energy aware protocols for wireless sensor networks. J Commun 8(3):192–205

14. Pal V, Yogita, Singh G, Yadav RP (2015) Cluster head selection optimization based on genetic algorithm to prolong lifetime of wireless sensor networks. Procedia Comput Sci 57:1417–1423

15. Gupta V, Sharma SK (2015) Cluster head selection using modified ACO. In: Proceedings of fourth international conference on soft computing for problem solving, advances in intelligent systems and computing, pp 11–20

16. Das S, Wagh S (2015) Prolonging the lifetime of the wireless sensor network based on blending of genetic algorithm and ant colony optimization. J Green Eng 4:245–260

17. Singh B, Lobiyal DK (2012) A novel energy-aware cluster head selection based on particle swarm optimization for wireless sensor networks. Human Centric Comput Inf Sci 2(1):2–13

18. Rao PCS, Banka H, Jana PK (2016) PSO-based multiple-sink placement algorithm for protracting the lifetime of wireless sensor networks. In: Proceedings of the second international conference on computer and communication technologies, pp 605–616

19. Guru SM, Halgamuge SK, Fernando S (2005) Particle swarm optimizers for cluster formation in wireless sensor networks. In: Proceedings of the IEEE international conference on intelligent sensors, sensor networks and information processing (ISSNIP'05), pp 319–324. Melbourne, Australia, Dec (2005)

20. Lati NMA, Tsimenidis CC, Sharif BS (2007) Energy-aware clustering for wireless sensor networks using particle swarm optimization. In: Proceedings of 18th IEEE international symposium on personal, indoor and mobile radio communications, pp 1–5

21. Davidovic T, Teodorovic D, Selmic M (2015) Bee colony optimization part I: the algorithm overview. Yugoslav J Oper Res 25(1):33–36

22. Gao W, Liu S, Huang L (2012) A global best artificial bee colony algorithm for global optimization. J Comput Appl Math 236(11):2741–2753

23. Bhardwaj M, Chandrakasan AP (2001) Upper bounds on the lifetime of wireless sensor networks. In: Proceedings of IEEE international conference on communications (ICC), vol. 3, pp 785–790

Modified Krill Herd Algorithm for Global Numerical Optimization Problems

Laith Mohammad Abualigah, Ahamad Tajudin Khader and Essam Said Hanandeh

Abstract For the purpose of improving the search strategy of the krill herd algorithm (KHA), an improved robust approach is proposed to address the function optimization problems, namely, modified krill herd algorithm (MKHA). In MKHA method, the modification of krill herd algorithm focuses on genetic operators (GOs) and it occurs in the ordering of procedures of the basic krill herd algorithm, where the crossover and mutation operators are employed after the updating process of the krill individuals position, the krill herd (KH) motion calculations, is finished. This modification is conducted because the genetic operators insignificantly exploit to enhance the global exploration search in the basic krill herd algorithm so as to speed up convergence. Several versions of benchmark functions are applied to verify the proposed method (MKHA) and it is showed that, in most cases, the proposed algorithm (MKHA) obtained better results in comparison with the basic KHA and other comparative methods.

Keywords Global optimization problem · Modified krill herd algorithm Genetic operators · Optimization techniques

1 Introduction

Generally, several versions of metaheuristic optimization algorithms have been applied to solve optimization problems [1–3]. The process of optimization is how to find an optimal solution by searching through the search space of the given problem.

L. M. Abualigah (✉) · A. T. Khader
School of Computer Sciences, Universiti Sains Malaysia (USM), 11800 Pinang, Malaysia
e-mail: laythdyabat@ymail.com

L. M. Abualigah
Faculty of Computer Sciences and Informatics, Amman Arab University, Amman, Jordan

E. S. Hanandeh
Department of Computer Information System, Zarqa University, Zarqa 13132, Jordan

© Springer International Publishing AG, part of Springer Nature 2019
S. K. Shandilya et al. (eds.), *Advances in Nature-Inspired Computing and Applications*, EAI/Springer Innovations in Communication and Computing, https://doi.org/10.1007/978-3-319-96451-5_9

All of the possible values are candidate solutions to solve the problem and the final obtained best value is the optimal solution [4].

In the metaheuristic optimization algorithms, randomization works a very important function in both exploitation and exploration search strategies. Based on this reality, many randomization procedures such as Markov chains, Gaussian random number, Levy flights, and numerous new procedures have been used in metaheuristic algorithms to improve its performance. In general, all metaheuristic optimization algorithms, which have been adapted to solve any optimization algorithm obtained, result in less computational time to find an optimum global solution, trapped in local optima, and fast convergence [5].

A common simple classification form for optimization algorithms is based on the nature of the algorithms. Optimization algorithms have been divided into two prime categories: (1) Deterministic algorithms: it used the gradient operator (a direction to move) such as hill-climbing technique. This technique has a strict movement and it almost generates the same solution if the iterations start with the same initial point. (2) Stochastic algorithms: it does not use the gradient operator, and it often generates several different solutions even with the same initial value [4]. In general, stochastic algorithms are divided into two main types: heuristic and metaheuristic as in [6].

Recently, the main trend of optimization is to enhance the performance of metaheuristic algorithms by combining with chaos theory, adaptive randomization technique, levy flights strategy, evolutionary boundary handling scheme, local search strategies, and genetic operators (i.e., crossover and mutation) [7]. Popular genetic operators have been already used in basic KHA [8], which can fine-tuning its global convergence speed, but the current basic version of KHA suffers from accelerating in the global search, which means that the genetic operators insignificantly exploited to improve the performance of the basic KHA.

In this paper, we propose a modified krill herd algorithm (MKHA) to improve the performance of the basic algorithm. In MKHA, the genetic operators are applied to significantly improve the performance of the basic KHA. This modification occurs in the order of KHA operators where the crossover and mutation processes are invoked after updating the krill positions because the nature of the search space of most optimization problems is ragged and deep. The proposed algorithm was applied to 14 benchmark function problems from the literature. The experimental results show that the performance of MKHA is superior to basic KHA and other comparative algorithms.

The organization of this paper is as follows. First, a literature review of several basic, modified, hybridized metaheuristic algorithms is given in Sect. 2. A detailed presentation of the basic KHA is provided in Sect. 3. Overview of MKHA is described in Sect. 4. The performance of the MKHA is verified using 14 benchmark functions in Sect. 5. Finally, a conclusion of the present work is represented in Sect. 6.

2 Literature Review

Several basic, modified, hybridized metaheuristic algorithms have been successfully used in the literature to solve many optimization problems such as text document clustering problem [9, 10], information retrieval [11], date clustering problem [12], unsupervised feature selection problem [13–15], benchmark function problems [16–19], etc.

The performance of the basic KHA is improved in [5] by connecting an adaptive technique to solve benchmark optimization functions, namely, adaptive KH (AKH). The results revealed that implementation of AKH algorithm reduces the computational times of the basic KHA for solving benchmark problems.

Firefly algorithm (FA) is a powerful algorithm in local exploitation search but at sometimes it may trap during the local search. This affects that it cannot perform global exploration search well. The search strategy of the FA depends totally on random walks, and therefore a fast or a quick convergence cannot be guaranteed. In the paper [4], the main development was by adding the handling of top fireflies to the FA, namely, improved firefly algorithm (IFA), which is produced to speed the convergence up. Thus, this new process makes the approach more achievable for the wide range problems while maintaining the attractive characteristics of the basic version of the FA. The proposed algorithm (IFA) is assessed on ten benchmarking function problems. The results showed that the proposed algorithm (IFA) performed more efficiently in comparison with the other basic algorithms (i.e., FA, BBO, PSO, and DE).

Bat algorithm (BA) is a powerful optimization algorithm in local exploitation search, but at sometimes it traps in local optima [20]. This means that it hard to perform global search fine. In the BA, the search strategy depends entirely on random walks. Thus, a fast convergence cannot be reached. In this paper [6], pitch adjustment operation of the harmony search (HS) algorithm is added to the BA with the purpose of speeding its convergence up (i.e., HS/BA). This approach improves the diversity of the population for BA. , as well as, making the proposed approach more possible for a wide range of functional applications while maintaining the attractive properties of the basic BA. The proposed approach is assessed on 14 benchmark functions that have been applied to test the performance of algorithms for continuous problems. The results showed that the proposed approach (HS/BA) performed more accurately in comparison with basic BA, ACO, BBO, DE, ES, GA, HS, PSO, and SGA.

In order to improve the diversity of the population for KH algorithm, the main development is of adding HS operator to the KH with the purpose of speeding its convergence up, thus making the proposed approach more achievable for a wide range of functional applications while maintaining the attractive properties of the basic KH. In this paper [21], two approaches are combined to propose a new hybrid algorithm according to the origin of HS and KH algorithms (HS/KH), and then an improved KH algorithm (HS/KH) is employed to find the optimal value of the objective function. The proposed algorithm is evaluated on 14 standard functions that have been used to test optimization algorithms for continuous optimization problems. The results demonstrated that the proposed (HS/KH) performed more efficiently than basic KH, ACO, BBO, DE, ES, GA, HS, PSO, and SGA.

The text document clustering is an appropriate technique used to partition huge text documents into clusters. The documents' size affects the clustering process by reducing its performance. Subsequently, text documents include sparse and un-informative features, which decrease the performance of the underlying clustering algorithm and reduce the computational time. Feature selection is a primary unsuper-vised learning technique used to choose a new subset of more informative features to develop the performance of the clustering and reduce its computational time. In this paper [22], a hybrid of particle swarm optimization (PSO) algorithm with genetic operators (GOs), called H-FSPSOTC, is proposed for solving the feature selection problem. The experiments were conducted on eight common datasets with alternative characteristics. The results showed that the proposed hybrid algorithm (H-FSPSOTC) improved the performance of the clustering algorithm by creating a new subset of informative features. The proposed is compared with the other comparative optimization algorithms.

In this paper [23], the modification of cuckoo search algorithm (CSA) is proposed, called MCSA. It includes replacing the tournament selection system with random selection system. The computational results revealed that the proposed MCSA got faster convergence in comparison with the basic CSA, and is more powerful during the search for the optimal value of the objective function. The performance of the proposed algorithm (MCSA) is assessed on 13 benchmark numerical functions neatly selected from the literature. The results illustrated that the performance of the MCSA was better compared with the basic CSA in almost all cases.

In this paper [24], chaotic particle swarm krill herd-based methods are introduced, called CPKHs. In these methods, the authors used different 1-D chaotic maps instead of the parameter of the method. Therefore, by utilizing various chaotic maps as alternatives to pseudorandom sequences, several methods have been introduced. The CPKH method is assessed on 23 benchmark function problems and other problem (a gear train design problem). Experimental results indicated that the proposed method (CPKH) worked more effectively than KH, ABC, DE, ES, HS, PBIL, and PSO. The results showed that the new methods for the CPKH improved the results.

3   The Basic Krill Herd Algorithm

KH is a recent metaheuristic population-based algorithm that interdicted by Gandomi and Alavi in 2012 [8]. The inspiration of the KH algorithm is the herding behavior of krill individuals when searching for the nearest food; KH with high density based on communication with each other [25, 26]. The KH algorithm follows the Lagrangian model for efficacious search, which is calculated using Eq. (1) based on three factors as follows [21, 27, 28]:

1. Movement by other individual krill.
2. Foraging action.
3. Physical diffusion.

$$\frac{dx_i}{dt} = N_i + F_i + D_i, \tag{1}$$

where for krill i, N_i is first part, which indicates to the motion induced by other krill individuals, F_i indicates the forging motion, and D_i indicates the physical diffusion of the ith krill individual [8]. The H-KHA factors discussed below.

3.1 Motion Calculation

3.1.1 Movement Induced by Other Krill Individuals

Based on certain theoretical arguments, each krill individual attempts to maintain a high density and closeness to the nearest food. The direction of the induced motion is derived from the local effect of each solution density, a target effect of the density of the individuals, and a repulsive effect of the individuals [29, 30]. Equation (2) is used to calculate the motion induced by other krill individuals.

$$N_i^{new} = N^{max}\alpha_i + \omega_n N_i^{old}, \tag{2}$$

where

$$\alpha_i = \alpha_i^{local} + \alpha_i^{target}, \tag{3}$$

N^{max} is the parameter used to tune the part of the induced motion, ω_n is the array of random values in the range [0, 1], and N^{old} is the current motion induced. For more details, refer [31]. In this study, the effect of the individual in the krill movement is calculated by Eq. (4).

$$\alpha_i^{local} = \sum_{j=1}^{N} \widehat{K}_{i,j}\widehat{x}_{i,j}, \tag{4}$$

where

$$\widehat{x}_{i,j} = \frac{x_j - x_i}{\|x_j - x_i\| + \epsilon}, \tag{5}$$

$$\widehat{K}_{i,j} = \frac{K_i - K_j}{K^{worst} - K^{best}}, \tag{6}$$

K^{best} and K^{worst} represent the best and worst fitness function values, respectively, among all krill individuals. $\widehat{x}_{i,j}$ represents the difference among position x_i and x_j ($j = 1, 2, \ldots, N$), K_i represents the objective function value of the current krill individual, K_j represents the objective function of the jth individual where ($j = 1, 2, \ldots, N$), N is the number of krill element, $x_{i,j}$ represents the jth position in solution i, and ϵ is a small positive number [27, 32].

$$\alpha_i^{target} = C^{best} \, \widehat{K}_{i,best} \widehat{x}_{i,best}, \tag{7}$$

where

$$C^{best} = 2\left(rand + \frac{I}{I_{max}}\right). \tag{8}$$

C^{best} is the coefficient of individuals, $\widehat{K}_{i,best}$ is the best objective function of the ith krill individual, $\widehat{x}_{i,best}$ is the best position of the ith krill individual, $rand$ is a random number between [0, 1] for improving the local exploration, I is the current iteration number, and I_{max} is the maximum number of iterations [8]. In this model, the neighbors are chosen by Eq. (9). The movement of krill individuals and their neighbors are illustrated in Fig. 1.

$$de_i = \frac{1}{5N} \sum_{j=1}^{N} \|x_i - x_j\|. \tag{9}$$

Equation (9) is used to find the neighbors of the ith krill individual by finding the distance between them, and the known goal for each krill individual is to achieve the highest fitness function, which is obtained by Eq. (10). This procedure allows the solution to move toward the current best solution.

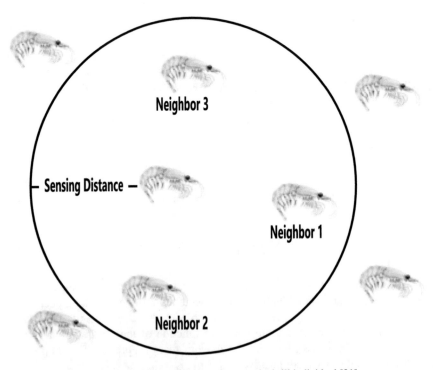

Fig. 1 A schematic represents the sensing domain around a krill individual [31]

$$\alpha_i^{target} = C^{best} \, \widehat{K}_{i,best} \widehat{x}_{i,best}, \tag{10}$$

where

$$C^{best} = 2 \left(rand + \frac{I}{I_{max}} \right), \tag{11}$$

C^{best} is the effective coefficient and α_i^{target} leads the solutions to the global optima values, which should be nearer to the optimal solution. The $rand$ is a random number between [0, 1], which is used to improve the exploration. I is the current iteration number and I_{max} is the maximum number of iterations [8].

3.1.2   Foraging Motion

This factor has two affected parameters: the first is the food location and the second is the old food location. This action can be expressed for the ith krill individual by Eq. (12).

$$F_i = V_f \beta_i + \omega_f F_i^{old}, \tag{12}$$

where

$$\beta_i = \beta_i^{food} + \beta_i^{best}, \tag{13}$$

V_f is the forging speed, ω_f is the intra-weight used to balance the local exploitation and global exploration for each individual, β_i^{food} is the food attraction, and β_i^{best} is the best food attraction so far.

$$x^{food} = \frac{\sum_{i=1}^{N} \frac{1}{K_i} x_i}{\sum_{i=1}^{N} \frac{1}{K_i}}, \tag{14}$$

where

$$\beta_i^{food} = C^{food} \, \widehat{K}_{i,food} \widehat{x}_{i,food}, \tag{15}$$

C^{food} is the food coefficient which affected the krill herding by decreasing during the execution time. The food coefficient is calculated by Eq. (16).

$$C^{food} = 2 \left(1 - \frac{I}{I_{max}} \right), \tag{16}$$

where

$$\beta_i^{best} = \widehat{K}_{i,ibest} \widehat{x}_{i,ibest}. \tag{17}$$

Equation (17) is used in order to handle the best fitness function of the ith krill individual, where K_{ibest} represents the best-visited location.

3.1.3 Physical Diffusion

In this factor, the krill individual is estimated as the random process which used two terms to express the physical diffusion: the first is the maximum diffusion speed and the second is the random directional vector [8]. The physical diffusion is determined by Eq. (18).

$$D_i = D^{max}\left(1 - \frac{I}{I_{max}}\right)\delta, \tag{18}$$

where D^{max} is the maximum diffusion speed and δ is the random values of the vector which has arrays containing random values between $[-1, 1]$. This action decreased the speed value of the krill individual [31].

3.1.4 Motion Process of the KH Algorithm

The motion-inducing and foraging motion contained two local and two global strategies. These strategies work in parallel to obtain a powerful algorithm. The physical diffusion generates random vectors [8]. KH algorithm parameters are effective during the algorithm acts. The positions of krill individuals are updated in each iteration using the Langranging model by Eq. (19).

$$x_i(I + 1) = x_i(I) + \Delta t \frac{dx_i}{dt}, \tag{19}$$

where

$$\Delta t = C_t \sum_{j=1}^{N}(UB_j - LB_j). \tag{20}$$

x_i is the position i in the search space; $(I + 1)$ is the next iteration; Δt is an important and more sensitive constant computed by Eq. (20); and N represents the total number of variables, the lower bounds LB_j, and the upper bounds UB_j of the ith variables $(J = 1, 2, \ldots, N)$, respectively. C_t is a constant value between $[0, 2]$ [31].

3.2 Genetic Operators

Reproduction procedures are incorporated into H-KHA algorithm to improve its performance. Crossover and mutation operators are inspired from the classical differential evolutionary algorithm. For more details, refer [8].

3.2.1   Crossover Operator of MKHA

The crossover is used in the genetic algorithm as an effective operator for reaching the global solution [8]. The crossover probability Cr is used as a control on the crossover operator by generating a uniformly distributed random value between [0, 1] [32]. The mth component of $x_{i,m}$ is determined by Eq. (21).

$$x_{i,m} = \begin{cases} x_{r,m}, & if\, rand < Cr \\ x_{i,m}, & else \end{cases} \tag{21}$$

$$Cr = 0.2\widehat{K}_{i,best}, \tag{22}$$

where $r \in \{1, 2, \ldots, i-1, i+1, \ldots, N\}$ is the new crossover probability for global best, which will increase with decreasing the fitness function.

3.2.2   Mutation Operator of MKHA

The mutation operator plays a beneficial role in evolutionary algorithms. The mutation probability Mu is used as a control on the mutation operator [8]. The mth component of $x_{i,m}$ is determined by Eq. (23).

$$x_{i,m} = \begin{cases} x_{gbest,m} + \mu(x_{p,m} - x_{q,m}), & if\, rand < Mu \\ x_{i,m}, & else \end{cases} \tag{23}$$

$$Mu = 0.05/\widehat{K}_{i,best}, \tag{24}$$

where $p, q \in \{1, 2, , i1, i+1, , S\}$, S represent the number of all solutions, μ is the value between [0, 1]. This mutation probability is considered new for global best, which determines the interconnection based on the increases in decreasing the fitness function [8].

4   Overview of Modified Krill Herd Algorithm

The previous section thoroughly presented the basic KHA through six main steps that have been procedurally described. The results obviously showed that the basic KHA has the potential to generate viable solutions in comparison with the other well-known algorithms, but not as impressive as those described in the literature.

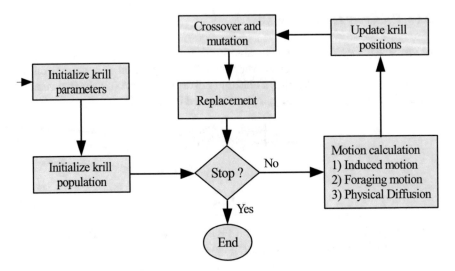

Fig. 2 The flowchart of the modified krill herd algorithm

The basic KHA needs to focus on exploration rather than exploitation to obtain preferable results. Thus, the basic KHA fails to determine the optimal solution in that area. Subsequently, the MKHA is proposed to overcome weaknesses of the basic KHA. The sequence of procedures of the MKHA is shown in Fig. 2.

This section addresses and attempts to overcome the weaknesses of the basic KHA by modifying the basic KH operators to achieve high-quality clustering solutions. The proposed method shown in this section is called MKHA. The MKHA is compared with the basic KHA and the other well-known algorithms, which used the same benchmark functions. Modification occurs in the order of KH operators where the crossover and mutation processes are invoked after updating the krill positions because the nature of the search space of optimization problems is ragged and deep. The pseudocode of the MKHA is shown in Algorithm 1.

However, the basic KHA is sometimes trapped in the local optimal solution because of its poor capability in exploration and exploitation. The MKHA is invented to improve the TD clustering technique by enhancing the procedures of the basic KHA. Algorithm 1 shows further details on the KH modification. This modification is performed by updating the current solutions by comparing them with the generated solutions of the motion calculations, applying the genetic operators, and comparing the introduced solutions of the genetic operators with the current solutions (best solutions so far). The MKHA enhances the global searchability by applying the genetic operators on the high-quality solutions introduced by the motion calculations.

Algorithm 1 Modified krill herd algorithm (MKHA)

1: Initialization the krill parameters: N^{max}, D^{max}, and $etc.$
2: **for** $i = 1$ to S **do**
3:    **for** $j = 1$ to n **do**
4:       $x_{i,j} = 1 + rand$ mod K, Initialization of KH memory.
5:    **end for**
6:    Compute the fitness function f_i Evaluate each krill
7: **end for**
8: Sort the krill and find x^{best}, where $best \in (1, 2, ..., S)$
9: **while** $I \neq I_{max}$ **do**
10:    **for** $i = 1$ to S **do**
11:       Perform the three motion calculation.
12:       $x_i(I + 1) = x_i(I) + \delta I \frac{dx_i}{dI}$
13:       Evaluate each solution using fitness.
14:    **end for**
15:    **for** $i = 1$ to S **do**
16:       Applying KH operators on the KH memory.
17:       Crossover within dynamic probability
18:       Mutation within dynamic probability
19:    **end for**
20:    Replace the worst krill with the best solution
21:    Sort the solution and find x^{best}, where $best \in (1, 2, ..., S)$
22:    $I = I + 1$
23: **end while**
24: Return x^{best}

---

5 Experimental Results

This section presents various experiments on optimization benchmark function problems to verify the performance of the proposed MKHA. In order to make a fair comparison, all the experiments are conducted using the same PC with the detailed settings (i.e., Windows 7 environment using MATLAB software (7.10.0) computer programming with 8 RAM).

Fourteen different functions are used to evaluate the proposed algorithm (MCA). The definitions and characteristics of these functions can be presented in Fig. 3. For further information, refer to [29]. For the purpose of showing the superiority of the proposed algorithm (i.e., MKHA), its performance is analyzed using function optimization problems with seven algorithms, which are CS [28], BA [4], ACO [6], HS [33], GA [21], PSO [29], and KHA [8].

For all comparative algorithms used in this research, population size NP is 50, an elitism parameter Keep is 2, and maximum generation is 50. The optimal results achieved by the algorithms for each function, in the following sections, are explained. The dimension of the functions used in this research is 20 (i.e., d = 20).

No.	Name	Definition
F01	Ackley	$f(\vec{x}) = 20 + e - 20 \cdot e^{-0.2 \cdot \sqrt{\frac{1}{n}\sum_{i=1}^{n} x_i^2}} - e^{\frac{1}{n}\sum_{i=1}^{n}\cos(2\pi x_i)}$
F02	Fletcher–Powell	$f(\vec{x}) = \sum_{i=1}^{n}(A_i - B_i)^2, \quad A_i = \sum_{j=1}^{n}(a_{ij}\sin\alpha_j + b_{ij}\cos\alpha_j) \quad B_i = \sum_{j=1}^{n}(a_{ij}\sin x_j + b_{ij}\cos x_j)$
F03	Griewank	$f(\vec{x}) = \sum_{i=1}^{n}\frac{x_i^2}{4000} - \prod_{i=1}^{n}\cos\left(\frac{x_i}{\sqrt{i}}\right) + 1$
F04	Penalty #1	$f(\vec{x}) = \frac{\pi}{n}\left\{10\sin^2(\pi y_1) + \sum_{i=1}^{n-1}(y_i - 1)^2 \cdot [1 + 10\sin^2(\pi y_{i+1})] + (y_n - 1)^2\right\} + \sum_{i=1}^{n}u(x_i, 10, 100, 4), \quad y_i = 1 + 0.25(x_i + 1)$
F05	Penalty #2	$f(\vec{x}) = 0.1\left\{\sin^2(3\pi x_1) + \sum_{i=1}^{n-1}(x_i - 1)^2 \cdot [1 + \sin^2(3\pi x_{i+1})] + (x_n - 1)^2[1 + \sin^2(2\pi x_n)]\right\} + \sum_{i=1}^{n}u(x_i, 5, 100, 4)$
F06	Quartic with noise	$f(\vec{x}) = \sum_{i=1}^{n}(i \cdot x_i^4 + U(0,1))$
F07	Rastrigin	$f(\vec{x}) = 10 \cdot n + \sum_{i=1}^{n}(x_i^2 - 10 \cdot \cos(2\pi x_i))$
F08	Rosenbrock	$f(\vec{x}) = \sum_{i=1}^{n-1}[100(x_{i+1} - x_i^2)^2 + (x_i - 1)^2]$
F09	Schwefel 2.26	$f(\vec{x}) = 418.9829 \times D - \sum_{i=1}^{D}x_i\sin(|x_i|^{1/2})$
F10	Schwefel 1.2	$f(\vec{x}) = \sum_{i=1}^{n}\left(\sum_{j=1}^{i}x_j\right)^2$
F11	Schwefel 2.22	$f(\vec{x}) = \sum_{i=1}^{n}|x_i| + \prod_{i=1}^{n}|x_i|$
F12	Schwefel 2.21	$f(\vec{x}) = \max_i\{|x_i|, 1 \leq i \leq n\}$
F13	Sphere	$f(\vec{x}) = \sum_{i=1}^{n}x_i^2$
F14	Step	$f(\vec{x}) = 6 \cdot n + \sum_{i=1}^{n}\lfloor x_i \rfloor$

In benchmark function F02, the matrix elements $a_{n \times n}, b_{n \times n} \in (-100, 100), \alpha_{n \times 1} \in (-\pi, \pi)$ are draw from uniform distribution.
In benchmark functions F04 and F05, the definition of the function $u(x_i, a, k, m)$ is as follows:

$$u(x_i, a, k, m) = \begin{cases} k(x_i - a)^m, & x_i > a \\ 0, & -a \leq x_i \leq a \\ k(-x_i - a)^m, & x_i < a \end{cases}$$

Properties of benchmark functions, *lb* denotes lower bound, *ub* denotes upper bound, *opt* denotes optimum point.

No.	Function	lb	ub	opt	Continuity	Modality
F01	Ackley	−32.768	32.768	0	Continuous	Multimodal
F02	Fletcher–Powell	−π	π	0	Continuous	Multimodal
F03	Griewangk	−600	600	0	Continuous	Multimodal
F04	Penalty #1	−50	50	0	Continuous	Multimodal
F05	Penalty #2	−50	50	0	Continuous	Multimodal
F06	Quartic with noise	−1.28	1.28	1	Continuous	Multimodal
F07	Rastrigin	−5.12	5.12	0	Continuous	Multimodal
F08	Rosenbrock	−2.048	2.048	0	Continuous	Unimodal
F09	Schwefel 2.26	−512	512	0	Continuous	Multimodal
F10	Schwefel 1.2	−100	100	0	Continuous	Unimodal
F11	Schwefel 2.22	−10	10	0	Continuous	Unimodal
F12	Schwefel 2.21	−100	100	0	Continuous	Unimodal
F13	Sphere	−5.12	5.12	0	Continuous	Unimodal
F14	Step	−5.12	5.12	0	Discontinuous	Unimodal

Fig. 3 Benchmark functions

Table 1 report the average results from 100 runs for each comparative algorithm. According to the average results, MKHA performed better than other comparative algorithms on 14 benchmark functions (i.e., F01, F02, F04, F05, F06, F07, F08, F09, F10, F12, F13, and F14) when searching for minimizing objective values. KHA and ACO are the second most productive algorithms according to the average values across the 14 benchmarks (F13, and F11, respectively). From Table 1, it can be observed that MKHA performed better in comparison with the basic KHA almost in the 14 benchmarks (i.e., F01, F02, F04, F05, F06, F07, F08, F09, F10, F11, F12, F13, and F14).

Moreover, the process of the proposed optimization algorithms (KHA, and MKHA) is given in Figs. 4, 5, 6 and 7. The values in these figures are presented the best optimum value achieved through 100 runs. Hither, all the taken values are true function values. Furthermore, it should be remarked that the global best of the

Table 1 The values are normalized so that the minimum in each row is 1.00. These are not the absolute minima found by each algorithm, but the average minima found by each algorithm

Function No.	CS [28]	BA [4]	ACO [6]	HS [33]	GA [21]	PSO [29]	KHA [8]	MKHA
F01	05.26	04.49	02.31	02.75	06.82	03.70	01.20	**01.15**
F02	06.95	07.51	24.58	12.87	04.26	02.27	01.06	**01.05**
F03	66.85	50.73	316.0	68.69	38.53	14.28	**06.10**	07.14
F04	1.4E6	1.9E7	3.8E3	9.6E6	1.6E5	545.3	435.2	**430.2**
F05	6.5E3	2.6E4	1.1E3	5.8E5	3.2E5	131.4	101.8	**99.51**
F06	422.4	1.4E3	489.0	3.0E3	1.4E3	24.12	24.48	**24.10**
F07	07.97	07.35	08.09	09.25	06.41	01.85	02.02	**01.80**
F08	14.47	54.64	42.25	21.80	16.36	04.28	05.02	**04.79**
F09	06.32	04.39	03.17	14.72	02.47	03.52	04.80	**02.35**
F10	04.04	18.68	01.75	03.06	03.78	01.64	05.14	**04.69**
F11	16.61	12.80	**01.05**	14.83	25.73	02.23	8.4E1	13.15
F12	04.87	06.95	01.86	02.53	08.00	06.55	01.73	**01.69**
F13	207.5	66.67	98.30	113.3	485.4	16.90	07.04	**07.00**
F14	53.29	42.69	07.73	85.79	140.9	15.78	06.43	**06.21**

benchmark functions are illustrated like convergence plots (F01 in Fig. 4, F02 in Fig. 5, F03 in Fig. 6, and F04 in Fig. 7).

Figure 4 shows the best values achieved by the KH algorithms on the function number one (i.e., F01). From Fig. 4, MKHA performance is superior to the basic KHA in the optimization process, while KHA performed the second best in this benchmark overall comparative algorithms. Moreover, all the proposed algorithms have almost the same initial values. Figure 5 shows the obtained optimization values for the function number two (i.e., F02). In this case, the figure clearly displays that MKHA performance differs from the other basic algorithm. From Figs. 6 and 7, MKHA has the best performance for these two functions as well. Figure 6 presents that the MKHA obtained the best solution in this benchmark function.

6 Conclusion

In the current work, the genetic operators are modified with the basic KH to propose an enhanced modified krill herd algorithm for solving optimization problems, called MKHA.

The results of the MKHA reveal that the proposed MKHA can produce the best-recorded results for all benchmark functions used compared with the other comparative algorithms from the literature. Thus, the KH algorithm with crossover and

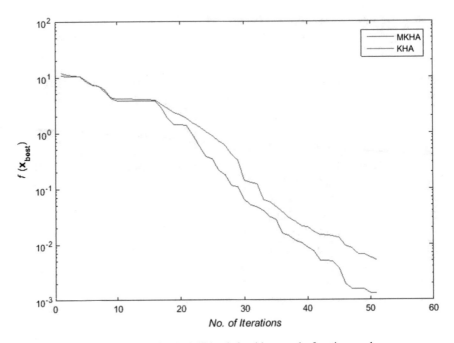

Fig. 4 The best values achieved by the krill herd algorithms on the function number one

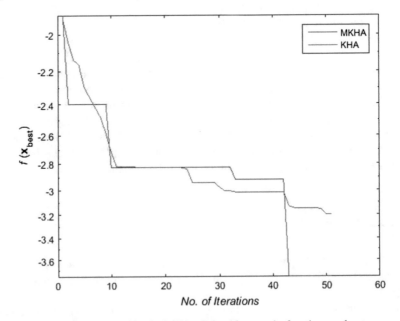

Fig. 5 The best values achieved by the krill herd algorithms on the function number two

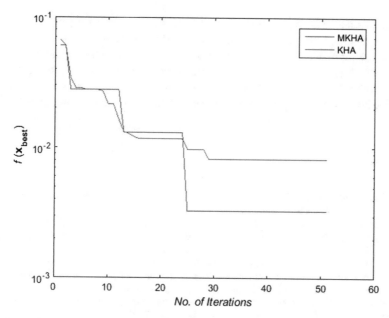

Fig. 6 The best values achieved by the krill herd algorithms on the function number three

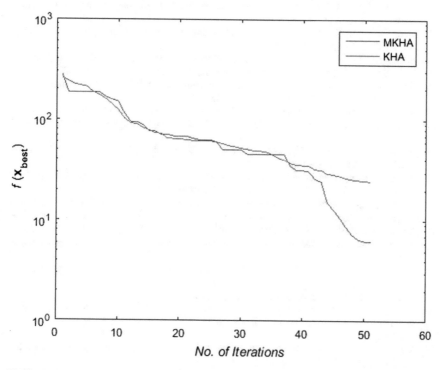

Fig. 7 The best values achieved by the krill herd algorithms on the function number four

mutation after the updating process is an effective and superior method for, indicating that the modified versions significantly exploit the genetic operators.

For the future works, first, MKHA can be used to solve other optimization problems such as text clustering and feature selection. Second, can be combined MKHA with another local search method to develop a new hybrid algorithm to improve the local exploitation search ability of the modified krill herd algorithm (MKHA).

References

1. Ghanem WAHM, Jantan A (2017) An enhanced Bat algorithm with mutation operator for numerical optimization problems. Neural Comput Appl 1–35
2. Alomari OA, Khader AT, Mohammed A, Abualigah LM, Nugroho H, Chandra GR et al (2017) Mrmr Ba: a hybrid gene selection algorithm for cancer classification. J Theor Appl Inf Technol 95(12):1
3. Shehab M, Khader AT, Al-Betar MA, Abualigah LM (2017) Hybridizing cuckoo search algorithm with hill climbing for numerical optimization problems. In: 2017 8th international conference on information technology (ICIT). IEEE, pp 36–43
4. Wang GG, Guo L, Duan H, Wang H (2014) A new improved firefly algorithm for global numerical optimization. J Comput Theor Nanosci 11(2):477–485
5. Trivedi IN, Gandomi AH, Jangir P, Kumar A, Jangir N, Totlani R (2017) Adaptive krill herd algorithm for global numerical optimization. In: Advances in computer and computational sciences. Springer, pp 517–525
6. Wang G, Guo L (2013) A novel hybrid bat algorithm with harmony search for global numerical optimization. J Appl Math 2013:21
7. Abualigah LM, Khader AT, Hanandeh ES, Gandomi AH (2017) A novel hybridization strategy for krill herd algorithm applied to clustering techniques. Appl Soft Comput 60:423–435
8. Gandomi AH, Alavi AH (2012) Krill herd: a new bio-inspired optimization algorithm. Commun Nonlinear Sci Numer Simul 17(12):4831–4845
9. Abualigah LM, Khader AT, Hanandeh ES (2018) A combination of objective functions and hybrid Krill herd algorithm for text document clustering analysis. Eng Appl Artif Intell 73:111–125
10. Abualigah LM, Khader AT, Hanandeh ES (2018) Hybrid clustering analysis using improved krill herd algorithm. Appl Intell 48:1–25
11. Abualigah LMQ, Hanandeh ES (2015) Applying genetic algorithms to information retrieval using vector space model. Int J Comput Sci, Eng Appl 5(1):19
12. Abualigah LM, Khader AT, AlBetar MA, Hanandeh ES (2017) A new hybridization strategy for krill herd algorithm and harmony search algorithm applied to improve the data clustering
13. Abualigah LM, Khader AT, Al-Betar MA, Alomari OA (2017) Text feature selection with a robust weight scheme and dynamic dimension reduction to text document clustering. Expert Syst Appl 84:24–36
14. Abualigah LM, Khader AT, Al-Betar MA. Unsupervised feature selection technique based on genetic algorithm for improving the text clustering. In: 2016 7th international conference on computer science and information technology (CSIT), pp. 1–6. IEEE
15. Abualigah LM, Khader AT, AlBetar MA, Hanandeh ES (2017) Unsupervised text feature selection technique based on particle swarm optimization algorithm for improving the text clustering. EAI
16. Xinchao Z (2010) A perturbed particle swarm algorithm for numerical optimization. Appl Soft Comput 10(1):119–124
17. Alatas B (2010) Chaotic bee colony algorithms for global numerical optimization. Expert Syst Appl 37(8):5682–5687

18. Chuang LY, Tsai SW, Yang CH (2011) Chaotic catfish particle swarm optimization for solving global numerical optimization problems. Appl Math Comput 217(16):6900–6916
19. Ghanem WA, Jantan A (2016) Hybridizing artificial bee colony with monarch butterfly optimization for numerical optimization problems. Neural Comput Appl 1–19
20. Sakib N, Kabir MWU, Subbir M, Alam S (2014) A comparative study of flower pollination algorithm and bat algorithm on continuous optimization problems. Int J Soft Comput Eng 4(3):13–19
21. Wang G, Guo L, Wang H, Duan H, Liu L, Li J (2014) Incorporating mutation scheme into krill herd algorithm for global numerical optimization. Neural Comput Appl 24(3–4):853–871
22. Abualigah LM, Khader AT (2017) Unsupervised text feature selection technique based on hybrid particle swarm optimization algorithm with genetic operators for the text clustering. J Supercomput 1–23
23. Shehab M, Khader AT, Laouchedi M (2017) Modified cuckoo search algorithm for solving global optimization problems. In: International conference of reliable information and communication technology, pp. 561–570. Springer
24. Wang GG, Hossein Gandomi A, Hossein Alavi A (2013) A chaotic particle-swarm krill herd algorithm for global numerical optimization. Kybernetes 42(6):962–978
25. Guo L, Wang GG, Gandomi AH, Alavi AH, Duan H (2014) A new improved krill herd algorithm for global numerical optimization. Neurocomputing 138:392–402
26. Wang GG, Gandomi AH, Alavi AH, Hao GS (2014) Hybrid krill herd algorithm with differential evolution for global numerical optimization. Neural Comput Appl 25(2):297–308
27. Abualigah LM, Khader AT, Al-Betar MA, Awadallah MA (2016) A krill herd algorithm for efficient text documents clustering. In: 2016 IEEE symposium on computer applications & industrial electronics (ISCAIE), pp. 67–72. IEEE
28. Wang GG, Gandomi AH, Alavi AH (2014) An effective krill herd algorithm with migration operator in biogeography-based optimization. Appl Math Modell 38(9):2454–2462
29. Wang GG, Guo L, Gandomi AH, Hao GS, Wang H (2014) Chaotic krill herd algorithm. Inf Sci 274:17–34
30. Abualigah LM, Khader AT, Hanandeh ES (2018) A hybrid strategy for krill herd algorithm with harmony search algorithm to improve the data clustering. Intell Decis Technol 12:1–12
31. Bolaji AL, Al-Betar MA, Awadallah MA, Khader AT, Abualigah LM (2016) A comprehensive review: Krill Herd algorithm (KH) and its applications. Applied Soft Computing. 49:437–446
32. Jensi R, Jiji GW (2016) An improved krill herd algorithm with global exploration capability for solving numerical function optimization problems and its application to data clustering. Applied Soft Computing. 46:230–245
33. Wang GG, Gandomi AH, Zhao X, Chu HCE (2016) Hybridizing harmony search algorithm with cuckoo search for global numerical optimization. Soft Computing. 20(1):273–285

Application of Nature-Inspired Optimization Techniques in Vessel Traffic Control

Ž. Kanović, V. Bugarski, T. Bačkalić and F. Kulić

Abstract This chapter aims to present the analysis and comparison of some well-known nature-inspired global optimization techniques applied to an expert system controlling a ship locking process. A ship lock zone represents a specific area on waterway, and control of the ship lockage process requires a comprehensive approach. The initially proposed Fuzzy Expert System (FES) was developed using suggestions obtained from lockmasters (ship lock operators) with extensive experience. Further optimization of the membership function parameters of the input variables was performed to achieve better results in the local distribution of vessel arrivals. The purpose of the analysis and comparison is to find the best algorithm for optimization of membership functions parameters of FES for the ship lock control. The initially proposed FES is optimized (fine-tuned) with three global optimization algorithms from the group of evolutionary and swarm intelligence algorithms, in order to achieve the best value of the economic criterion defined as a linear combination of two opposed criteria: minimal average waiting time per vessel and minimal number of empty lockages (lockages without a vessel in a chamber). Besides the well known and widely applied Genetic Algorithm (*GA*), two relatively new but very promising global optimization techniques were used: Particle Swarm Optimization (*PSO*), the technique based on behavior of animals living in swarms and Artificial Bee Colony (*ABC*) algorithm, inspired by social organization of honey bees. Although all these algorithms have been widely applied and showed a great

Ž. Kanović · V. Bugarski · F. Kulić
Computing and Control Department, Faculty of Technical Sciences,
University of Novi Sad, Novi Sad, Serbia
e-mail: kanovic@uns.ac.rs

V. Bugarski
e-mail: bugarski@uns.ac.rs

F. Kulić
e-mail: kulic@uns.ac.rs

T. Bačkalić (✉)
Department of Traffic Engineering, Faculty of Technical Sciences,
University of Novi Sad, Novi Sad, Serbia
e-mail: tosa@uns.ac.rs

© Springer International Publishing AG, part of Springer Nature 2019
S. K. Shandilya et al. (eds.), *Advances in Nature-Inspired Computing and Applications*, EAI/Springer Innovations in Communication and Computing,
https://doi.org/10.1007/978-3-319-96451-5_10

potential in engineering applications in general, their application in ship lock control and similar transportation problems is not so common. However, this chapter will present that all three algorithms may obtain the significant improvement of the adopted economic criterion value and succeed to find its (possibly global) optimum. Furthermore, the performances of these algorithms in FES parameters optimization are compared and some conclusions are adopted on their applicability, efficiency, and effectiveness in similar systems. The developed fuzzy algorithm is a rare application of artificial intelligence in navigable canals and significantly improves the performance of the ship lockage process. This adaptable FES is designed to be used as a support in decision-making processes or for the direct control of ship lock operations.

Keywords Ship lock control · Fuzzy expert system · Genetic algorithm · Artificial bee colony · Particle Swarm Optimization

1 Introduction

The history of humankind is a history of human endeavors and efforts in encountering, analysis, and overcoming obstacles. The overcoming of the difference in water levels on inland waterways, caused by the construction of dams as artificial obstacles, is one of the challenges that are most often solved using ship locks. The beginning of the implementation of ship locks dates back to the distant past. Today, ship locks are the most commonly used hydrotechnical structures that enable vessels to overcome the difference in water levels easily and safely. Ship locks are designed to enable vessels to overcome rises in the water level and help to maintain navigation on inland waterways [46, 39].

Although inland waterway transport is perceived to have considerable societal importance in achieving sustainable mobility, it is growing at only a modest rate [67]. Because this potential requires new concepts to be realized, attention should be paid to innovations that can improve vessel traffic management. Intelligent transportation systems have been developed in the field of road transportation and hence the term "intelligent infrastructure" typically refers to that transportation mode. Recent research has been increasingly directed towards intelligent infrastructure development and control structure design [42]. Unlike other transport modes, the use of computational intelligence in inland water transportation is still in its infancy [68], particularly in regard to replace humans in the decision-making process in real time. Transportation systems control is gaining importance, and in the marine systems field, the focus of research is on the swarming behavior of vessels [31]. The aim of this research is to emphasize the potential application of artificial intelligence as a control tool in vessel traffic management on inland waterways.

Campbell et al. [9] presented decision tools for reducing congestion at ship locks on the upper Mississippi river. Bugarski et al. [8], proposed a decision support system for a ship lock control based on the fuzzy logic. The fuzzy logic is chosen as a control approach that does not require a precise mathematical model of the controlled system [27] and as the most suitable mathematical method for addressing uncertainty, subjectivity, polysemy, and indefiniteness [34]. Other authors [15, 44] have also used the fuzzy logic in decision support processes. Teodorović and Vukadinović [62] successfully applied fuzzy reasoning and artificial intelligence in traffic control.

Although expert systems have been successfully used in the design of large structures [2] such as ship locks, the proposed analysis focuses on the implementation of a Fuzzy Expert System (FES) designed to assist ship lock operators (lockmasters) in the decision-making process. Adeli [1] published an article extolling the advantages of expert systems based on artificial intelligence implemented in construction engineering and management. Developing an expert system for ship lock control raises two specific challenges: gathering expert knowledge and adapting to changes in control criteria priorities.

The proposed model is based on previous research. The initial research on designing a control algorithm based on artificial intelligence for ship lock control was published by [8]. The control algorithm relied on fuzzy logic and was designed solely on the basis of operator's experience. Kanović et al. [25] published a paper where three different global optimization techniques were tested for possible implementation in a fuzzy expert system for a ship lock control. The present research is performed on the ship locks on navigable canals in Serbia, where it is possible to implement a Supervisory Control And Data Acquisition (SCADA) system with an FES [6].

The essence of the proposed approach is the optimization (fine tuning) of the previously developed FES in accordance with the given economic criterion. The proposed economic criterion is defined as a linear combination of two opposing norms—a minimum number of empty lockages (lockages without a vessel) and a minimal waiting time (vessel's delay). In order to improve FES for the control of the ship locking process and for finding the best optimization technique, three well-known nature-inspired optimization algorithms have been used as optimization tools: Particle Swarm Optimization (*PSO*), Artificial Bee Colony Optimization (*ABC*), and Genetic Algorithm (*GA*) [19, 23, 63]. The results achieved in presented research confirmed that chosen nature-inspired global optimization algorithms could be successfully applied in problems regarding transportation performance improvement and optimization.

The presented FES is intended to be a decision support system implemented in an existing Programmable Logic Controller (PLC) and SCADA system in a ship lock control room. There are existing examples of improving PLC and SCADA control logic in irrigation canals [17], but the presented model is a rare application of artificial intelligence in navigable canals. Today, around the world, actively operates a large number of ship locks, which are different in size, age, area of navigation, ways of functioning and organization of traffic. From a wide range of types of ship locks, the choice was limited to a system that is usually applied on European navigable canals on inland waterways, i.e., a single-chamber ship lock that operates only with one ship. Observed ship lock could be described as a single-channel queuing system with two independent, stochastic streams of arrivals from two opposite directions. Even though the proposed model was established and tested in the particular real system, the principle of generality could be established. With minor modifications in the design of the basic FES, the presented model could be extended to any other ship lock from the observed category.

The chapter is organized as follows: Sect. 2 describes the problem of vessel traffic control in the ship lock zone and fuzzy expert system as a decision support for ship lock control. Section 3 discusses the proposed methodology and the nature-inspired optimization techniques. In Sect. 4, the application of nature-inspired optimization techniques for vessel traffic control in ship lock zone is described. In Sect. 5, the results of different optimization techniques are compared and discussed. Finally, Sect. 6 contains the concluding remarks and directions of further research.

2 Vessel Traffic Control in the Ship Lock Zone

2.1 Problem Description

The ship lock or navigation lock is a hydraulic structure that consists of an enclosed chamber with watertight gates at each end. The water level difference is overcome by filling or emptying of the lock chamber. Namely, the ship lock operates on the simple buoyancy principle that any vessel, no matter what proportion, will float upon a large enough volume of water. By raising or lowering the level of water within the ship lock chamber, the vessel itself goes up or down correspondingly. As hydraulic engineering systems, ship locks are designed to enable vessels to surmount obstacles (rapids, weirs, or dams) on the waterway [46]. Lockmasters (ship lock operators) regularly make effort to fill or empty the ship lock chamber in the fastest time possible with a minimum of turbulence. The vessel traffic organization in the zone of the ship lock is a compromise between a reasonable usage of the lock and minimizing a vessel's delay while waiting to pass the ship lock zone. Basic elements of the classic ship lock are presented in [5, 59].

Although there is an extensive range of types of ship locks, each ship lock consists of three basic elements:

- The lock chamber is the basic and largest element of the ship lock. It connects the upper and lower approach channels of the ship lock and is sufficiently large to accommodate one or more vessels. The position of the lock chamber is fixed but water level inside it can vary.
- The lock gates are the most important movable elements, and they are constructed to be watertight. Mitre gates are mostly used lock gates and represent a pair of half gates at each end of the lock chamber (Fig. 1). In addition, there are also single pivot gates, wing gates, tainter gates, rolling gates, lift gates, and drop gates. The ship lock gates keep water level in the lock chamber and the possibility of entering or exiting of a vessel depends on their current state (open or closed). Possible states of the ship lock are (a) one gate is open and the second is closed and (b) both of gates are closed [39].
- Intake and discharge system.

From among the wide variety of ship locks, the focus was narrowed to one specific type of ship lock, but there is no loss of generality and the presented model is applicable to any other type of ship lock. This study focused on a specific choice: a single-channel two-way lock (Fig. 1). The analyzed system as a single queuing node utilizes a First–In, First-Out (FIFO) queuing discipline, and vessel arrivals are random from both levels [41, 59]. A multi-trajectory approach of vessels from the same direction is possible, but only up to a reference point. The reference point is the first pre-signal. Overtaking is forbidden after the reference point, and vessels form a queue according to the FIFO principle. The primary objective in controlling a ship lock and managing vessel transitions is evaluating and reducing traffic delays [30] while minimizing the consumption of water and energy [8, 9, 64]. The owners of the ship lock prefer fewer empty lockages (change of level in the lock chamber without vessel) because such lockages reduce the ship lock efficiency (both in terms of the total number of lockages and in terms of energy consumption). However, shippers prefer to wait as little as possible for lockage. In the case that more vessels are approaching to the ship lock from the same direction, operators have to change the level of the water in the lock chamber without vessel to reduce vessels waiting times, which increase the costs of operating the ship lock.

The lockmasters, as part of the inland waterways tradition, are responsible for the proper functioning of the lock and vessel traffic control in the lock zone. Their

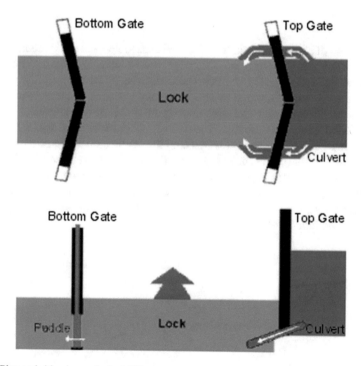

Fig. 1 Plan and side views of a lock [8]

vessel traffic control decisions are based on estimates under conditions of uncertainty. Therefore, the experience of the lockmaster plays a significant role in the decision-making process. The essentiality of the decision-making problem in the ship lock control is in the choice (*LC*) which vessel will be served first: vessel which is far from the ship lock and comes from the level at which the gate is open (*LGO*) or the vessel which is closer to the ship lock, but it is at the level at which the gate is currently closed (*LGC*) (Fig. 2). The lockmaster often faces a decision-making dilemma as to whether to prioritize saving water and energy or reducing the waiting time of a vessel. The lockmaster's dilemma rarely appears in situations where the traffic density is significantly lower than the ship lock capacity or close to the limit of the ship lock capacity.

The lockmaster must simultaneously control lock operations and vessel traffic in the ship lock zone. Similar to a Vessel Traffic Service (VTS) operator in maritime, experience is necessary in both ship lock control and quality decision support [51]. An adaptive expert system for ship lock control based on human experience can provide the necessary decision support. A fuzzy expert system was thus developed and applied to describe and solve the lockmaster's problem.

2.2 Fuzzy Expert System as a Decision Support for Ship Lock Control

Achievement of high-quality control of an observed system normally requires knowledge of nature and principles of the process, and a suitable control algorithm for realization the desired goals or system performance. In cases where the system to be controlled is complex, variable, or it is difficult to define a precise mathematical model, a control approach that does not require a precise mathematical model of the controlled system is desirable. The fuzzy logic is the method that fulfills these requirements [27]. Fuzzy control is based on the application of the fuzzy set theory and the fuzzy logic. Using fuzzy logic and fuzzy inference systems [58], one can collect knowledge from experts in a specific field and implement it in the control algorithm to realize the required control of an observed system. The fuzzy set theory is the most suitable mathematical approach to focus on uncertainty, subjectivity, polysemy, and indefiniteness [34].

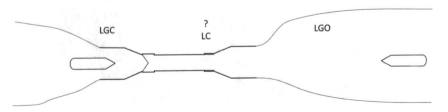

Fig. 2 A dilemma situation in control of a single-channel two-way lock

The concept of fuzzy logic was originated by Lotfi Zadeh, who proposed it not as a control methodology but as a method of processing data by allowing partial membership in a set instead of simply belonging or not belonging to a set [32]. The fuzzy logic approach to control problems mimics the process by which humans would make decisions, only much faster. Fuzzy logic is a multivalued logic that allows values to be defined between the values associated with traditional attitudes: yes/no, true/false, black and white, for example [60]. Fuzzy logic uses human experience, experts in the form of linguistic IF-THEN rules and approximate reasoning mechanisms and calculates control inputs for the particular case [38]. Zadeh [72] introduced the concept of linguistic variables and approximate reasoning. Very common applications of fuzzy logic include modeling of complex systems, where it is very difficult to determine relationships between variables using some other method [16], and group clustering [21]. Teodorović and Vukadinović [62] successfully applied fuzzy logic in traffic control. Successful applications of expert systems based on fuzzy logic may be found in [11, 71]. The fuzzy logic approach attempts to mimic the process of human decision-making, only at a much faster rate.

The primary goal when building the basic FES was the incorporation of descriptive estimations from the lockmaster that were based on experience. Required data for the basic FES design were collected through interviews with lockmasters and during observations in field research. The interviews consisted of answer requests to a given question about the decision made depending on distances of a vessel from the level at which the gate is currently open and from the level at which the gate is currently closed. The lockmaster makes decisions based on subjective estimations of vessel distances from the lock. The distance between a vessel in motion and the ship lock cannot be precisely defined. Therefore, a narrow zone around the vessel can be considered as a vessel domain [49, 66]. It is important to note that the proposed hypothesis that the distance of a vessel from the ship lock represents an input fuzzy variable is in conformity with research that defined a vessel's domain as a fuzzy value [48].

To form the model for ship lock control, one of the ship locks from Serbia (ship lock "Kucura" on the DTD Hydro system) was chosen as the example system. This ship lock is a one-channel queuing system with two independent, stochastic flows of demand (demand for lockage) that are from opposite directions; in the other words, from different water levels. The basic problem during the control of the ship locking process is the achievement of a compromise between minimizing the costs of the ship lock operations and water consumption as the interest of the ship lock owner on the one hand, and on the other hand minimizing vessels' delays during the transit process as the interest of vessel owners. For example, in cases of more consecutive arrivals from the same direction, it is necessary to make empty lockages, i.e., water level changes in an empty ship lock. This action decreases vessel delays but increases operating costs of the ship lock. The lockmaster controls the complete ship locking process, relying on subjective estimations. The goal is to have the least possible number of empty lockages (level changes without a vessel in the lock chamber), on the demand that vessel delays in the ship lock zone are within acceptable limits. Qualitative and descriptive (i.e., subjective) estimations by the lockmaster (based on experience) are represented here by fuzzy reasoning. Bugarski et al. [8] composed basic membership functions of the fuzzy expert system for ship lock control. Regarding the guidelines and recommendations for river information services of the PIANC [47], FES is designed to be compatible with the River Information Service (RIS). FES collects information from RIS such as data on speed, distance and the direction from which the vessel is coming [68]. The process of construction a Fuzzy Inference System (FIS) is arranged through next steps:

Step 1: Definition of input and output variables.

Based on the state of the lock (lower or upper gate is open), the operator must consider two main variables: the *"Distance of the vessel from the lock on the level where the gate is open"* (*LGO*) and the *"Distance of the vessel from the lock on the level where the gate is closed"* (*LGC*). Three categories related to the distance from the ship lock (small, medium, and large) are subjectively assessed and they define fuzzy sets for both fuzzy input variables, *LGO* (Fig. 3) and *LGC* (Fig. 4). Distances of vessels from the ship lock (input variables) are expressed in minutes. The output variable represents the control variable *"Change of the lock condition"* (*LC*) (Fig. 5), which is expressed in three categories: change, no change, and indefinite. Output value after defuzzification is given in universal units and after comparison to the limit value gives a binary decision ("change" or "no change"). The lockmaster decides whether to change the present state of the ship lock according to the vessel distances at both levels.

Two main variables are implemented as inputs to the FES (distances of vessels from the ship lock on both sides—open gate (*LGO*) and closed gate (*LGC*)) (Fig. 1). Both input variables are described with three linguistic values (small, medium, and large) and they are represented by corresponding membership functions (Figs. 2 and 3). These distance measures can be calculated from data obtained from river

Fig. 3 Membership functions of input fuzzy variable *LGO—Level where the Gate is Open* (*"Distance of the vessel from the lock on the level where the gate is open"*)

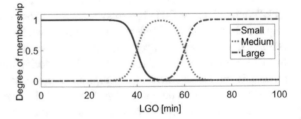

Fig. 4 Membership functions of input fuzzy variable *LGC—Level where the Gate is Closed* (*"Distance of the vessel from the lock on the level where the gate is closed"*)

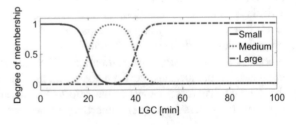

Fig. 5 Membership functions of output fuzzy variable *LC—Lock Condition* ("*Change of the lock condition*")

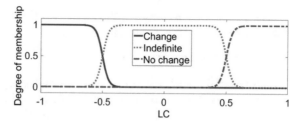

Table 1 Fuzzy rules

LGO	LGC		
	Small	Medium	Large
Small	**No change**	**No change**	**No change**
Medium	*Indefinite*	**No change**	**No change**
Large	***Change***	*Indefinite*	**No change**

information services or predicted [73]. The output variable is the "*Change of the lock condition*" (*LC*) described with *Change, No Change* or *Indefinite* (Fig. 4). The distances from the ship lock (input variables) are expressed in minutes required to reach the ship lock, and the output value after defuzzification is given in universal units. This universal value is compared with the limit value (zero in this research) and generates a binary decision (to change or not).

Step 2: Fuzzification

During the control of the ship locking process, the lockmaster chooses between two opposite decisions, "yes" or "no", concerning a change of the present state of the ship lock. His/her subjective estimations are quantified by fuzzy sets. In the presented research, the relation between the membership function of the fuzzy set and the observed variable is described by a logistic curve (e.g., S-curve or sigmoid function), as shown by several authors on the basis of experiments and research [10, 59, 70, 74]. The logistic curve has the mathematical form:

$$f(x, \sigma, a) = e^{\frac{-(x-a)^2}{2\sigma^2}} \tag{1}$$

where

　　a—position of the center of the peak;
　　σ—standard deviation of function.

Step 3: Control rules design

The third step is the design of the fuzzy control rules; i.e., determining which rules applies under which conditions [74]. In our case, we have three fuzzy sets for both input variables, and therefore we should have $3 \times 3 = 9$ fuzzy rules (of type AND) in a well-defined fuzzy inference system. Table 1 presents these nine fuzzy rules.

Fuzzy rules are presented in Table 1 and are not part of the optimization process. They are based on the work of [8].

Step 4: Defuzzification

To make exact decisions (control actions), it is necessary to choose one of the control variable values. Approximate reasoning in the fuzzy inference system is performed in several phases: fuzzification, "AND" phase, implication, aggregation, and defuzzification [22, 43]. The methods chosen for each of the mentioned phases influence the output results of the fuzzy inference system. The choice of the functions that will implement these phases has a significant impact on the results and the algorithm speed [36]. For that reason, most of the practical applications of the fuzzy logic are based on Takagi-Sugeno type, but in presented case, speed of the algorithm is not important because of slow nature of inland navigation and the ship locking process. This research covered 19 experiments with test sub-datasets and with different methods for the abovementioned phases. The experiments enclosed the Minimum and Product method for the implication phase, the Maximum, Sum, and Probabilistic OR for the aggregation phase and five different methods for the defuzzification phase: Center of gravity, Bisection of area, Mean of maximum, Largest of maximum and Smallest of maximum. The best results were acquired with the combination of methods presented in Table 2. This combination is very common in implementation of the fuzzy inference systems. However, other combinations of methods are taken into consideration and presented combination provided the best results in all of the conducted simulation tests.

If we use this algorithm to calculate the output values in many different coordinates, we can graph the control surface (see Fig. 6).

3 Nature-Inspired Optimization Techniques

3.1 Introduction

As are previously mentioned, the lock operation involves opposing interests from the lock owners and the shippers. Two opposing criteria describe these interests: the minimal Average Waiting Time per vessel ($AWTpV$) and the minimal Number of Empty Lockages ($NoEL$), as introduced by [8]. This is a trade-off situation designed

Table 2 Selected methods of approximate reasoning

Phase	Method
"AND" phase	Minimum
Implication	Minimum
Aggregation	Maximum
Defuzzification	Center of gravity

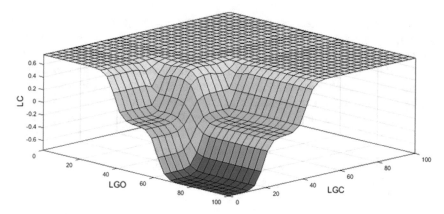

Fig. 6 Control surface calculated with fuzzy algorithm

to provide efficiency between these two criteria. The primary goal is to maximize the profit made by the lock company, but not at the expense of the vessel owners.

Three popular global optimization algorithms were applied in this research: Genetic Algorithm (*GA*), Artificial Bee Colony Optimization (*ABC*), and Particle Swarm Optimization (*PSO*), with objective to find the best optimization technique for presented expert system controlling a ship lock process. The Particle Swarm Optimization (*PSO*) method is selected because it converges faster and much more smoothly to optimal values of the membership function parameters. Additionally, certain complexities can arise during execution of the *GA* with multiple optima [4]. Unlike the *GA*, the *PSO* is a population-based stochastic optimization technique that operates on the principle inspired by the social behavior of flocks of birds or schools of fish [3, 28, 45]. Although it is a relatively new optimization algorithm, the PSO has been confirmed as advantageous for multiple objective fuzzy optimization scenarios [45, 61]. *ABC* algorithm, as a relatively new technique, was also involved in this research, since it already showed some good results in transportation problems [63].

3.2 Genetic Algorithm (GA)

Genetic algorithm is an evolutionary optimization technique inspired by Darwin's theory of natural evolution of the species. It was proposed in 1970s by John Holland [20] and improved during the years by numerous other researchers (Michalewicz [40]). It is the most common optimization technique used in decision support problems as it is outlined in [65]. In this technique, a population of candidate solutions (called individuals) to an optimization problem is evolved toward better solutions. Each candidate solution has a set of properties (its genotype, which represents the values of problem variables) which can be mutated and altered.

The evolution is an iterative process which starts from a population of randomly generated individuals. The population in each iteration is called a generation. The value of the objective function, called fitness, of every individual in the population is evaluated in each generation. The more fit individuals are selected from the current population, and each individual's genome is modified (recombined and possibly randomly mutated) to form a new generation. The new generation of candidate solutions is then used in the next iteration of the algorithm. When forming the new population, evolution mechanisms are used, such as selection, crossover and mutation, and imitating the process of natural evolution. Commonly, the algorithm terminates when either a maximum number of generations has been produced, or a satisfactory fitness level has been reached for the population.

3.3 Artificial Bee Colony (ABC)

Artificial bee colony (ABC) algorithm, also known as Bee Colony Optimization (BCO) is another, relatively novel swarm-based numerical optimization algorithm, based on simulation of the foraging behavior of honey bee swarm [26, 63]. In this algorithm, the food source position represents a possible solution of the optimization problem and the optimization criterion is defined as the nectar amount of a food source in that solution. The colony consists of three groups of bees: employed bees, onlookers, and scouts. The number of the bees, employed, or onlooker, is equal to the number of solutions in the population. At the first step, a randomly distributed initial population $P(G=0)$ of SN solutions (food source positions) is being generated, where SN denotes the size of population. Each solution x_i ($i=1, 2, ..., $ SN) is a D-dimensional vector, with D being the number of variables in optimization criterion. After initialization, the population of the positions (solutions) is subjected to repeated cycles, $C=1, 2, ..., C_{max}$, of the search processes. An employed or onlooker bee, using probability, produces a modification on the position (solution) in her memory in order to find a new food source and then tests the nectar amount of the new source. In ABC model, the artificial bees randomly select a food source position and produce a modification on the one existing in their memory, using the expression

$$v_{ij} = x_{ij} + \phi_{ij}\left(x_{ij} - x_{kj}\right) \tag{2}$$

where $k \in \{1, 2, ..., BN\}$ and $j \in \{1, 2, ..., D\}$ are randomly chosen indexes, BN is the number of employed bees, and ϕ is a random number in the range $[-1, 1]$.

If the nectar amount of the new source is higher than that of the previous one, the new position of the bee is being memorized, and the old one is being forgotten. Otherwise, she keeps the position of the previous one. When all employed bees complete the search, they share the information on nectar of the food sources and their position with the onlooker bees on the dance area. An onlooker bee evaluates the information on nectar taken from all employed bees and accordingly chooses a food source with a probability related to its nectar amount. As in the case of the employed bee, she produces a modification on the position in her memory and checks the nectar amount of the candidate source. Providing that its nectar amount is higher than that of the previous one, the bee memorizes the new position and forgets the old one.

An onlooker bee chooses a food source concerning the probability value associated with that food source, p_i, calculated by the following expression:

$$p_i = \frac{f(x_i)}{\sum\limits_{n=1}^{SN} f(x_n)} \tag{3}$$

where $f(x_i)$ is the value of the optimization criterion for the solution i evaluated by its employed bee, and SN is the number of food sources which is equal to the number of employed bees (BN). In this way, the employed bees exchange their information with the onlookers.

Employed bees whose solutions cannot be improved through a predetermined number of trials, called limit, become scouts and their solutions are abandoned. Then, the scouts start to search for new solutions, randomly. Hence, those sources that are initially poor or have been made poor by exploitation are abandoned.

Described search process is conducted until a termination criterion is satisfied, for example, a maximum cycle number or a maximum CPU time.

3.4 Particle Swarm Optimization (PSO)

The Particle Swarm Optimization (PSO) represents a sociological system of simple individuals who interact with other individuals and the environment. Several variants of optimization algorithms based on a swarm (cluster) of particles exist, including insects (bees and ants with pheromones), arthropods, and water drops. The generally accepted name of these systems is "swarm intelligence". The PSO algorithm represents a swarm-based technique, inspired by the social behavior of animals that live and move in large groups, such as birds [28].

The particles (potential solutions) move throughout the search space by following the current best particles [33]. PSO has been successfully used to solve various types of problems, including optimization functions [56], the training of artificial neural networks, and fuzzy classification systems [12]. One of the advantages of PSO is that it does not use derivatives in determining an optimum (Ren et al. [54], though combining PSO with gradient algorithms [50] or differential evolutions is not rare [55]. PSO is preferable to optimization algorithms when addressing multi-objective optimization problems [69] and is frequently used in combination with fuzzy logic [37]. A few variations exist in the PSO algorithms proposed by different authors. For example, [35] proposed a "co-evolutionary" PSO with a Gaussian probability distribution of accelerating coefficients.

Each particle is associated with its position (x) and velocity (v). The position of the particle represents a potential solution; the best position ever achieved by each particle during the entire optimization process is memorized (p). The best position ever achieved by any of the particles is memorized by the swarm as a whole (g). In the k-th iteration, the position and the velocity of each particle are updated as

$$v[k + 1] = w \cdot v[k] + cp \cdot rp[k] \cdot (p[k] - x[k]) + cg \cdot rg[k] \cdot (g[k] - x[k])$$
$$x[k + 1] = x[k] + v[k + 1] \tag{4}$$

The relative impact of the personal (local) and common (global) knowledge on the movement of each particle is controlled by acceleration factors cp and cg. Inertia factor w keeps the swarm together and prevents it from diversifying excessively, diminishing thereby *PSO* into a pure random search. Factors rp and rg represents mutually independent random numbers uniformly distributed on the range [0, 1].

Many modifications of *PSO* algorithm have already been presented in the literature. Early concept of *PSO* algorithm employes constant parameter set (cp, cg, w), while some more recent modifications introduce variable parameter set, improving the overall performance of the algorithm [13, 23, 53]. An example of tuning a fuzzy controller using *PSO* was proposed by [7], and [18] successfully applied swarm intelligence in traffic control.

The version of the algorithm selected for application in presented research was introduced by [53] and generalized in Kanović et al. [23, 24].

4 Application of Nature-Inspired Optimization Techniques for Vessel Traffic Control in Ship Lock Zone

4.1 Description of the Real System

The proposed optimization and testing of the FES were carried out based on the generated dataset of vessel traffic densities. The dataset was chosen to correspond to actual traffic conditions and was formed on the basis of simulation experiments that could describe possible states of the system. Simulation models that closely described the complex process of vessel traffic were developed, verified and validated from research on navigable canal capacity [5].

In Serbia, there is a complex system of Danube–Tisa–Danube navigable canals with a total length of approximately 600 km. In this system, 12 ship locks are in use and can be classified into three characteristic groups. All of them are designed for the same vessel category but differ in some technical details. Three relevant representatives of each group were selected (locks with the largest traffic density). Although they differ in some technical details, the rules of navigation and order of control operations are identical for all analyzed locks. The ship locks "Kucura" and "Sombor" (Figs. 7 and 8) were observed as actual representative systems. They can be described as a single-channel servicing systems with two independent, stochastic flow requirements for the two opposing directions. Time intervals for the lockage (i.e., passage) were defined experimentally, measuring the time and interviewing lock operators. The average time for the lockage of a vessel is set to 25 min, and the time interval for the level change in the chamber without a vessel is set to 15 min. It is also assumed that only a single vessel can be held by the lock chamber. Based on these time intervals, two possible situations can occur. The first is a "regular lockage", where the vessel enters the end where the gate is open and it does not have to wait for the water level to change in the chamber. This lockage type lasts 25 min. The second case is an "empty lockage" when the lock gate for the approaching vessel is closed, and the water level in the chamber must be changed before lockage can take place. An empty lockage followed by the regular lockage lasts 40 min.

The main objective in the ship lock control problem is to achieve a compromise between minimizing the waiting time for the lockage, preferred by the ship owners, and minimizing the energy and water consumption for operating the lock, preferred by the owners of the lock [64]. If more vessels are approaching the lock from the same direction, empty lockage(s) must be performed by lock operators, in order to reduce waiting times, and this increases the costs of lock operation.

Set of vessel arrivals is generated for the simulation based on statistic data altered with stochastic parameters. It can be considered as a vessel traffic database. In the observed case, there is an annual navigation break during the winter (from 21 December to 21 March), which is included in the construction of the arrivals set. On other days, the average of the traffic load is 10 vessels per day. There are a total of 2786 generated arrivals at the lock (Table 3).

Fig. 7 The ship lock "Kucura"

Fig. 8 The ship lock "Sombor"

4.2 Optimization of Membership Functions

The principle of fuzzy inference systems is that they try to mimic the human way of reasoning. That is why the original FES is designed based on suggestive estimations of lock operators. Without diminishing the importance and significance of human logic in decision-making, one question can arise: Did we find the optimal tactic (the best choice) in controlling and decision-making? Can we take the advantage of computers that can process much more information for a short period than the human brain? What if some optimization algorithm can find a better solution? Maybe some changes in membership functions parameters can upgrade the performances of an expert system?

Table 3 Summary of vessel arrivals per month generated for simulation

Month	Total number of vessels	Arrivals at upper gate of lock	Arrivals at lower gate of lock	Ratio of arrivals up/down
March	106	50	56	1.12
April	311	147	164	1.11
May	322	167	155	0.93
June	306	161	145	0.90
July	289	143	146	1.02
August	313	161	152	0.94
September	297	153	144	0.94
October	294	166	128	0.77
November	325	152	173	1.14
December	223	118	105	0.89

Earlier was mentioned that interests of lock managers and shippers conflict with each other. When a vessel approaches to ship lock zone, the most important thing to shippers is to complete the lockage as soon as possible. That means that their goal is to spend the least possible time while waiting for the lockage to occur. The situation, when a vessel is approaching a chamber with open gates, means that the lockage will start immediately upon arrival and that is the most favorable situation for shippers. Based on this particular goal of shippers, it is possible to construct an assessment criterion that will reflect the shippers' interests—a *Minimum Waiting Time (MWT)* criterion. On the other side, there is the interest of lock owners and workers. They try to perform minimal number of lockages, as possible. The situation, where empty lockage is necessary, is not a favorable one. Energy is needed to run the pumps and water to fill the chamber. If an empty lockage is performed (without a vessel in the chamber), both energy and water are spent. Based on the described goal of lock owners, it is possible to construct an assessment criterion which will reflect their interests—a minimum number of empty lockages performed, or simpler, a *Minimum Number of Lockages (MNL)* criterion. Two defined assessment criteria are extremely opposite.

In real situations, lockmasters find a compromise (between two extremes). The advantage of FES running on a PC is that its history is saved and its actions can be further analyzed. Analysis has shown that in most real cases, only two of nine fuzzy rules were truly activated. It was the signal that triggered the following actions on adjustments in fuzzy control logic. Idea is to optimize some fuzzy parameters. Which optimality criterion to use? With goals to minimize both waiting times and number of lockages, there are two objectives present. Is it a multi-objective optimization [14, 52]? How to construct a criterion that can be used in optimization algorithm?

FES rules were designed during an interview with lockmasters. There was not a doubt in rules logic (Table 2). But, it is very difficult to determine the right values for membership function parameters. A small shift in parameter value can influence the number of activated rules in a particular case and at the end the final decision of an expert system.

A universal criterion as an assessment factor for comparison of different set of FES parameters is needed. If two earlier described criteria are taken in consideration, logically is to find an "economic" criterion somewhere in the middle. Optimality criterion (Eq. 5) is chosen to be a weighted sum of average waiting time per vessel (*AWTpV*) and the number of empty lockages (*NoEL*).

$$E = A * NoEL + B * AWTpV \tag{5}$$

where

E—the optimality criterion, *A* and *B* are the weight coefficients, *NoEL*—number of empty lockages, *AWTpV*—average waiting time per vessel.

The purpose of weight coefficients *A* and *B* in Eq. 5 is to describe the individual importance of both parts in optimality criterion. Which is more expensive? To perform one more empty lockage and spend extra water and energy or to increase an average waiting time of all vessels by one minute? Coefficients A and B gives us opportunity to weigh the two expenses.

There is a question—how to choose the right values for A and B? Logically, the first guess is to set both coefficients to equal value (usually 1). In practice, it would mean that an extra empty lockage has similar weight as one minute of average vessel waiting. Idea is that lockmaster set the values of weight coefficients to achieve the desired strategy in accordance with current conditions of the ship lock. Every time the relation between the two coefficients is changed, it is necessary to perform optimization once more with a new optimality criterion. That would result in new optimal values of membership function parameters.

The FES is designed to have two input variables and one output variable. All variables are defined with three linguistic values. Differential sigmoid membership function (Eq. 6) is used to describe the linguistic values of input variables.

$$\mu(x) = \frac{1}{1 + e^{-a_1(x-c_1)}} - \frac{1}{1 + e^{-a_2(x-c_2)}} \tag{6}$$

The parameters a_1 and a_2 in Eq. 6 represent the left and right tilt of the function, while the parameters c_1 and $c2$ represent its transient points.

The structure of the particles (variables to be determined) is such that they are encoded with four values (coordinates in the solution space). These values are, in fact, parameters that uniquely determine the positions (transient points) of the sigmoidal functions of the set of input variables of the FES.

The structures of the individual are X_{LGO}, Y_{LGO}, X_{LGC}, and Y_{LGC}.

The first two variables, X_{LGO} and Y_{LGO}, determine the positions of the membership functions belonging to the input fuzzy variable LGO (the distance of the closest vessel from the gateway to the open gate side). X and Y are two variables that uniquely determine the parameters of all the membership functions belonging to one input fuzzy variable. The other two variables, X_{LGC} and Y_{LGC}, determine the shape and position of the functions of the fuzzy input variable LGC (the distance of the closest vessel from the chamber on the closed gate side).

As shown earlier in Figs. 3 and 4, the input fuzzy variables LGO and LGC are defined with three sigmoid functions. Each slope is, as a rule, defined by two parameters (the intensity of the slope and its transient point). The membership function "*Medium*" contains in its definition two slopes, while the other two membership functions ("*Small*" and "*Large*") are defined with only one slope since they are borderline functions. Therefore, overall, there are four slopes to be defined with $4 \times 2 = 8$ parameters. In the presented case, the tilt intensities are fixed to the predetermined equal values for all input fuzzy sets, so the number of parameters is reduced to four (the transient points of the sigmoid or crossings) by each input variable. Fuzzy sets are defined, in our case, in such a way that they overlap and complement each other to their full membership. That implies it is sufficient to define the position of the "*Medium*" set with two parameters and by automatism these two same parameters will be used for the positions of the other sets of given input variable, because their membership functions can be observed as inverse (complementary) functions. In this way, with two variable parameters, one can uniquely determine the positions and shape of all of three membership functions belonging to one variable. Since in our FES we have two input variables, all of their fuzzy sets can be described with four parameters: X_{LGO}, Y_{LGO}, X_{LGC}, and Y_{LGC}. Therefore, the number of unknown parameters in optimization process is four. That number is reduced to achieve quality in optimization process. The following assumptions are taken into account:

(a) the final decision of FES is not significantly affected with output fuzzy variable;
(b) tilts (slopes) of sigmoid functions are fixed to a priori value;
(c) transient points of adjacent membership functions are interconnected to achieve full overlapping.

Figure 9 presents the steps of optimization process:

Three optimization algorithms (*GA*, *ABC*, and *PSO*) are used to fine-tune the FES parameters. Population size is set to 30 individuals (or particles or solutions, depending on the algorithm) for all 3 algorithms, due to fair comparisons. This is relatively large number for population size, which caused quick convergence to a certain optimal solution (even after nine or ten iterations).

The parameters X_{LGO}, Y_{LGO}, X_{LGC}, and Y_{LGC} that represent the structure of the individual are limited in the range of real numbers [0, 100], which corresponds to the range of the input fuzzy variables in minutes. When generating the initial population (generation, colony) they are randomly generated in the range [0, 100] in order to achieve a uniform distribution within the search area.

4.3 Optimization Results

The results related to the economic criterion were obtained in the case of the weight factors $A = 1$ and $B = 1$ (see Eq. 5). Additional cases are considered with different values of weight factors. Values are changed to give more or less importance to the two parts of economic criterion: *NoEL* and *AWTpV*. The first case is the economic criterion with weight factors $A = 2$ and $B = 0.5$ and the second case with $A = 0.5$ and $B = 2$. The first case is an economic criterion in which it is more important to reduce the number of empty lockages rather than the average waiting time. The second case addresses the opposite situation, that is, the *AWTpV* is more important to minimize than *NoEL*. In the initial economic criterion, the ratio between A and B is 1:1 (FES 1:1), while in the new two cases this ratio is 4:1 (FES 4:1) for the first case, and 1:4 (FES 1:4) for the second case.

After the application of the optimization algorithms with the new optimality criteria (with new coefficients A and B), new membership function parameters were obtained that satisfy the newly created criteria. Tables 4, 5, and 6 show the values of these parameters in the original FES, then FES optimized according to the first variant of the economic criterion, and at the end of the FES optimized in two new cases of economic criteria in which the coefficients A and B are different from 1, for three variants of optimization algorithm: *GA*, *ABC*, and *PSO*, respectively.

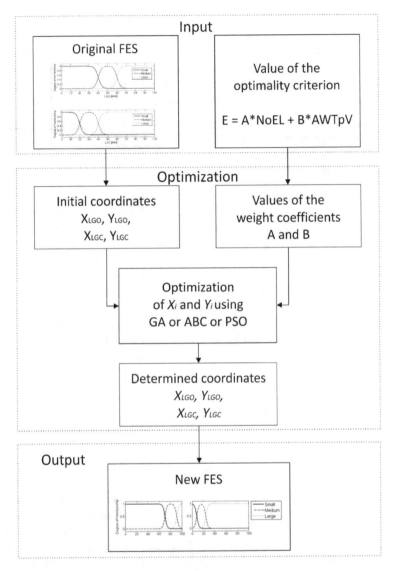

Fig. 9 Schematic of the optimization procedure of the FES for control of ship lockage process

The membership functions of input fuzzy variables constructed based on the values of the parameters from the last two columns in Table 6 (achieved with PSO algorithm) are shown in Fig. 10.

The top two graphics (Fig. 10a and b) shows the membership functions of *LGO* (level where the gate is open) and *LGC* (level where the gate is closed) fuzzy variables in FES optimized with the coefficients $A = 2$ and $B = 0.5$ in the optimality criterion. The lower two graphics (Fig. 10c and d) shows the membership functions of input

Table 4 FES parameter values obtained by different forms of economic criteria—GA

FES parameter	Original FES	Optimized FES 1:1	Optimized FES 4:1	Optimized FES 1:4
X_{LGO}	40	48.10	70.55	32.11
Y_{LGO}	60	69.82	91.31	59.91
X_{LGC}	20	17.21	8.21	69.86
Y_{LGC}	40	46.62	25.04	89.72

Table 5 FES parameter values obtained by different forms of economic criteria—ABC

FES parameter	Original FES	Optimized FES 1:1	Optimized FES 4:1	Optimized FES 1:4
X_{LGO}	40	49.28	71.22	33.65
Y_{LGO}	60	69.27	90.24	61.88
X_{LGC}	20	18.23	9.21	66.28
Y_{LGC}	40	47.43	25.32	89.73

Table 6 FES parameter values obtained by different forms of economic criteria—PSO

FES parameter	Original FES	Optimized FES 1:1	Optimized FES 4:1	Optimized FES 1:4
X_{LGO}	40	49.94	70.11	31.92
Y_{LGO}	60	69.94	91.03	60.23
X_{LGC}	20	16.00	8.10	68.54
Y_{LGC}	40	48.32	24.00	88.31

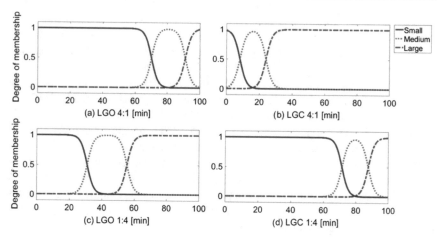

Fig. 10 Optimized input fuzzy variables based on the parameters obtained with different economic criteria (*PSO*) **a** *LGO* 4:1, **b** *LGC* 4:1, **c** *LGO* 1:4, **d** *LGC* 1:4

fuzzy variables in the FES optimized with the weight factors $A = 0.5$ and $B = 2$.

From Fig. 10, it can be seen that there are significant differences in the positions of membership functions in relation to the original FES (Figs. 2 and 3). The rule table (Table 1) was not part of the optimization and is the same as in the original FES.

In the first case ($A = 2$ and $B = 0.5$), a greater part of the economic criterion is related to the number of empty lockages. An empty lockage occurs as a result of only one rule out of nine in the fuzzy rule base, when the LGO is "*Large*" and LGC is "*Small*" (see the rules in Table 1). In this case, judging by the above two graphics (Fig. 10a and b), this rule refers to the situation when a vessel approaches to open gate side of lock over a distance of more than 90 min, while the vessel approaching from the side of closed gates at a distance of less than 10 min. The difference of these times (80 min) is two times greater than the difference in the same parameters in the basic FES, which tells us that according to this criterion, the translation will only occur in extreme cases, which was the goal of setting such parameters in the economic criterion. This reduces the number of empty lockages because conditions that are more stringent are applied. In the second case ($A = 0.5$ and $B = 2$), the average waiting time for lockage is more important. From the lower graphics (Fig. 10c and d), it can be seen that the positions of the fuzzy sets (the transient points of the membership functions) "*Large*" for LGO and "*Small*" for LGC are very close.

5 Comparison of Results

Population for every algorithm consisted of totally 30 individuals (particles, bees). Optimization process was conducted in 15 iterations (generations), which was sufficient for all 3 algorithms to converge to the optimal solution. Each individual (particle, bee) consists of four values (coordinates). These values are parameters that define the shape and position of sigmoid membership functions in fuzzy sets. First two variables X_{LGO} and Y_{LGO} determine the shape and the position of membership functions for fuzzy variable LGO. The other two variables X_{LGC} and Y_{LGC} determine the shape and the position of membership functions for fuzzy variable LGC. As shown in Figs. 3 and 4, input fuzzy variables LGO and LGC consist of three sigmoid functions. "Medium" sigmoid function is defined with two values X and Y, and other two sigmoid functions can be observed as inverse functions.

Three variants of economic criterion were considered, with different values of coefficients A and B (see Eq. 5). In the first variant, both coefficients were equal to one, giving equal significance to both number of empty lockages and average waiting time per vessel. The second criterion variant had values of $A=2$ and $B=0.5$, favoring the number of empty lockages in ratio 4:1, and the third one of $A=0.5$ and $B=2$, favoring the average waiting time per vessel, in ratio 1:4.

The values of optimal FES parameters obtained using all three optimization algorithms are shown in Table 7. One can notice that all three algorithms converged to similar parameter values for each criterion variant, which implies that obtained solution is close to global optimum. Also, it should be emphasized that the shape of membership function varies significantly for different criterion variant, implying that fuzzy parameters depend to a large degree on desired system behavior.

Table 8 depicts the optimal values of different variants of economic optimization criterion. Best results for each criterion variant are typed in bold font. In the case of equal significance of number of empty lockages and average waiting time (variant 1:1), *PSO* provided the best criterion value, while the values obtained using *GA* and *ABC* were almost the same. In criterion variant 4:1, which favors number of empty lockages, *GA* and *PSO* converged to the same, probably global optimal value, while *ABC* obtained slightly worse result. However, in the third variant, favoring average waiting time (1:4), *ABC* showed the best performance, while *GA* and *PSO* followed with the same criterion value. Thus, it is not possible to distinct the best algorithm for universal application. We can only conclude that all three algorithms showed similar performance, implying the global nature of the obtained results.

Table 9 shows the simulation results obtained using the criteria of *MWT* and *MNL*, the original fuzzy system and new optimized fuzzy systems, obtained using all three optimization algorithms. When comparing number of empty lockages obtained using original and optimal fuzzy systems, one can notice a significant improvement with optimized fuzzy systems, particularly in the criterion variant 4:1, favoring this parameter (670, i.e., 676 compared to 746). In the case of average waiting time, it is also distinctive that optimal fuzzy systems provide a significant improvement in the criterion variant 1:4 (114 and 115 min, compared to 137 with the original fuzzy system).

Both FESs (basic and optimized) give similar outputs in the control process when the vessel traffic densities are low (less than 30% of lock capacity) or high (more than 70% of lock capacity). At low traffic densities, empty lockages are frequent and vessel delays are minimal, regardless of the defined criteria of the optimality. Similarly, empty lockages at higher vessel traffic densities are very rare because a vessel is almost always waiting for the lockage on the other side. The greatest improvements are achieved in the traffic density interval between 30 and 70% of capacity because the operator's dilemma mostly appears in these cases.

In practice, a ship lock management or a lockmaster must choose coefficients *A* and *B*, and this raises questions about how to analyze, assess and establish this relation. First, a lockmaster must analyze which is more important: water and energy consumption or vessel waiting times in the ship lock zone. The solution to this problem demands in-depth analysis because it is subject to many unpredictable conditions. Energy consumption for operating the lock strictly depends on the number of lockages. Water consumption depends on several different factors (location: river or canal; water supply: free flow or pumps; consumers: settlements, industry, and irrigation; evaporation: low or high temperatures; economics: energy and water prices). All of these factors require a very complex analysis as a basis for the new assessment of the relation between the two coefficients; then the new optimization process can be performed with a new objective function. The results of the optimization will return two values to the operator: the estimated number of empty lockages and the average waiting time in minutes per vessel. If the operator is satisfied with the provided results, he can accept the new parameters of the membership functions of the

Table 7 FES parameter values obtained by different forms of economic criteria

FES parameter	Original FES	Optimized FES 1:1			Optimized FES 4:1			Optimized FES 1:4		
		GA	ABC	PSO	GA	ABC	PSO	GA	ABC	PSO
X_{LGO}	40	48.10	49.28	49.94	70.55	71.22	70.11	32.11	33.65	31.92
Y_{LGO}	60	69.82	69.27	69.94	91.31	90.24	91.03	59.91	61.88	60.23
X_{LGC}	20	17.21	18.23	16.00	8.21	9.21	8.10	69.86	66.28	68.54
Y_{LGC}	40	46.62	47.43	48.32	25.04	25.32	24.00	89.72	89.73	88.31

Table 8 Economic criteria optimal values obtained by *GA*, *PSO*, and *ABC*

Optimization algorithm	Optimization criterion variant		
	1:1	4:1	1:4
GA	863.91	**1412.09**	640.33
ABC	863.79	1424.22	632.89
PSO	**860.53**	**1412.09**	640.13

Table 9 Comparative presentation of simulation results, for different FESs and economic criteria

Evaluation model		Number of empty lockages			Average waiting time [min]		
MWT		1410			4.18		
MNL		50			3090.85		
Original FES		746			137.3		
Optimized FES	Criterion	1:1	4:1	1:4	1:1	4:1	1:4
	GA	726	670	824	137.9	144.19	114.17
	ABC	726	676	802	137.79	144.44	115.94
	PSO	720	670	822	140.53	144.19	114.56

FES; if the results are not satisfactory, the optimization process should be repeated with new coefficients *A* and *B*. If the management of the ship lock is able to define the costs for each part of the criterion, then the optimal values for *A* and *B* can be determined and the proposed FES can be optimized to that specific goal. Moreover, operators can expect to achieve better FES performance.

6 Conclusions and Future Research

A ship lock zone represents a specific area on waterway, and control of the ship lockage process requires a comprehensive approach. In reality, ship lock control is mostly based on the subjective estimations and the experience of a lockmaster (ship lock operator). The development of an expert system for the control of ship locks that relies on a fuzzy logic is the first step in design process that must be done before the optimization.

Research presented in this chapter emphasizes the potential application of nature-inspired optimization techniques in vessel traffic control on inland waterways. Appropriate optimization techniques used in the proposed method for creating a fuzzy expert system can be used as a support in decision-making or in training the ship lock operators. This new approach in the field of vessel traffic control in the ship lock zone represents a rare use of artificial intelligence in water transport. The basic characteristics of the proposed system were adaptability and flexibility.

The parameters of such a system were optimized using three popular global optimization procedures in order to minimize three different variants of economic optimization criterion. The presented results show that all algorithms, with slight variations in criterion results, provide performance improvement, i.e., the number of empty lockages and average waiting time decrease, compared to originally proposed fuzzy expert system. Thus, we can conclude that all these algorithms can be successfully applied in this kind of transportation planning and control problems.

An FES was designed for a single-channel two-way lock (single chamber—single vessel). Future research should consider the development of a control algorithm for a multichannel lock (operating in series or parallel), which is not rare in actual systems. This may require more input variables and/or more complex fuzzy rules. The authors hope that including vessel priorities (military, commercial, etc.) in the proposed system could significantly improve the results. Lockmasters could use the proposed system as a valuable aid in making decisions, particularly if many vessels with different priorities request lockages over a short time span.

Based on the obtained results, it can be concluded that a good selection of economic criteria parameters can significantly improve the FES performance. Thus, the optimization approach was proven to be satisfactory. It should be noted that the obtained results largely depended on traffic density. The developed FES was designed according to existing navigation rules (allowed navigation speed) and the technical characteristics of a ship lock (duration of the chamber filling/emptying and duration of additional operations in the lock zone). The expert system must be redesigned if any characteristic of the current situation is changed. In addition, the proposed expert system can be applied in two modes. In semiautomatic mode, it is designed to be used as a support in the decision-making process. In the more automated variant, the FES can directly control ship lock operations, thus eliminating the need for human operators and decreasing the probability of errors caused by human factors.

The final research results confirm that all these procedures show similar output and provide an overall improvement of ship lock operation performance compared to an originally proposed fuzzy expert system. Presented research speaks in favor of their application in similar transportation problems optimization.

Further research can proceed in the direction of a greater complexity in lock functionality. Attention should be given to different disciplines of queues and more complex cases (i.e., multiple arrivals and multi-trajectory arrivals on both sides of a ship lock). Instead of the single-channel lock, a multichannel lock can be considered. Such systems would need to have fuzzy input variables and fuzzy rules that are more complex. Moreover, military, service, commercial and private vessels in actual systems do not have the same priority; some classes of vessels have a higher priority for using the lock than others. This could introduce additional fuzzy rules. A well-designed fuzzy expert system could serve as a valuable aid in the choice of the control action when there are more requests for lockage by a number of vessels with different priorities.

In addition, decision-making in cases of a multi-trajectory approach differs from the proposed model. A multi-trajectory approach occurs in ship locks whose chambers are designed to accommodate more vessels simultaneously. In these cases, the total ship lockage process takes significantly longer and the problem exhibits completely different characteristics and sequences of activities.

Acknowledgements This research was supported by the Ministry of Education, Science and Technological Development (Government of the Republic of Serbia) under Grant Number TR 36007.

References

1. Adeli H (1988) Expert systems in construction and structural engineering. Chapman and Hall, New York
2. Adeli H, Balasubramanyam KV (1988) A novel approach to expert systems for design of large structures. AI Mag 9(4):54–63
3. Ankur M, Debanjan D, Mehta SP, Shalivahan S, Bhattacharya, BB (2011) PSO vs. GA vs. VFSA: A comparison of performance, accuracy and resolution with respect to inversion of SP data. In: Proceedings of Japan Geoscience Union Meeting 2011, Makuhari, Chiba, Japan
4. Aytug H, Koehler GJ (2007) The effect of multiple optima on the simple GA run-time complexity. Eur J Oper Res 178(1):27–45
5. Bačkalić T (2001) Traffic control on artificial waterways of limited dimensions in function of its throughput capacity (Upravljanje saobraćajem na veštačkim plovnim putevima ograničenih dimenzija u funkciji njihove propusne sposobnosti - in original). PhD thesis, University of Novi Sad, Serbia
6. Bačkalić T, Bugarski V, Kulić F, Kanović Ž (2016) Adaptable fuzzy expert system for ship lock control support. J Navig 69(6):1341–1856
7. Bouallegue S, Haggege J, Ayadi M, Benrejeb M (2012) PID-type fuzzy logic controller tuning based on particle swarm optimisation. Eng Appl Artif Intell 25(3):484–493
8. Bugarski V, Bačkalić T, Kuzmanov U (2013) Fuzzy decision support system for ship lock control. Expert Syst Appl 40(10):3953–3960
9. Campbell JF, Smith LD, Sweeney II, Mundy R, Nauss, RM (2007) Decision tools for reducing congestion at locks on the upper Mississippi river. In: 40th annual Hawaii international conference on system sciences, Hawaii, USA
10. Camps-Valls G, Martín-Guerrero JD, Rojo-Alvarez JL, Soria-Olivas E (2004) Fuzzy sigmoid kernel for support vector classifiers. Neurocomputing 62:501–506
11. Castanho MJP, Hernandes F, De Re AM, Rautenberg S, Billis A (2013) Fuzzy expert system for predicting pathological stage of prostate cancer. Expert Syst Appl 40(2):466–470. https://doi.org/10.1016/j.eswa.2012.07.046
12. Chen CC (2006) Design of PSO-based fuzzy classification systems. Tamkang J Sci Eng 9(1):63–70
13. Clerc M, Kennedy J (2002) The particle swarm: explosion, stability and convergence in a multidimensional complex space. IEEE Trans Evol Comput 6(1):58–73
14. Collette Y, Siarry P (2004) Multiobjective optimization: principles and case studies. Springer. ISBN 3-540-40182-2
15. Comes T, Hiete M, Wijngaards N, Schultmann F (2011) Decision Maps: a framework for multi-criteria decision support under severe uncertainty. Decis Support Syst 52(1):108–118
16. Feng G (2006) A survey on analysis and design of model-based fuzzy control systems. IEEE Trans Fuzzy Syst 14(5):676–697
17. Figueiredo J, Botto MA, Rijo M (2013) SCADA system with predictive controller applied to irrigation canals. Control Eng Pract 21(6):870–886
18. Garcia-Nieto J, Alba E, Olivera AC (2012) Swarm intelligence for traffic light scheduling: application to real urban areas. Eng Appl Artif Intell 25(2):274–283
19. He Q, Wang L (2007) An effective co-evolutionary particle swarm optimization for constrained engineering design problems. Eng Appl Artif Intell 20:89–99

20. Holland J (1975) Adaptation in natural and artificial systems. University of Michigan Press
21. Hsiao B, Chern C-C, Yu M-M (2012) Measuring the relative efficiency of IC design firms using the directional distance function and a meta-frontier approach. Decis Support Syst 53(4):881–891
22. Jantzen J (2007). Foundations of fuzzy control. Wiley
23. Kanović Ž, Rapaić M, Jeličić Z (2011) Generalized particle swarm optimisation algorithm—Theoretical and empirical analysis with application in fault detection. Appl Math Comput 217(24):10175–10186
24. Kanović Ž, Rapaić M, Jeličić Z (2013) The generalized particle swarm optimization algorithm: idea, analysis, and engineering applications. In: Swarm intelligence for electric and electronic engineering. Book News Inc, Portland, OR, pp 237–258
25. Kanović Ž, Bugarski V, Bačkalić T (2014) Ship lock control system optimization using GA, PSO and ABC: a comparative review. PROMET—Traffic&Transp 26(1):23–31
26. Karaboga D (2005) An idea based on honey bee swarm for numerical optimization. Technical Report- TR06. Erciyes University, Kayseri, Turkey
27. Kecman V (2001). Learning and Soft Computing: Support Vector Machines, Neural Networks, and Fuzzy Logic Models, Massachusetts Institute of Technology
28. Kennedy J, Eberhart RC (1995) Particle swarm optimization. In: Proceedings of IEEE international conference on neural networks, Perth, Australia, pp 1942–1948
29. Kennedy J, Eberhart RC (2001) Swarm intelligence. Morgan Kaufmann Publishers, San Francisco
30. Khisty CJ (1996) Waterway traffic analysis of the Chicago River and lock. Marit Policy and Manag 23(3):261–270
31. Kiencke U, Nielsen L, Sutton R, Schilling K, Papageorgiou M, Asama H (2006) The impact of automatic control on recent developments in transportation and vehicle systems. Annual Rev Control 30(1):81–89
32. Klir GJ, Yuan B (1996) Fuzzy sets, fuzzy logic, and fuzzy systems: selected papers by Lotfi A. Zadeh. World Scientific Publishing Co Pte Ltd, Singapore
33. Kordon AK (2010) Applying computational intelligence: how to create value. Springer, Berlin
34. Kosko B (1993) Fuzzy thinking—The new science of fuzzy logic. Hyperion, New York
35. Krohling RA, dos Santos Coelho L (2006) Co-evolutionary particle swarm optimisation using Gaussian distribution for solving constrained optimisation problems. IEEE Trans Syst Man Cybern B Cybern 36(6):1407–1416
36. Lancaster S (2008) Fuzzy logic controllers. Maseeh College of Engineering and Computer Science at PSU
37. Li J-q., Pan, Y-x. (2013) A hybrid discrete particle swarm optimisation algorithm for solving fuzzy job shop scheduling problem. Int J Adv Manuf Technol 66(1–4):583–596
38. Mamdani EH (1974) Application of fuzzy algorithms for the control of a simple dynamic plant. In: Proceedings of the institution of electrical engineers, 121(12), pp 1585–1588, ISSN: 0020-3270, https://doi.org/10.1049/piee.1974.0328
39. McCartney BL, George J, Lee BK, Lindgren M, Neilson F (1998) Inland navigation: locks, dams, and channels, ASCE Manuals No 94, ASCE
40. Michalewicz Z (1999) Genetic algorithms + data structures = evolution programming, 3rd edn. Springer
41. Mundy RA, Campbell JF (2005) Management systems for inland waterway traffic control, volume II: vessel tracking for managing traffic on the Upper Mississippi River, Technical report, Final Report, Midwest Transportation Consortium, MTC Project 2004-003
42. Negenborn RR, Lukszo Z, Hellendoorn H (2010) Intelligent infrastructures. Springer, Dordrecht
43. Nguyen HT, Sugeno M (1998) Fuzzy systems—Modeling and control. Kluwer Academic Publishers
44. Onieva E, Milanes V, Villagra J, Perez J, Godoy J (2012) Genetic optimization of a vehicle fuzzy decision system for intersections. Expert Syst Appl 39(18):13148–13157. https://doi.org/10.1016/j.eswa.2012.05.087

45. Panigrahi BK, Shi Y, Lim M-H (2011) Handbook of swarm intelligence: concepts, principles and applications (adaptation, learning, and optimisation). Springer, Berlin
46. Partenscky HW (1986) Inland waterways: ship lock installations, (Binnenverkehrswasserbau: Schleusenanlagen – in original). Springer, Berlin
47. PIANC (2011) Guidelines and recommendations for river information services. The World Association for Waterborne Transport Infrastructure
48. Pietrzykowski Z (2008) Ship's fuzzy domain—A criterion for navigational safety in narrow fairways. J Navig 61:499–514
49. Pietrzykowski Z, Uriasz J (2009) The ship domain—A criterion of navigational safety assessment in an open sea area. J Navig 62:93–108
50. Plevris V, Papadrakakis M (2011) A hybrid particle swarm—gradient algorithm for global structural optimisation. Comput Aided Civil and Infrastruct Eng 26(1):48–68
51. Praetorius G, Lützhöft M (2012) Decision support for vessel traffic service (VTS): user needs for dynamic risk management in the VTS. Work: J Prevention Assess and Rehabil 41:4866–4872
52. Rao RV, Patel V (2013) Multi-objective optimization of two stage thermoelectric cooler using a modified teaching–learning-based optimization algorithm. Eng App Artif Intell 26(1):430–445. Elsevier, ISSN 0952-1976, http://dx.doi.org/10.1016/j.engappai.2012.02.016
53. Rapaić M, Kanović Ž (2009) Time-varying PSO—convergence analysis convergence related parameterization and new parameter adjustment schemes. Inf Process Lett 109(1):548–552
54. Ren Y, Cao G-Y, Zhu X-J (2006) Particle swarm optimisation based predictive control of proton exchange membrane fuel cell (PEMFC). J Zhejiang Univ Sci A 7(3):458–462
55. Sedki A, Ouazar D (2012) Hybrid particle swarm optimisation and differential evolution for optimal design of water distribution systems. Adv Eng Inform 26(3):582–591
56. Shafahi Y, Bagherian M (2013) A customized particle swarm method to solve highway alignment optimisation problem. Comput Aided Civil and Infrastruct Eng 28(1):52–67
57. Shi Y, Eberhart RC (1999) Empirical study of particle swarm optimization. In: Proceedings of IEEE international congress on evolutionary computation, 3:101–106
58. Siddique N, Adeli H (2013) Computational intelligence: synergies of fuzzy logic, neural networks and evolutionary computing. Wiley, London
59. Smith LD, Sweeney DC II, Campbell JF (2009) Simulation of alternative approaches to relieving congestion at locks in a river transportation system. J Oper Res Soc 60(4):519–533
60. Subašić P (1997) Fuzzy logic and neural networks (Fazi logika i neuronske mreže – in original). Tehnička knjiga, Beograd
61. Tapkan P, Ozbakir L, Baykasoglu A (2013) Solving fuzzy multiple objective generalized assignment problems directly via bees algorithm and fuzzy ranking. Expert Syst Appl 40(3):892–898
62. Teodorović D, Vukadinović K (1998) Traffic control and transport planning: a fuzzy sets and neural networks approach. Kluwer Academia Publishers, Norwel, MA
63. Teodorović D, Dell'Orco M (2005) Bee colony optimization–a cooperative learning approach to complex transportation problems. Advanced OR and AI Methods in Transportation. In: Proceedings of 16th Mini–EURO Conference and 10th Meeting of EWGT, September 2005, pp 51–60
64. Ting CJ, Schonfeld P (2001) Control alternatives at a waterway lock. J Waterw Port Coast Ocean Eng 127(2):89–96
65. Tsou MC, Kao SL, Su CM (2010) Decision support from genetic algorithms for ship collision avoidance route planning and alerts. J Navig 63(01):167–182
66. Wang N, Meng X, Xu Q, Wang Z (2009) A unified analytical framework for ship domains. J Navig 62:643–655
67. Wiegmans BW, Konings R (2007) Strategies and innovations to improve the performance of barge transport. Eur J Transport Infrastruct Res EJTIR 7(2):145–161
68. Willems C, Schmorak N 2010 River information services on the way to maturity. In: Proceedings on 32nd PIANC international navigation congress, Liverpool, UK
69. Xu G, Yang Z-T, Long G-D (2012) Multi-objective optimisation of MIMO plastic injection molding process conditions based on particle swarm optimisation. Int J Adv Manuf Technol 58(5–8):521–531

70. Yager RR, Filev DP (1994) Essentials of fuzzy modelling and control. Wiley, New York
71. Yunusoglu MG, Selim H (2013) A fuzzy rule based expert system for stock evaluation and port-folio construction: an application to Istanbul Stock Exchange. Expert Syst Appl 40(3):908–920. https://doi.org/10.1016/j.eswa.2012.05.047
72. Zadeh LA (1975) The concept of a linguistic variable and its application to approximate rea-soning. Inf Sci 8:199–249
73. Zhang Y, Ge H (2013) Freeway travel time prediction using takagi-sugeno-kang fuzzy neural network. Comput Aided Civil and Infrastruct Eng 28(8):594–603
74. Zimmermann HJ (2001) Fuzzy set theory and its applications, 4th edn. Kluwer Academic Publishers, ISBN 0-7923-7435-5

Enhanced Throughput and Accelerated Detection of Network Attacks Using a Membrane Computing Model Implemented on a GPU

Rufai Kazeem Idowu and Ravie Chandren Muniyandi

Abstract Membrane computing (MC) is a versatile, nondeterministic, and maximally parallel computing model. We explore the advantages of MC parallelism to flag intrusive connection records in a set of network traffic using a graphic processing unit (GPU) that built on a parallelism platform with a single-program multiple data (SPMD) feature. We build a P system model for attack detection by combining some of the features of a recognizer P system and a tissue-like P system with symport rules. Most previous implementations for intrusion detection have been performed on sequential or minimally low parallel machines called a central processing unit (CPU), so the issue of large data handling has always been a major challenge. Using a massively parallel NVIDIA CUDA architecture, we were able to overcome this problem. Comparison of processing on a GPU and a CPU reveals an increase in average throughput of 50,000 packets/s and more than fivefold acceleration for the detection rate.

Keywords Membrane computing · Attack detection · Graphic processing unit Cybersecurity · Data parallelism · P system

1 Introduction

Membrane computing (MC) , otherwise called P systems, was introduced by Gheorghe Păun more than a decade ago [18, 19]. Since then, MC has been applied in several fields because of its great parallelism, which reduces computational time complexity.

R. K. Idowu
Computer Science Department, College of Science & Information Technology,
Ijagun, Ogun State, Nigeria
e-mail: rufaiki@tasued.edu.ng

R. C. Muniyandi (✉)
Faculty of Information Science & Technology, Centre for Software Technology & Management,
Universiti Kebangsaan Malaysia, Bangi, Malaysia
e-mail: ravie@ukm.edu.my

© Springer International Publishing AG, part of Springer Nature 2019
S. K. Shandilya et al. (eds.), *Advances in Nature-Inspired Computing
and Applications*, EAI/Springer Innovations in Communication and Computing,
https://doi.org/10.1007/978-3-319-96451-5_11

MC comprises three distinct features that mimic the structure and functionality of biological cells: a membrane structure; objects found within the membranes; and operational rules that guide activities within the membrane. There are many variants of P systems, including cell-like, neural-like, and tissue-like systems. A cell-like P system has a hierarchically arranged set of membranes that can be described by a tree, whereas in a tissue-like system the membranes are located at the nodes of arbitrary graph. A neural-like P system has neurons (cells) linked by a specific set of synapses [19]. In this study, we focus on a tissue-like P system with an embedded recognizer P system that has a total Boolean function over a halting computation $\prod$ [11].

An intrusion (or attack) detection system (IDS) is a security measure used on a network or host-based system to check for activities symptomatic of an attack. An IDS is frequently used to oversee a networked environment to flag and report events capable of (i) compromising the system's integrity, (ii) denying its availability, and (iii) rendering its performance inefficient [8, 25, 28]. Ways of handling intrusion problems within a classification domain include ensuring that the processing throughput and detection acceleration are kept high.

According to the literature, most detection systems have been implemented using conventional CPUs, which become overwhelmed and eventually drop packets when their throughput can no longer cope with the increasingly large data found within extremely high-speed networks [3, 17, 21, 26, 27]. This deficiency leads to the problem of packet dropping and eventually to defective detection and high false alarm rates.

Myriad detection methods are available, but relatively few researchers have explored the parallelization offered by GPUs, and no study has investigated how the inherent advantages of MC could be used in this context. Here, we describe a novel approach in the use of MC for attack detection on GPU. The model is based on a recognizer tissue P system with evolution and symport communication rules for objects contained within membrane regions to ensure load balancing among GPU processors. The implementation combines the parallelism advantages of GPU and MC to enhance IDS performance, and yields sustainable and greatly increased throughput and accelerated detection under worst-case network traffic scenarios.

The remainder of the paper is organized as follows. Section 2 briefly discusses related work on MC applications on GPUs and parallelized IDS. Section 3 outlines techniques used for intrusion detection, while Sect. 4 describes GPU machines with a specific focus on the Nvidia GeForce GTX 680. In Sect. 5, we present our P system model for attack detection on a GPU. Section 6 discusses the implementation and other experimental details. In Sect. 7, we present and discuss our results. Section 8 concludes.

2   Related Work

MC variants have been applied in various fields and have yielded highly efficient results. Ishdorj et al. [10] used an MC variant to identify a deterministic solution to two well-known PSPACE-complete problems: QSAT and Q3SAT. For QSAT they hypothesized that the answer to any instance of the problem is computable in a time that is linear with respect to both the number of Boolean variables and the number of clauses that compose the instance. For Q3SAT, they postulated that the answer is computable in a time that is at most cubic in the number of Boolean variables.

Díaz-Pernil et al. [7] developed a new method for segmenting images via gradient-based edge detection by adopting an MC parallelism paradigm. They implemented a tissue P system algorithm in a compute unified device architecture (CUDA) environment.

Zhang et al. [32] tackled an NP-complete combinatorial optimization problem, the traveling salesman problem, using a membrane-inspired approximate algorithm. To generate their qualitative solutions, they combined the communication advantage of the hierarchical structure of P systems with ant colony optimization algorithms.

While studying GPUs as an alternative architecture for parallelism, Cecilia et al. [5] proposed that P systems could yield an efficient and uniform solution in the satisfiability (SAT) problem. They achieved this by adapting an initially developed simulator to the GPU architecture idiosyncrasies. In a previous study using the same GPU platform, the authors demonstrated that the N-Queens problem could also be solved by applying P system tools [4].

Despite MC successes in various areas, there have been few studies on the impact of MC in the security field. The first investigation of an MC security application was by Leporati and Ferretti [12], who modeled and analyzed firewalls using tissue-like P systems.

Rufai et al. [22] applied MC to the Bee algorithm used for an anomaly-based IDS to minimize redundant features that adversely affect the detection rate. Their approach yielded high detection rates and reasonably decreased false positives and negatives. Zaher et al. [30] adopted the parallel computing of a membrane environment to enhance the encryption/decryption processing time of the Rivest–Shamir–Adleman public key protocol in cryptography.

Despite a few studies on IDSs implemented on GPUs [3, 21, 29, [31], no research has combined the inherent parallelism benefits of both MC and GPUs for detection of network attacks.

3　Intrusion Detection Techniques

Every network system is vulnerable to attack, so IDS introduction is an important security measure in curbing or reducing intrusions. An intrusion is a security threat that is deliberately committed to access and compromise the integrity and confidentiality of a resource and to render an information system unreliable or unusable. An IDS is a device that checks a system to (i) flag unusual activities or actions that exhibit traits similar to an attack and (ii) report these activities to an administrator [1,

6, 8]. A dependable IDS should be able to (i) guarantee network integrity, (ii) make system services available whenever needed, and (iii) ensure that service delivery is efficient.

There are two main IDS techniques: anomaly-based detection and signature-based detection.

3.1 Anomaly-Based Detection

Anomaly-based detection (also called behavior- or heuristic-based detection) mainly involves identification of network behaviors symptomatic of an attack. Every user within an information system environment has a unique pattern. An action that seems to deviate from this unique pattern can often be flagged as an anomaly. Methods used to identify anomalies include statistical modeling, data mining, and the hidden Markov approach, for which specific metrics are used to determine breaches of a threshold created from a database of activity profiles that are legitimate [15]. Self-learning (automated or online) and programmed learning (manual or offline) have been identified as appropriate tools for learning what constitutes normal behavior in anomaly-based intrusion detection [2]. Any behavior that is adjudged to be normal is given free passage, while behaviors failing to conform to the predefined standard are rejected, thereby triggering an event in the anomaly detection system.

The widely acknowledged strength of anomaly-based detection is that it has very high potential for tracking novel attacks. It also has good scalability and minimizes resource usage. However, evaluation via anomaly-based detection may be inaccurate owing to the scarcity of adequate datasets [24]. Further weaknesses include (i) excessive time for retraining of behavior profiles, (ii) high false negative rates, and (iii) high computational costs if many metrics are involved. Anomaly-based IDSs can be further categorized as protocol-based or application payload-based anomalies.

3.2 Signature-Based Detection

Signature-based detection is also called misuse detection. In this approach, it is assumed that every known attack has a unique signature by which it can be identified in a network. Such signatures must be predefined by the network administrator before any vulnerability is considered intrusive. Methods commonly used in signature-based intrusion detection include application of rules and string or pattern matching.

The major advantage of signature-based intrusion detection is its high precision and accuracy in flagging known attacks. It also has very low levels of false negatives and development of signatures is easy. However, signature-based intrusion detection has the following disadvantages:

(i)   It leaves systems highly vulnerable because before the signature of a new attack is studied and rules developed, the attack might have caused havoc in the system.

(ii) It is excessively demanding because signatures have to be developed for new attacks to keep the system safe.

(iii) The database may become enormous as every signature pattern has to be kept.

(iv) Resource consumption is high, leading to slow throughput and inefficiency.

4   GPUs

GPUs are suitable for high-performance computing and were introduced in 1999 with the launch of the NVIDIA GeForce 256. GPUs were primarily designed to handle processes relating only to graphics such as texture mapping, lighting, and transformation of vertices and polygons, among others. However, later research showed that GPUs could also be used for other scientific calculations [20]. A GPU machine is distinguished from a traditional CPU system by the number of processor cores. While a CPU has multiple cores, a GPU has hundreds of cores (Fig. 1).

A GPU is a specialized, massively parallel computing device with several inbuilt processors. An application is considered to be suitable for a GPU if it is computationally intensive and can be parallelized such that there is little or no dependence or communication between tasks. For example, because of their inherent data parallelism, GPUs can be used to rapidly solve large problems relatively easily [29]. Some GPUs can deliver hundreds of billions of highly repetitive operations per second. Powerful and flexible programming languages for the design of accelerated GPU applications include C/C++, Python, and MATLAB.

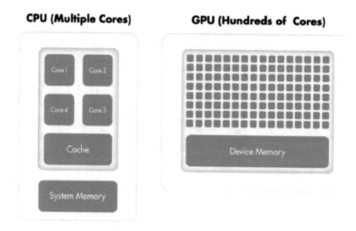

Fig. 1   Comparison of CPU and GPU systems [33]

4.1 Typical GPU Activities

A typical GPU machine performs three distinct functions: data input and issuance of instruction(s), execution based on instructions and available data, and data/information storage and output. Figure 2 shows a simple representation of these activities in a GPU.

4.2 Nvidia GPU GeForce GTX 680 Architecture

The Nvidia GPU GeForce GTX 680 is one of the fastest and most efficient GPU machines built. It comprises four graphics processing clusters (GPCs), eight next-generation streaming multiprocessors (SMs), and four memory controllers [16]. The eight SMs have 192 CUDA cores each, resulting in a total of 1536 CUDA cores. The machine properties are listed in Table 1.

Fig. 2 Simplified view of activities in a GPU

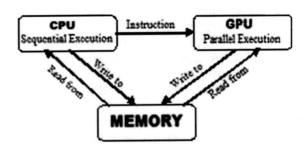

Table 1 Features of GeForce GTX 680

Parameter	Value
ComputeCapability	3.0
MaxThreadsPerBlock	1024
MaxShmemPerBlock	49,152
MaxThreadBlockSize	[1024 1024 64]
MaxGridSize	[2.1475e + 09 65535 65535]
SIMDWidth	32
TotalMemory	2 GB
MultiprocessorCount	8
ClockRate	1 GHz

5   P System Model for Attack Detection

A tissue P system is a powerful machine that processes multisets of symbol impulses in a maximally parallel manner within a net of cells. Since these cells are elementary and coexist in a single environment, a premium is attached to the communication between the cells and/or with the environment that holds the computation results [14, 19].

Since we are purely considering data parallelism, objects inside the membranes do not necessarily need to communicate with one another, but instead with the environment where the results (intrusive connection records) are obtained. Hence, the principle of membrane permeability was invoked to allow intrusive connections to leave the membranes.

In this model, referred to as an attack detection P system ($\prod_{AD\_P}$), network connection information is modeled as a multiset of objects and the parameters of the P system defined in this respect. The model is configured using a system with the structure $\prod_{AD\_P} = [\ ]_1 [\ ]_2 [\ ]_3 \cdots [\ ]_m$, where m is the number of connection records. Computation in this system allows for process convergence because classified intrusive connection records are obtained in the environment.

We formally define our attack detection P system model as a system of degree $m \geq 1$ of the form

$$\prod_{AD\_P} = (O, \gamma_1, \ldots, \gamma_m, r, \beta, l),$$

where

- O is a set of objects. An object represents connection records, $\epsilon\ O$, where $O \mid \epsilon\ [0, 4898430]$;
- $\gamma_1, \ldots, \gamma_m$ are membranes (cells) representing the zones of the network;
- r are evolution/symport rules used for classification;
- β is the environment/zone. This external membrane environment is where the results of computation are obtained. They do not hold any rule. This stage signifies the end of computation–final configuration; and
- $l \subseteq \{1, 2, \ldots, m\} \times \{\beta\}$, which is a link (also known as channel or synapse) between the membranes and the environment β.

It important to note that this model functions on a single level of membrane hierarchy in which each membrane acts as an elementary or skin membrane. The results (intrusive attack connections) generated by these membranes are sent directly to the external environment. Figure 3 shows how membranes are assigned to threads on the GPU machine.

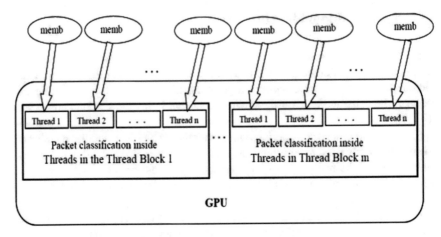

Fig. 3 Assigning membranes to GPU threads

5.1 Rule Application

The rules are called transition or maximally parallel rules. The evolution and symport rules with guards used here follow Ipate et al. [9]. Rules with guards have the property whereby all possible rule assignments must take place during every communication step. Therefore, in this scenario, these rules are applied at the classification stage.

In each membrane, multisets are initialized according to

$$j_1 = O_1 \text{ defined by } \gamma^{1,1}\gamma^{2,1} \ldots \gamma^{41,1}$$
$$j_2 = O_2 \text{ defined by } \gamma^{1,2}\gamma^{2,2} \ldots \gamma^{41,2}$$
$$j_3 = O_3 \text{ defined by } \gamma^{1,3}\gamma^{2,3} \ldots \gamma^{41,3}$$
$$\vdots$$
$$j_m = O_m \text{ defined by } \gamma^{1,m}\gamma^{2,m} \ldots \gamma^{41,m},$$

where j_i are multisets for the ith membrane, $O_1, \ldots, O_m$ are objects (connection records), and $(\gamma^{1,1}\gamma^{2,1} \ldots \gamma^{41,1})$, $(\gamma^{1,2}\gamma^{2,2} \ldots \gamma^{41,2})$, $(\gamma^{1,3}\gamma^{2,3} \ldots \gamma^{41,3})$, ..., $(\gamma^{1,m}\gamma^{2,m} \ldots \gamma^{41,m})$ are features of the m connection records.

Definition (i) Let $\beta = \{j_1, j_2, \ldots, j_n\}$ be a finite set of connection packets in traffic such that n $\leq$ 4,898,430.

Definition (ii) Let $\theta = \{j_1, j_2, \ldots, j_k\}$ be a group of intrusive connection packets such that θ is a subset of β ($\theta \subseteq \beta$ and $\beta > \theta$).

The 41 features of the KDD Cup dataset were used to formulate 111 rules with 110 conditioned guards. In line with Ipate et al. [9], a particular rule with a guard is only applicable when the conditioned guard (g) is evaluated as true. Thus, using features of the connection records for the $\prod_{AD\_P}$ model, some of the conditions used to generate the guards in the rules are as follows:

$$\text{Cond}_1 : \text{If } \gamma_{23} < 76.5, \text{ then node 2 elseif } \gamma_{23} \geq 76.5 \text{ then node 3 else 0;} \quad (1.1)$$

$$\text{Cond}_2 : \text{If } \gamma_{37} < 0.495, \text{ then node 4 elseif } \gamma_{37} \geq 0.495 \text{ then node 5 else 0;} \quad (1.2)$$

6 Implementation

We used a computer with an Intel Core-i7-3820, 3.60 GHz CPU with 8 GB RAM and a NVIDIA GTX 680 GPU for implementation. As typical with almost every (if not all) GPUs, execution of a project involves some processes as simplified in Fig. 2.

6.1 Experimental Setup

We use the KDD Cup'99 dataset, which has 41 attributes (Table 2) that can be discrete, continuous, or symbolic. This is a very large dataset that uses Transmission Control Protocol/Internet Protocol (TCP/IP) level information and has nearly five million (4,898,430 labeled) connection records. The KDD Cup dataset was chosen because it remains the benchmark dataset for intrusion-related studies by the security research community.

The first step in the implementation is preprocessing of the data to ensure that textual data are converted to numeric form.

Table 2 Attributes of the KDD cup dataset

No. Attribute name	No. Attribute name	No. Attribute name
01 duration	15 su_attempted	29 same_srv_rate
02 protocol_type	16 num_root	30 diff_srv_rate
03 service	17 num_file_creations	31 srv_diff_host_rate
04 flag	18 num_shells	32 dst_host_count
05 src_byte	19 num_access_files	33 dst_host_srv_count
06 dst_byte	20 num_outbound_cmds	34 dst_host_same_srv_rate
07 land	21 is_host_login	35 dst_host_diff_srv_rate
08 wrong_fragment	22 is_guest_login	36 dst_host_same_src_port_rate
09 urgent	23 count	37 dst_host_srv_diff_host_rate
10 hot	24 srv_count	38 dst_host_serror_rate
11 num_failed_login	25 serror_rate	39 dst_host_srv_serror_rate
12 logged_in	26 srv_serror_rate	40 dst_host_rerror_rate
13 num_compromise	27 rerror_rate	41 dst_host_srv_rerror_rate
14 root_shell	28 srv_rerror_rate	

6.2   Execution Processes

We adopted the principle of recognizer P systems [11] to make it possible for all regions to evolve concurrently by sending intrusive connections to the environment as output. Since maximization of the utilization of functional GPU units depends on thread-level parallelism, we designed our data parallelism for several connection records and subsequent application of the rules in a parallel manner.

A recognizer P system is often applied in decision-making scenario-based problems in which only one option is applicable, such as True/False, Yes/No, and On/Off. In Fig. 4, the object in each membrane of the embedded recognizer P system is either a normal or anomalous connection. Thus, its output produces "1" values that are released to the environment, thereby satisfying the acceptance computational condition. Hence, the computation halts, at which point our $\prod_{AD\_P}$ system flags intrusive connection records.

Since constant data transfer between the CPU and GPU has a negative effect on performance, we vectorized our data on the device for computational efficiency. We used the GPU-enabled function *gather* to transfer results back to the CPU host.

Execution of the $\prod_{AD\_P}$ model involves the following procedural steps:

Step 1: Downloading the dataset and preprocessing to accommodate data with symbolic and continuous features. The KDD Cup dataset was downloaded from http://kdd.ics.uci.edu/databases/kddcup99/kddcup99.html.

Step 2: Creating/compiling *.cu* and *.ptx* files using the NVIDIA CUDA compiler (NVCC). This step takes care of non-CUDA compilation handled via the C++ compiler, which is an instance of Microsoft Visual Studio supported by NVCC.

Step 3: Vectorization. The dataset was vectorized to avoid any deterioration in the required GPU acceleration.

Step 4: Assigning membranes to threads. Since this model uses the data parallelism of a tissue P system, the membrane/thread (m/n) ratio was 1:1. Furthermore,

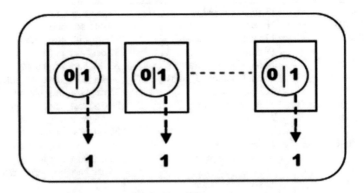

Fig. 4 Membrane structure of the embedded recognizer P system

the membranes used here can be said to be independent, so assigning each membrane to a thread does not lead to unnecessary communication between threads, which could decrease the performance [13]. Since maximizing the utilization of GPU functional units depends on thread-level parallelism, the same *kernel* that operates on every thread was launched. Meanwhile, shared memory instead of global memory was used because it is faster and easily accessible by all threads within a thread block [16].

Step 5:   Application of the evolution rules via kernels to classify the packet data. Once the membranes have been appropriately assigned to the threads with the *kernel*, then rules of the following type are invoked to distinguish intrusive from nonintrusive connection records:

$$R_1 = \gamma_{i23}\gamma_{i6}\gamma_{i27}\gamma_{i5} \rightarrow s_{i1}; (\gamma_{i23} > 76.5 \text{ and } \gamma_{i6} \geq 40.5 \text{ and } \gamma_{i27} > 0.45 \text{ and}$$
$$\gamma_{i5} > 0.495); \quad 1 \leq i \leq \text{MaxPac} \tag{1.3}$$

$$R_2 = \gamma_{i34}\gamma_{i37}\gamma_{i13}\gamma_{i40} \rightarrow s_{i0}; (\gamma_{i34} \geq 0.015 \text{ and } \gamma_{i37} < 0.495 \text{ and } \gamma_{i13} < 0.5 \text{ and}$$
$$\gamma_{i40} \geq 0.89); \quad 1 \leq i \leq \text{MaxPac}, \tag{1.4}$$

where

$$(\gamma_{i23} > 76.5 \text{ and } \gamma_{i6} \geq 40.5 \text{ and } \gamma_{i27} > 0.45 \text{ and } \gamma_{i5} > 0.495 \text{ and}$$
$$\gamma_{i34} \geq 0.015 \text{ and } \gamma_{i37} < 0.495 \text{ and } \gamma_{i13} < 0.5 \text{ and } \gamma_{i40} \geq 0.89)$$

represent some of the conditional guards derived from a classification tree, and $s_i = \{1,0\}$ denotes the status of connection record O_i, which may be either intrusive (0) or nonintrusive (1) according to features 23, 6, 27, 5, 34, 37, 13, and 40.

Step 6:   Running on both sequential (CPU) and parallel (GPU) platforms. To measure and compare the acceleration and throughput, the model was run on both the GPU and CPU for varied numbers of membranes.

Step 7:   Output stage. In line with the principle of recognizer P systems [11], the model was designed so that all regions can evolve concurrently by sending the right answer (intrusive connection records) to the environment (β) as output. Thus, the rules are formulated so that intrusive connections enjoy permeability through the membranes according to

$$O_i \rightarrow (\text{anomaly}, \beta), \tag{1.5}$$

where i denotes the membranes that release anomalous connection records into the environment (β) after invocation and execution of the evolution rules for classification. The symport rules can thus be succinctly expressed using the format

$$\text{Rule}_{\text{symp}(i)} : [O^{\text{anomaly}}]_i \rightarrow [\,]_i O^{\text{anomaly}}. \tag{1.6}$$

So, the output produces "1" values that are released to the environment, thereby satisfying the acceptance computational condition. Hence, the computation halts, at which point the $\prod_{AD\_P}$ system flags anomalous packets.

The execution steps outlined above are depicted in Fig. 5. Since the objects used here are independent, assigning each to a thread does not lead to unnecessary communication between threads and thus avoids any decrease in performance [13].

In the implementation of our P system model, membranes are represented as vectors in the GPU and connection records are the objects. This approach ensures that there is efficient use of GPU memory. For this work, we use GPU-enabled functions in the MATLAB parallel computing toolbox.

7 Results and Discussion

We applied a P system model for attack detection on a GPU to flag intrusive attacks and to handle large data from the KDD Cup dataset. Our approach yielded an increase in average throughput of 50,000 packets/s and processing acceleration of more than fivefold for detection.

7.1 Throughput

Table 3 lists throughput data for the GPU and CPU, which were obtained by dividing the packet size by the processing time. A key performance metric for a network IDS is sustainable throughput [23], for which the $\prod_{AD\text{-}P}$ model achieves very good results.

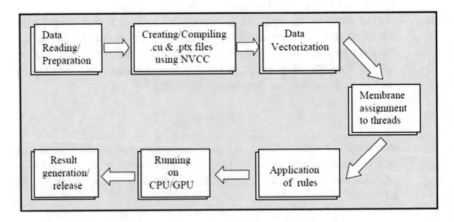

Fig. 5 Execution processes of the $\prod_{AD\_P}$ system

Table 3 Performance of the $\prod_{AD\_P}$ model on GPU and CPU

All membranes (test)	Number of membranes	GPU time (s)	CPU time (s)	GPU/CPU acceleration	GPU throughput (packets/s)	CPU throughput (packets/s)
314572	2	323.7	34.7	0.1	971.7	9039.8
314572	4	167.4	34.7	0.2	1879.5	9039.8
314572	8	85.7	34.7	0.4	3669	9039.8
314572	16	44.8	34.7	0.8	7014.3	9039.8
314572	32	23.1	34.7	1.5	13,606.4	9039.8
314572	64	13.8	34.7	2.5	22,814.5	9039.8
314572	128	9.3	34.7	3.7	33,696.4	9039.8
314572	256	7.5	34.7	4.6	41,853.9	9039.8
314572	512	6.6	34.7	5.3	47,580.2	9039.8
314572	1024	6.2	34.7	5.6	50,945	9039.8
314572	2048	5.9	34.7	5.9	53,102.6	9039.8
314572	4096	6.2	34.7	5.6	50,652.4	9039.8

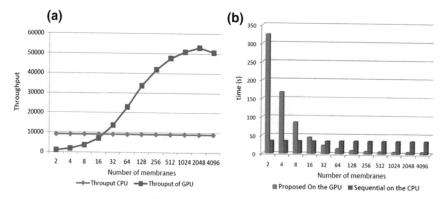

Fig. 6 Comparison of CPU and GPU in terms of **a** throughput and **b** acceleration

Figure 6a shows that for 2, 4, 8, or 16 membranes, the CPU has better throughput performance than the GPU. This is because of ineffective utilization of GPU resources when the number of membranes is small. The highest GPU throughput (53,102.6) was achieved for 2048 membranes. The decrease in throughput to 50,652.4 for 4096 membranes might be related to other loads on the machine. In general, the average GPU throughput for the $\prod_{AD\_P}$ model is sufficient to check for packet drop/loss in an IDS. This is closely related to the higher multiprocessor occupancy of the GPU, which ultimately improves system efficiency.

7.2 Acceleration

Table 3 and Fig. 6b show the detection acceleration obtained for 2–4096 membranes with appropriate connection records in the $\prod_{AD\_P}$ model. Similar to the through-put results, the GPU acceleration increases with the number of membranes [13]. This implies that the model would continue to perform well on a GPU when the multiprocessor occupancy and load balancing are adequate.

Figure 6b compares the GPU and CPU detection times for varying numbers of membranes in the $\prod_{AD\_P}$ model. As already pointed out, the model can classify the content of membranes as either intrusive or nonintrusive at high speed once there are sufficient numbers of membranes to check for underutilization of resources. For instance, classification of the content of 2048 membranes as either normal or anomalous took 34.7 s on the CPU and 5.9 s on the GPU, representing more than fivefold acceleration compared to the CPU.

8 Conclusions

We described a GPU IDS model based on a tissue P system and presented an architectural overview. This approach has not been explored to date.

While timing is of the essence in network security, the ability to handle ever-increasing connection records is equally as important. We established that a combination of the highly parallelized structures of both P systems and GPUs yields good synchronization and hence could be very advantageous in IDS. An example using the large data in the KDD Cup dataset (approximately five million connection records) confirmed the suitability of our approach. We used 100% of the KDD Cup dataset rather than the 10% typically used.

Acknowledgements This work has been supported by Fundamental Research Grant of Ministry of Higher Education of Malaysia (Grant Code : FRGS/1/2015/ICT04/UKM/02/3).

References

1. Alomari O, Othman ZA (2012) Bees algorithm for feature selection in network anomaly detection. J Appl Sci Res 8(3):1748–1756
2. Axelsson S (2000) Intrusion detection systems: a taxonomy and survey. Chalmers University of Technology, Sweden, Tech. Report
3. Bul'ajoul W, James A, Pannu M (2015) Improving network intrusion detection system performance through quality of service configuration and parallel technology. J Comput Syst Sci 81(6):981–999
4. Cecilia JM, García JM, Guerrero GD, Martínez–del–Amor MA, Pérez–Hurtado I, Pérez–Jiménez MJ (2009) Implementing P systems parallelism by means of GPUs. In: Membrane computing, Springer, Berlin, pp 227–241

5. Cecilia JM, García JM, Guerrero GD, Martínez-del-Amor MA, Pérez-Hurtado I, Pérez-Jiménez MJ (2010) Simulating a P system based efficient solution to SAT by using GPUs. J Logic Algebraic Program 79(6):317–325
6. Dartigue C, Jang HI, Zeng W (2009) A new data-mining based approach for network intrusion detection. In: Communication Networks and Services Research Conference, 2009. CNSR'09. Seventh Annual, pp 372–377
7. Díaz-Pernil D, Berciano A, Pena-Cantillana F, Gutierrez-Naranjo MA (2013) Segmenting images with gradient-based edge detection using membrane computing. Pattern Recogn Lett 34(8):846–855
8. Folorunso O, Akande OO, Ogunde AO, Vincent OR (2010) ID-SOMGA: a self organizing migrating genetic algorithm-based solution for intrusion detection. Comput Inform Sci 3(4):80–92
9. Ipate F, Dragomir C, Lefticaru R, Mierla L, Pérez-Jiménez MDJ (2012) Using a kernel P system to solve the 3-col problem. In: Proceedings of the 13th international conference on membrane computing. Computer and Automation Research Institute, Hungarian Academy of Sciences, pp 243–258
10. Ishdorj TO, Leporati A, Pan L, Zeng X, Zhang X (2010) Deterministic solutions to QSAT and Q3SAT by spiking neural P systems with pre-computed resources. Theoret Comput Sci 411(25):2345–2358
11. Jiménez MJP, Jiménez ÁR, Caparrini FS (2003) Complexity classes in models of cellular computing with membranes. Nat Comput 2(3):265–285
12. Leporati A, Ferretti C (2010) Modeling and analysis of firewalls by (tissue-like) P systems. Rom J Inf Sci Technol 13(2):169–180
13. Maroosi A, Ravie CM, Elankovan S, Abdullah MZ (2014) Parallel and distributed computing models on a graphics processing unit to accelerate simulation of membrane systems. Simul Model Pract Theory 47:60–78
14. Martí C, Păun G, Pazos J (2003) Tissue P systems. Theoret Comput Sci 296(2):295–326
15. Modi C, Patel D, Borisaniya B, Patel H, Patel A, Rajarajan M (2013) A survey of intrusion detection techniques in cloud. J Netw Comput Appl 36(1):42–57
16. Nvidia (2012) Whitepaper on NVIDIA GeForce GTX 680
17. Papadogiannakis A, Polychronakis M, Markatos EP (2010) Improving the accuracy of network intrusion detection systems under load using selective packet discarding. In: Proceedings of the third European workshop on system security, pp 15–21
18. Păun Gh, Rozenberg G (2002) A guide to membrane computing. Theoret Comput Sci 287:73–100
19. Păun Gh (2006) "Introduction to membrane computing," Applications of membrane computing. Springer, Berlin, pp 1–42. ISBN 978-3-540-29937-0
20. Reese J, Zaranek S (2012) GPU programming in matlab. MathWorks News & Notes. The MathWorks Inc, Natick
21. Rietz R, Vogel M, Schuster F, König H (2014) Parallelization of network intrusion detection systems under attack conditions. In: Detection of intrusions and malware, and vulnerability assessment. Springer, Berlin, pp 172–191
22. Rufai KI, Ravie CM, Othman ZA (2014) Improving bee algorithm based feature selection in intrusion detection system using membrane computing. J Netw 9(3):523–529
23. Schaelicke L, Freeland JC (2005) Characterizing sources and remedies for packet loss in network intrusion detection systems. In Workload characterization symposium, proceedings of the IEEE international, pp 188–196
24. Shiravi A, Shiravi H, Tavallaee M, Ghorbani AA (2012) Toward developing a systematic approach to generate benchmark datasets for intrusion detection. Comput Secur 31(3):357–374
25. Uma M, Padmavathi G (2013) A survey on various cyber attacks and their classification. Int J Netw Secur 15(5):390–396
26. Vasiliadis G, Michalis P, Sotiris I (2011) MIDeA: a multi-parallel intrusion detection architecture. In: Proceedings of the 18th ACM conference on computer and communications security, ACM, pp 297–308

27. Vasiliadis G, Antonatos S, Michalis P, Markatos EP, Sotiris I (2008) "Gnort: high performance network intrusion detection using graphics processors," Recent advances in intrusion detection. Springer, Berlin, pp 116–134
28. Venter HS, Eloff JH (2003) A taxonomy for information security technologies. Comput Secur 22(4):299–307
29. Wu W, DeMar P, Holmgren D, Singh A (2011) G-NetMon: a GPU-accelerated network performance monitoring system. In: 2011 symposium on application accelerators in high-performance computing (SAAHPC), pp 76–79
30. Zaher S, Badr A, Farag I, Tarek Elmageed TA (2012) Using P system with GPU model to design and implement a public key cryptography. Int J Compt App 60(6):0975–8887
31. Zaher S, Badr A, Farag I (2012) Performance enhancement of RSA cryptography algorithm by membrane computing. Int J Adv Res Compt Sci Softw Eng 2(9)
32. Zhang GX, Cheng JX, Gheorghe M (2011) A membrane-inspired approximate algorithm for traveling salesman problems. Rom J Info Sci Technol 14(1):3–19
33. Reese J, Zaranek S (2012) GPU programming in Matlab. MathWorks News & Notes. The MathWorks Inc, Natick, MA, pp 22–5

A Computational Physics-Based Algorithm for Target Coverage Problems

Jordan Barry and Christopher Thron

Abstract The problem of optimally covering a set of point targets in a region with areas of a specific shape has several important applications in the fields of communications, remote sensing, and logistics. We consider the case where a target is covered when it falls within a coverage area (so-called "Boolean" coverage), and we specialize to the case of identical circular (or spherical) coverage areas. The problem has been shown to be NP-hard, and most practical algorithms use statistical methods to look for near-optimal solutions. Previous algorithms cannot guarantee 100% target coverage. In this chapter we demonstrate a physics-based algorithm (called the "nebular algorithm") that guarantees full coverage while seeking to minimize the number of coverage areas employed. This approach can generate solutions with reduced numbers of sensors for systems with thousands of targets within a few hours. The algorithm, its implementation, and simulation results are presented, as well as its potential applicability to other coverage problems such as area and/or probabilistic coverage.

Keywords Target coverage · Sensor placement · Boolean coverage · Physics-based · Algorithm · Stochastic optimization

1 Introduction

1.1 Practical Applications of Point Coverage

The term "point coverage problem" generally refers to a situation where a finite set of points within a region must be covered by sets, so as to minimize a cost function which depends on the sets included in the cover. The region may be 2-, 3-, or higher dimensional; and usually there are constraints on the size, shape, and location of the sets that may be included in the cover.

J. Barry · C. Thron (✉)
Texas A&M University–Central Texas, 1001 Leadership Place, Killeen, TX 76549, USA
e-mail: thron@tamuct.edu

© Springer International Publishing AG, part of Springer Nature 2019
S. K. Shandilya et al. (eds.), *Advances in Nature-Inspired Computing and Applications*, EAI/Springer Innovations in Communication and Computing,
https://doi.org/10.1007/978-3-319-96451-5_12

Applications of point coverage problems are readily apparent in several areas of modern technology. In wireless telecommunications, network designers may be tasked with configuring a set of transmit-receive nodes (which may be routers, relays, or base stations) which can optimally service dispersed clients within a region. In the field of logistics, a company may need to establish a warehouse infrastructure for distribution purposes. Each warehouse has a maximum delivery radius based on product spoilage; and naturally the company wants to minimize the number of warehouses built while still covering all delivery locations.

Some of the most prominent instances of coverage problems involve sensor networks. Sensors may be used to monitor environmental pollution and/or habitat change [8, 14]. In a military context, sensors may be used to monitor activity in battlefields and other sensitive environments [4]. In the following discussion we will focus primarily on sensor networks, while keeping in mind that our results also pertain to other applications as well. In accordance with our primary focus, we will refer to sets used in the cover as "sensors" and points to be covered as "targets".

The rest of the chapter is organized as follows. We first provide some mathematical background to the problem, including some estimates of computational complexity. Next, we briefly introduce previous approximate solution algorithms, many of which are nature-inspired. We then develop the physical roots of our own method. A description of the algorithm and its implementation is provided, followed by some simulation results. Finally we discuss generalizations and extensions.

2 Problem Statement and Mathematical Background

2.1 General Formulation of Set Cover Problem

As our benchmark scenario, we consider the problem of covering a set of points located in a region with area $A \gg 1$ with the smallest possible number of disks of fixed size (for simplicity, we take the disks to have area 1). This is a special case of the general problem of selecting a set of allowable covering sets with lowest total cost that contain a given set of points. This general problem may be formulated rigorously as follows.

Given an arbitrary set R, let $\mathcal{P}(R)$ be the set of all subsets of R. Let $\mathcal{S} \subset \mathcal{P}(R)$ be the set of possible covering sets (sensors), and let $c : \mathcal{S} \rightarrow [0, \infty)$ be a cost function defined for these sets. We shall define:

$$T = \{\mathbf{t}_1, \mathbf{t}_2, \ldots, \mathbf{t}_N\}$$

(elements of R are denoted by boldface) as the set of all points in R which are to be covered (i.e. targets). For any $S \in \mathcal{S}$ we may define its intersection with T:

$$I_S \equiv T \cap S.$$

The set of all $\{I_S\}_{s \in \mathcal{S}}$ is a subset of $\mathcal{P}(T)$, and hence finite. Let $I_1, I_2, \ldots, I_M$ be a listing of the elements of $\{I_S\}_{s \in \mathcal{S}}$. Let c_m be the cost of I_m, defined as follows:

$$c_m = \inf_{I_S = I_m} c(S)$$

Define characteristic functions $\chi_m: T \rightarrow \{0, 1\}$ as follows:

$$\chi_m(\mathbf{t_n}) = \begin{cases} 0, & \text{if } \mathbf{t_n} \notin I_m \\ 1, & \text{if } \mathbf{t_n} \in I_m \end{cases}$$

We must then find $a_1, a_2, \ldots, a_M \in \{0, 1\}$ that satisfy the following linear programming problem:

$$\text{Minimize } Z = \sum_{m=1}^{M} c_m a_m \text{ subject to } \sum_{m=1}^{M} a_m \chi_m(\mathbf{t_n}) > 0, \ n = 1, \ldots N.$$

This coverage problem is often referred to as "Boolean coverage" because the functions χ_m take the values 0 or 1: the problem may be generalized by allowing χ_m to take values between 0 and 1, corresponding to coverage probabilities. In the case where c_m is a constant independent of m (the so-called "unicost" case), the problem can be reformulated as a "hitting set problem" as follows. Let $J_1 \ldots J_{M'}$ be a listing of the maximal elements of $\{I_S\}_{s \in \mathcal{S}}$, i.e., those elements of I_S that are not proper subsets of any other element of I_S. For each $n \le N$, let $C_n \subset \{1, \ldots, M'\}$ be the set of all numbers m such that $\mathbf{t_n} \in J_m$. Then the minimization problem amounts to finding the smallest k for which there exists a subset of $\{1, \ldots, M'\}$ with cardinality k which intersects C_n for all n in $\{1, \ldots, N\}$. The general hitting set problem is known to be NP-hard, and no polynomial-time solution is known. This means that the complexity grows very rapidly with the size of the problem, which in our case corresponds to M'. In the 1-dimensional case (i.e. points are located on a line segment) by symmetry arguments one may obtain the estimate $E[M'] \ge N/2$; and the value should be somewhat higher in 2 dimensions.

Although there is no known polynomial-time algorithm for solving the point coverage problem, there are polynomial-time algorithms which give solutions which approach optimality to within any given tolerance. However, in practice these algorithms are impractically expensive. For example, the algorithm in Hochbaum and Maas [5], if applied to the problem of covering N points in a region with unit disks, has an estimated complexity of greater than $10^{2500} N^{1500}$ to guarantee optimality within 10%.

In view of difficulties posed by exact solution, we must turn to alternative strategies. One possibility is an exploratory approach, where promising solution candidates are sequentially generated and evaluated. The key to such a strategy is the generation of good solution candidates; and here is where we shall seek inspiration from physical systems. But first, we shall briefly review some basic concepts of Markov chains, and describe their usefulness in approximate optimization algorithms.

2.2 *Markov Chains and Stochastic Optimization*

We have already shown that there are too many possible solutions to the point coverage problem to examine exhaustively. Instead, we seek a guided, non-exhaustive search process to look for prospective optimum solutions. Such a process should have some randomness built in, for otherwise it may systematically avoid certain configurations which could be optimal. Markov chains exactly fit this description of guided but partially random processes; and indeed, Markov chains form the mathematical basis for most stochastic optimization algorithms, including genetic algorithms, swarm intelligence, and other nature-inspired methods. In this section we will give a brief overview of the concept of Markov chain and explain why it is suitable for point coverage problems.

For a system to be represented as a Markov chain, the system must change state according to an iterative process such that at each step of the process, the probability of transitioning from one state to another depends only on the two states involved in the transition (and possibly the iteration number). Figure 1 shows a graphical representation of a simple Markov chain with three states labeled 1, 2, 3. The labeled arrows in Fig. 1 show the transition probabilities of moving from one state to another during a single iteration of the process. For time-independent Markov chains these transition probabilities are constant; but in general they may vary from iteration to iteration, as long as they are calculated based only on the two states involved in the transition.

In an *irreducible, positive recurrent* Markov chain, a system started in any state will, with probability 1, eventually visit every other state if allowed to continue to run for a sufficiently large number of steps. In other words, an irreducible, positive recurrent Markov chain provides a method for "touring" all possible states of the system. Figure 2 shows an example of a time-independent Markov chain that is not irreducible positive recurrent: if the chain starts in either of states 2 or 3, it will never reach state 1.

Not all states of a system are good candidates for an optimal solution. The value of Markov chains in search problems lies in their ability to be "tuned" to heavily

Fig. 1 Graphical representation of a simple Markov chain

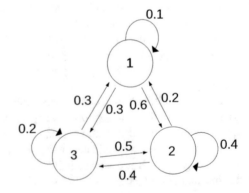

Fig. 2 Transition diagram
for a non-positive recurrent
Markov chain

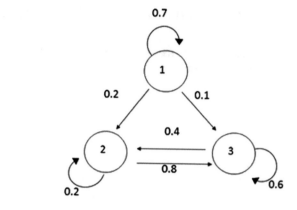

Fig. 3 Example Markov
chain trajectory tuned for
optimal search

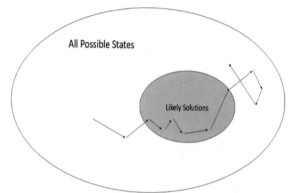

sample from among likely solutions and mostly avoid states with poor prospects (see
Fig. 3). This technique improves the chance of finding better solutions in cases where
it is not feasible to try all possible states.

Based on the above discussion, we may identify criteria for a "good" Markov
chain from the point of view of optimization:

- It should be easy to calculate the next state from the previous one, so many states
  can be investigated quickly;
- It should tend to remain near states that are likely optimal solutions;
- It should not get "stuck" in any one subset of states, even if that subset has many
  good candidates. (It is not necessary for the chain to be positive recurrent, because
  usually it is not practical to run the chain for a long enough time so that it reaches
  all states. Nonetheless, the chain should not be confined so as to systematically
  exclude possible states which may turn out to be optimal.)

Most common stochastic optimization approaches such as genetic algorithms
and algorithms based on swarm intelligence are Markov chains, so they may all be
evaluated on the basis of the above criteria.

3   Previous Nature-Inspired Algorithms for Point Coverage

In this section, we shall briefly survey previous point coverage algorithms which also draw inspiration from nature.

3.1   Biologically-Inspired Algorithms for Point Coverage

Due to the practical importance of point coverage problems, several researchers have tried various stochastic approaches for finding good (though not necessarily optimal) solutions. Many of these previous approaches are also nature-inspired. In this section, we briefly describe the application of two standard approaches with biological roots, namely genetic algorithms (GA) and swarm intelligence.

In order to apply a genetic algorithm, possible solutions must be encoded as "chromosomes". This may be done in a variety of ways. In Njoya et al. [13], a chromosome consists of a list of triples $\{(x_m, y_m, \sigma_m)\}, m = 1, \ldots, M$ where (x_m, y_m) is the mth potential sensor location and $\sigma_m = 1$ or 0 depending on whether or not a sensor at that location is used in the cover ([13] considers the case of multiple coverage by independent covers, so that σ_j may take additional values). In Xu and Yao [20], potential sensor locations are similarly enumerated, and a chromosome is an ordering of these locations. Covers are associated with chromosomes according to a "greedy" algorithm: the sensor locations are examined one by one in the order specified by the chromosome, and a sensor is placed at location m if the vicinity m is not sufficiently covered by previously-placed sensors from the set $\{1, \ldots m-1\}$. In Njoya et al. [14], a chromosome is a list of locations $\{(x_j, y_j)\}, j = 1 \ldots J$ of sensors used in the cover (meaning that covers are of fixed size J).

Besides a chromosome encoding, an objective function (also called "fitness function") must be specified. Xu and Yao [20] directly minimize number of sensors, while Njoya et al. [14] specify two objective functions: one reflecting the degree of coverage and the other the number of "free" sensors that cover no targets (so that for example if there are F free sensors, the actual cover size is J–F).

Once chromosome encoding and objective function(s) have been specified, the standard GA operations of crossover, mutation, and selection may be applied to "evolve" a population of chromosomes so as to raise the objective function level(s) in the surviving population. (Note that these operations produce a Markov chain on the set of possible chromosome populations.) Njoya et al. [14] use NSGA-II [3], a general-purpose multiobjective evolutionary algorithm.

GA methods are attractive in that they are general-purpose, and require very little specific information to apply them to different systems. On the flip side, this same non-specificity means that GA does not take advantage of particular system characteristics which can guide the search towards promising solutions. For example, if a candidate solution contains sensors that can be moved slightly to cover an additional targets, then GA is unable to detect this. The solution candidates generated in the crossover and mutation operations are agnostic to the objective function(s), so multitudes of chromosomes are produced that a more intelligent method would immediately reject. To perform selection, GA must compute the objective function(s) for every chromosome, which can be computationally intensive for large systems. These inefficiencies in GA operations take their toll in very large systems with thousands of sensors and/or targets. Indeed, in all systems studied in Njoya et al. [13, 14] and Xu and Yao [20] either the number of target points or the number of sensors is less than 200.

Besides GA, various flavors of "swarm intelligence" algorithms have been applied to the problem. Two prominent examples include "artificial bee colony" (ABC) and "particle swarm optimization", which we discuss briefly below.

The general methodology of "artificial bee colony" (ABC) optimization was introduced by D. Karaboga [7], (Intelligent Systems Research Group, n.d.): and the method has been applied to sensor network coverage problems by various authors [12, 16, 18]. These references consider slightly different optimization problems (e.g. maximization of network lifetimes and non-Boolean coverage), but their algorithms can be easily adapted to the Boolean coverage problem.

In ABC the optimization process is likened to a swarm of "bees" which explore the space of possible solutions. Three types of bees are used: employed bees, onlooker bees, and scout bees. Each employed bee has been assigned to a candidate solution, and explores the vicinity of that solution; onlooker bees choose worker bees randomly (with a distribution that more heavily weights workers associated with better solutions) and explore the neighborhoods of the chosen worker; and scout bees randomly seek out new solutions. A single iteration of the search process can be described as follows:

1. Loop through worker bees; locally perturb each worker bee solution, evaluate the fitness of the perturbed solution, and move the worker bee to a perturbed solution if it is better than the worker bee's current solution.
2. Loop through onlooker bees; each onlooker bee chooses a worker bee with probability proportional to the fitness function of the worker bee's solution; the onlooker bee then locally perturbs the chosen worker bee's solution and evaluates the fitness; and the worker bee is moved to the perturbed solution if it is better than the worker bee's current solution.
3. For each worker bee, a record is kept of the number of iterations since the worker bee has last moved. If this number surpasses a threshold (which is fixed by the programmer), then the worker bee's solution is "abandoned" and replaced by a randomly chosen solution (selected by a scout bee).

Since the perturbations in Steps 1 and 2 above are local, the fitness evaluations are not computationally costly. However, the number of local solutions increases exponentially as the number of sensors and/or targets increases. When there are so many local options, the algorithm spends enormous resources in trying multitudes of solutions without moving very far through the state space. This is an example of the so-called "curse of dimensionality" that affects many types of optimization problems [2].

In particle swarm optimization (PSO), a set of candidate solutions called "particles" undergo a probabilistic update process (particles are analogous to the "worker bees" in ABC). If we use P to denote the number of particles (a fixed parameter chosen by the user), then the Markov process is applied to states of the form $\{\{\mathbf{x}_p, \mathbf{v}_p, \mathbf{b}_p\}_{p=1, \ldots P}, \mathbf{g}\}$, where $\mathbf{x}_p$ represents the pth candidate sensor configuration (a.k.a. "particle"); $\mathbf{v}_p$ is the velocity of the pth particle; $\mathbf{b}_p$ is the best configuration obtained so far by the pth particle; and $\mathbf{g}$ is the global best configuration obtained so far by the system. The Markov process proceeds according to the following dynamical equations [9]:

$$\mathbf{v}_p \leftarrow w_0 \cdot \mathbf{v}_p + w_1 \cdot \psi_1 \cdot (\mathbf{b}_p - \mathbf{x}_p) + w_2 \cdot \psi_2 \cdot (\mathbf{g}_p - \mathbf{x}_p)$$
$$\mathbf{x}_p \leftarrow \mathbf{x}_p + \mathbf{v}_p$$

PSO (like ABC) is also subject to the "curse of dimensionality", and computation times increase rapidly with system size.

3.2 Previous Physics-Based Algorithms

In addition to biology-inspired algorithms, there are also coverage algorithms that draw inspiration from physics. Many naturally-occurring optimization problems are solved by statistical physical systems. Consider for example the shaking together of rice grains in a basket so that they settle into a smaller volume. Two interrelated quantities govern the evolution of statistical systems: entropy and energy. A system's entropy reflects the number of available states. If energy is added or removed from the system, the entropy correspondingly increases or decreases. If the system undergoes a process whereby energy is progressively lost to the environment, then the entropy will decrease and the system will settle into a minimum-energy state. In our rice-in-a-basket example, the falling rice grains lose potential energy, which is released to the environment as heat.

In the context of the set covering problem, we may treat the set covers as an evolving physical system and define the system energy as the cost of the current cover. It follows that lower-energy states of the system corresponds to a more optimal solution. Thus if we can define quasi-physical dynamics which tend to decrease energy, the dynamics will move the system towards a low-cost solution.

There is however a problem with this procedure. If the system dynamics keep decreasing the energy, we will eventually reach a solution that cannot be improved, but that does not mean that it is an overall best solution. Once again referring to our rice grain example, after each "shaking" the rice settles into a stable, low-energy configuration; but the settling may continue to improve after several shakings. Each shaking amounts to temporarily adding potential energy back into the system, thereby increasing the entropy and allowing exploration of more states.

In summary, a physics-based heuristic requires the following components:

- An energy function defined on sensor configurations, equal to the cost function that we are trying to minimize;
- A (possibly randomized) dynamics on the set of possible sensor configurations that tends towards lowering the energy;
- An entropy-increasing process which is periodically applied to "shake up" the system to give a chance to settle into a better low-energy state.

Typically both the dynamics and the entropy-increasing process specify Markovian transitions on the set of possible configurations.

In the case of the disk coverage problem, it is natural to make the energy function equal to the number of sensors. As far as dynamics, some options considered by previous researchers are described below.

Zou and Chakrabarty [21] define quasi-physical dynamics by treating the covering disks as "charged particles" with pairwise interactions between the particles. Particles that are too close repel each other, while those that are too far away attract. Each particle moves according to the "net force" which is the sum of pairwise interactions between that particle and all other particles. The repulsive interaction serves to inhibit coverage by multiple disks, while the attraction keeps the disks congregated in the area of interest. This dynamics is suitable for area coverage (the problem that Zou and Chakrabarty consider), but not for point target coverage because the target locations do not influence the dynamics. It seems plausible that the inclusion of an attractive force between targets and sensors could remedy this deficiency. Note also that Zou and Chakrabarty do not include an entropy-increasing process in their algorithm, so once the configuration settles there is no possibility of further improvements. Furthermore, the number of disks in the cover is fixed, so the algorithm can't be used directly to optimize the number of disks (unless the algorithm is run multiple times with different cover sizes, which is computationally expensive).

Lin and Chiu [11] rely on a very simple dynamics: choose one sensor in the cover at random, and replace it with another sensor that is not currently in the cover according to a probabilistic rule. The rule depends on the energy difference ΔE between the original cover and the cover with replacement as follows. If the $\Delta E < 0$, then the replacement is made; but if $\Delta E > 0$, the change is made with probability $e^{-\Delta E/T}$, where $T > 0$ is the "temperature" parameter. If the temperature is low, then energy-increasing changes are rarely accepted; but if T is raised, then energy-increasing changes become more common. The value of T is varied during the course of the dynamics according to predetermined rules: these rules constitute the "annealing schedule": and the model falls in the category of "simulated annealing" models. In this case, the energy-decreasing and entropy-increasing aspects are not implemented as separate dynamics, but rather are both included within a single probabilistic dynamics with a variable parameter.

4 Nebular Algorithm: Motivation and Description

In this section we give the physical background, flowchart, and functional description of our new algorithm which we call the "nebular algorithm". Two different MATLAB code implementations of the algorithm are available on GitHub [1].

4.1 Physical Basis of the Nebular Algorithm

We base our dynamics on a physical analogy with the accretion of matter in a nebula into stars. We may think of nebular matter as initially consisting of discrete particles. As the particles approach each other they merge, and the merged object grows in mass. Each eventual star consists of matter that is drawn from the surrounding region in space. The final positions of these stars reflect the original configuring of nebular matter from which they are formed. The fewer the stars the lower the gravitational energy, so the tendency of the process is towards larger and fewer stars. The process will steadily decrease the energy, unless another source of energy is introduced. In astronomical dynamics this source of energy comes from nuclear fusion: under some conditions, stars explode, their mass is released into space, and eventually is redistributed among the other stars.

In our analogy, the initial positions of particles represent targets, while stars formed from accreting particles correspond to sensor locations. In the following we alternatively refer to particles as "sensors" or "sensor-particles", with the understanding that the final location of these particles gives the sensor location solution produced by the algorithm.

We may express the system state mathematically as a triple (X, V, σ), where X and V are $N \times 2$ matrices whose jth rows gives the position and velocity coordinates (respectively) of the jth sensor-particle, and σ is an N-vector of indices whose jth entry gives the index of the sensor to which the jth target is currently assigned. The process is a Markov process on this state space.

In accordance with this nebular analogy, we may attribute a mutual attractive inverse-square force between particles within $2r_{\text{sensor}}$ of each other. We also introduce a velocity-dependent drag force on particles that dissipates the particles' kinetic energy, leading to an energy-decreasing process as described previously. If the location and velocity of the jth particle are given by $\mathbf{x}_j$, and $\mathbf{v}_j$ respectively, then the total force on the jth particle is given by:

$$\mathbf{F}_{j \text{ total}} = \sum_{k \neq j} f\left(\mathbf{x}_j, \mathbf{x}_k\right) - c\mathbf{v}_j,$$

where c is a system parameter and

$$f\left(\mathbf{x}_j, \mathbf{x}_k\right) = \begin{cases} \left(\dfrac{\mathbf{x}_j - \mathbf{x}_k}{\|\mathbf{x}_j - \mathbf{x}_k\|^2}\right), & \text{if } \|\mathbf{x}_j - \mathbf{x}_k\| < 2r_{\text{sensor}} \\ \qquad 0 & \text{otherwise} \end{cases}$$

We then use a simple Euler integration scheme to update $\mathbf{v}_j$ and $\mathbf{x}_j$:

$$\mathbf{v}_j \leftarrow \mathbf{v}_j + \left(\frac{dt}{m}\right) \cdot \mathbf{F}_{j\,\text{total}};$$

$$\mathbf{x}_j \leftarrow \mathbf{x}_j + dt \cdot \mathbf{v}_j$$

where dt and m are system parameters.

At the beginning of the process, each target is assigned to a separate particle. During the process, if particles approach sufficiently closely and consistency requirements are satisfied, then they merge: for example, a merger of 5 particles corresponds to a sensor that covers 5 targets. These sensor-particles continue to move under the same quasi-physical dynamics described above. The sensors that remain at the end of the process correspond to the solution produced by the algorithm. Thus on the one hand the number of sensors in the cover changes dynamically during the course of the process; and on the other hand, full coverage is guaranteed.

Once we have computed new locations for each sensor, we then must check if any sensors no longer cover all their assigned targets. When this happens, the sensor is pulled back along its velocity vector until all assigned targets are within the sensor's coverage area.

In analogy with supernovas, we also introduce a time-varying explosion probability for sensors. When a sensor explodes, it is replaced with the separate sensors for all targets covered by the sensor (with random velocities imparted to them). As the dynamical process continues, the newly-created sensor-particles are absorbed into other sensors. The time-varying explosion probability we used was given by:

$$P[\text{explode}] = 0.07e^{\frac{-\,\text{mod}\,(t,\,10\log_{10}(N^*))}{\log_{10}(N^*)}}.$$

where t is the iteration number, and N^* is the (current) number of sensors. Figure 4 shows the explosion probability as a function of time for various values of N^*. The explosion probability is designed so that the system will settle between periods of high explosion rate: a longer settling time is required for systems with more sensors.

4.2 Algorithm Flowchart and Functional Description

A flowchart for the algorithm is shown in Fig. 5.

The major code functions may be briefly described as follows:

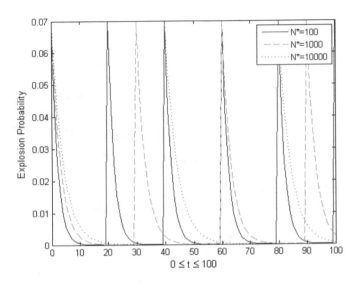

Fig. 4 Time-varying explosion probability

Initialization and preprocessing sets up data structures for storing the states, and to facilitate calculation. For this discussion, we are referring to the struct data type in the MATLAB language. These data types group data into structures with field names. Accessing data is done using the structName.fieldName notation. These structures include:

Target:

- target.loc is a $2 \times M$ matrix that is used to store the (x, y) coordinate pairs for target locations.
- target.number is a scalar value giving the total number of targets.

Sensor:

- sensor.loc is a $2 \times M$ matrix which stores (x, y) coordinate pairs for the location of sensors.
- sensor.vel is a $2 \times M$ matrix stores the sensors' velocities. The ith column of the matrix contains the ith sensor's velocity vector $[v_x, v_y]^T$.
- sensor.accel is a $2 \times M$ matrix that stores the sensors' accelerations. The ith column of the matrix contains the ith sensor's acceleration vector $[a_x, a_y]^T$.
- sensor.valid is a logical vector of length M. Each current sensor is indexed by one particular target which it contains, and sensor.valid indicates which targets are serving as indices in the current covering.
- sensor.pending is a logical vector of length M used to check whether a sensor is pending merger with another sensor. If the sensor is pending, it will change status from valid in the sensor.valid structure before the next iteration.

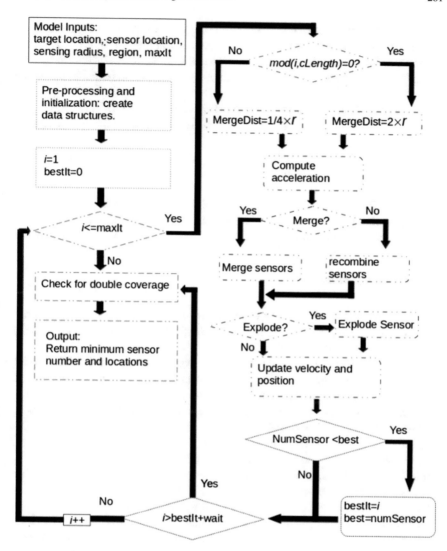

Fig. 5 Flowchart for the nebular algorithm

- sensor.mayMerge is a cell array data structure with M cells, in which the ith cell contains indexes of targets that are within $2r_{sensor}$ of sensor i. This means that sensor.mayMerge$\{i\}$ contains a list of all sensor indexes which can potentially merge with sensor i.

Sensor merge procedures

Check For Merges: Checks pairs of sensors for merge candidates. Two sensors are merge candidates if the distance between them is less than MergeDisk (a system parameter). Empirically it was found that the merge process was facilitated by periodically increasing MergeDisk to its maximum possible value (equal to the sensor diameter).

Compute Min Disk: Uses a recursive function [19] to compute the smallest disk which can contain all the targets from two sensors. From here we move on to execute the merge or recombine functions.

Execute Merge: If radius obtained from "compute min disk" is less than sensor coverage radius, merges sensors into a new sensor which contains all the targets of the previous two sensors.

Recombine: If radius obtained from "compute min disk" is greater than the sensor coverage radius, then recombines the targets into sensors in a way that preserves the greatest freedom of movement for at least one sensor by minimizing the number of targets for one of the sensors.

Sensor motion procedures

Compute Forces Between Sensors: Computes the attractive force between sensors.

Compute Velocity and Update Position: Uses computed forces to update the velocity and position using Euler's equations.

Check if New Position is in Bounds: Checks the position of the sensor relative to the targets it is supposed to cover. If any target lies outside the sensor's coverage area, the sensor location is pulled back towards the previous location.

Check for Best Solution: Compares the current number of sensors against the current best solution that has been found so far during the computation. (Initially, the best solution is taken to be one sensor per target.) If the current number is lower, replace the previous best configuration with the current configuration.

Explode sensors: Existing sensors are exploded with small probability (this probability is time-varying, as described above).

Check for Improvement in N Iterations: In this step, the program checks to see if there has been improvement after some given number of iterations. If there has been no noticeable improvement after the N iterations, the program will stop and move to the final stage, otherwise the program will go back to the beginning and go through again in hopes of improving the results. N is set during the initialization phase; it was found experimentally as a good approximation and not through rigorous analytic computation.

Check Double Coverage: Once the best configuration has been decided upon by the preceding functions, it is possible that extraneous sensors may be present. In this final step, the program checks for any unnecessary double coverage. For example, if all targets covered by a given sensor are also covered by other sensors, then the given sensor is not necessary. This routine identifies and eliminates such redundant sensors. In practice, checking the coverage does not typically yield a better solution, but experimentation revealed a need for such a function in some cases, due to occasional extraneous sensors in the algorithm's final solution.

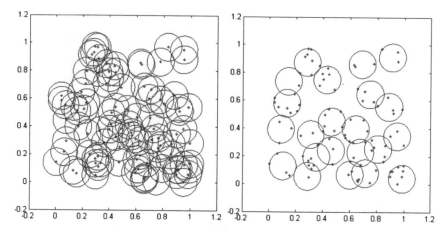

Fig. 6 Initial (left) and final (right) sensor configurations for random point coverage with 100 targets

5 Experimental Results

Figure 6 shows a typical initial and final configuration for the algorithm, for a small problem with 100 targets. As described above, the initial configuration places a sensor at every target.

Figure 7 shows the evolution of number of sensors as a function of iteration number during the Markov process. For a scenario with 170 initial uniformly random targets, we can see how the evolution is affected by explosions and variation in the MergeDisk parameter. The upward spikes correspond to spikes in the explosion probability (as shown in Fig. 4), while the sharp downward dips correspond to the temporary expansion of the MergeDisk parameter described in the "sensor merge procedures" subsection above. The asterisk plotted on the graph represents the least number of sensors obtained for all algorithm iterations.

We also wish to check how the explosions affect the final solutions. Table 1 shows final sensor numbers and execution times for several runs of the algorithm on the same configuration. We observe a 10% variation in the final sensor number: lower sensor numbers are correlated with longer execution times. Repeated runs like this can be used to tune algorithm parameters: in this case shown in Table 1, the 'wait' parameter is apparently too small and should be increased, because the system is not settling before the algorithm terminates.

6 Area Coverage

The nebular method can easily be adapted to area coverage. Figure 8 shows the final sensor configuration obtained by the nebular algorithm applied to grid points

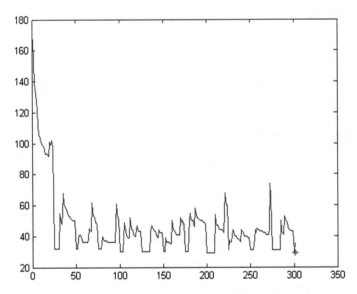

Fig. 7 Sensor number as a function of iteration number for 170 randomly-placed targets (initial number of sensors equals number of targets)

Table 1 Initial and final sensor number, execution time and iteration number for a single random configuration of 500 randomly-placed targets in a 20 × 20 square with sensors of radius 1, using a wait parameter of 100 iterations

Initial	Final	Execution time	Iteration
500	100	38.882	251
500	98	73.257	501
500	99	52.889	351
500	102	27.059	176
500	98	40.317	276
500	100	66.115	451
500	98	70.034	476
500	101	30.736	202
500	102	38.558	251
500	107	28.027	176
500	105	23.858	151
500	101	37.713	251
500	101	26.653	176
500	101	34.066	226

covering a map of Europe. The grid spacing may be empirically adjusted for optimal performance.

Another important measure of algorithm performance is the run time compared to the input size, which in this case is the number of targets. Figure 9 shows run time

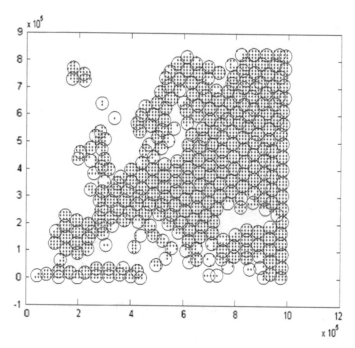

Fig. 8 Area coverage produced by nebular algorithm

(log scale) versus number of targets and target density for a range of target numbers. Run time data was obtained using a MATLAB code [1] executed on an Athlon II Quad Core processor with 8 GB of RAM. The graph shows that run time is nearly independent of target density. Also $\Delta\log_{10}$(run time)$/\Delta\log_{10}$(number of targets)≈ 1, indicating a nearly linear relationship between number of targets and run time.

7 Modifications and Extensions

The algorithm described above may be modified in various ways to apply to a variety of point coverage problems. In this section we indicate some of these possible modifications.

Higher-dimensional and ellipsoidal coverage
The nebular algorithm as described above may apply to points distributed in higher-dimensional space with minor changes. The dynamical equations are unchanged, and Welzl's algorithm works in higher dimensions as well as 2 dimensions [19]. Welzl's algorithm can also be modified to find smallest covering ellipsoids, so the nebular algorithm can easily be adapted to the case where coverage regions are ellipsoids.

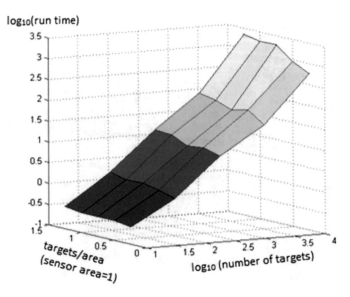

Fig. 9 Run time for nebular algorithm applied to randomly-distributed targets

Non-unicost coverage

In the algorithm we have described, all sensors have the same cost. If some sensors are more costly than others (e.g. depending on location or sensor size), we may introduce an explosion probability that is a function of cost, such that more costly sensors have a higher explosion probability.

Probabilistic (non-Boolean) coverage

So far we have assumed Boolean coverage, i.e. each target is fully covered by a single sensor. A more general problem is that of probabilistic coverage. For example, a target point that is sufficiently far from a sensor may have a probability between 0 and 1 of being detected. Generally, the detection probability is assumed to be a function of the distance between sensor and target. Various functions are cited in the literature (e.g. [10, 21], typically without justification. In the following we will derive a detection probability function based on realistic assumptions, and gauge its impact on performance and implementation of the nebular algorithm.

Sensors typically accumulate received power over a period of time. This implies that we may model both noise and signal power as Gaussian random variables. Without loss of generality, we may set the mean noise power as 1, leaving us with three parameters: signal power (at unit distance) denoted by s, signal relative standard deviation denoted by σ_s, and noise standard deviation denoted by σ_n. It is common in the wireless propagation literature to assume that the signal strength is proportional to $d^{-\alpha}$, where α is commonly referred to as the "path loss exponent", and typically takes values between 2 and 3 for electromagnetic signals depending on the environment [17]. Given these assumptions we may write:

Received signal power distribution $= (s/d^\alpha) \cdot (1 + \sigma_s Z_s)$;

Noise power distribution $= 1 + \sigma_n Z_n$,

where s/d^α is the expected signal at distance d, and Z_s and Z_n are both independent normal random variables with mean 0 and variance 1. The total received power is given by:

Received power $=$ signal power $+$ noise power $= 1 + (s/d^\alpha) + ((s/d^\alpha)^2 \cdot \sigma_s^2 + \sigma_n^2)^{1/2} Z$,

where Z is normal with mean 0 and variance 1. For detection a threshold θ must be set such that received power above θ is recognized as a detection: usually θ is set so as to control the false alarm probability to an acceptable level (which depends on the user's capabilities for dealing with false alarms). If we let ϕ be the acceptable false alarm probability, then the probability of detection is given as a function of d as

$$\text{Pr}[\text{detection of source}|\text{source} - \text{sensor distance} = d]$$
$$= \text{Pr}[Z > (\theta - 1 - s/d^\alpha) / ((s/d^\alpha)^2 \cdot \sigma_s^2 + \sigma_n^2)^{1/2}]$$
$$= \Phi((k\sigma_n - s/d^\alpha) / ((s/d^\alpha)^2 \cdot \sigma_s^2 + \sigma_n^2)^{1/2}),$$

where Φ is the standard normal cumulative distribution function, and $k = \Phi^{-1}(1-\phi)$. In the usual case, $10^{-8} < \phi < 10^{-5}$, $\sigma_s^2 \approx \sigma_n^2 \ll 1$, and the missed detection probability is between 10^{-6} and 10^{-12}.

Targets that are not sufficiently covered by a single sensor can be jointly covered by a pair of sensors. Assuming that detections by the two sensors are independent, then the probability of missed detection under joint coverage is equal to the product of the missed detection probabilities for the individual sensors. It follows that if p_d is the maximal acceptable missed detection probability, then each target must be covered by at least one sensor with missed detection probability less than $(p_d)^{1/2}$. This missed detection probability corresponds to an extended coverage radius r_{ext} which is the unique positive solution to:

$$1 - (p_d)^{1/2} = \Phi((k\sigma_n - r_{\text{ext}}^{-\alpha} s) / ((r_{\text{ext}}^{-\alpha} s \cdot \sigma_s)^2 + \sigma_n^2)^{1/2}),$$

while the single-sensor coverage radius r_{sensor} is the unique positive solution to:

$$1 - p_d = \Phi((k\sigma_n - r_{\text{sensor}}^{-\alpha} s) / ((r_{\text{sensor}}^{-\alpha} s \cdot \sigma_s)^2 + \sigma_n^2)^{1/2}).$$

Figure 10 shows $(r_{\text{ext}}/r_{\text{sensor}})^2$ for various parameter values: this ratio represents the maximum potential expansion in area coverage if joint coverage is used rather than single-sensor coverage. Potential coverage area increases range from 13% (for path loss exponent 3, false alarm probability 10^{-8} and missed detection probability 10^{-9}) up to about 25% (with path loss exponent 2, false alarm probability and missed detection probability 10^{-12}).

There are (at least) two possible ways to modify the nebular algorithm to effect joint coverage. First, the nebular algorithm may be run with coverage radius r_e, and an increased explosion probability may be assigned to sensors which cover targets that are not sufficiently jointly covered. Second, the nebular algorithm may be modified so that 2 sensors are initially assigned to each target. When a sensor merger takes place, either one or both of these sensors can be included in the merger, depending on the target's coverage by the merged sensor. If only one sensor is merged, the other sensor's dynamics may be defined so as to repel the merged sensors, and attract other sensors with which it can merge. This method has the advantage of guaranteeing full coverage—however, the implementation is more complicated.

8 Conclusions

We have provided a detailed overview of a new physics-based method for finding feasible, adequate solutions to target and area coverage problems. The algorithm is inspired by gravitational attraction in astronomical systems. Unlike previous approaches, our approach will also guarantee complete coverage for all targets in the system. The algorithm may be modified in various ways, to perform area coverage and non-Boolean, non-circular, and/or non-unicost coverage.

One of the highlights of our approach is reduced execution time. The nebular algorithm executes consistently within 2 h on systems as large as 10,000 targets; while other algorithms report require comparable runtimes on configurations with 100 times fewer targets [16]. Extensive performance tests of a modified version of the nebular algorithm programmed in C++ are given in [15].

Two MATLAB versions of the code may be downloaded from GitHub [1].

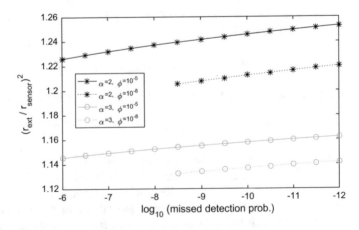

Fig. 10 Effective coverage area increase when joint coverage by two sensors is enabled

References

1. Barry J, Thron C (2017) Nebular algorithm code. Retrieved from https://github.com/jthomasb arry/nebular
2. Chávez E, Navarro G, Baeza-Yates R, Marroquín JL (2001) Searching in metric spaces. ACM Comput Surv (CSUR) 33(3):273–321
3. Deb K, Pratap A, Agarwal S, Meyarivan TAMT (2002) A fast and elitist multiobjective genetic algorithm: NSGA-II. IEEE Trans Evol Comput 6(2):182–197
4. Đurišić MP, Tafa Z, Dimić G, Milutinović V (2012) A survey of military applications of wireless sensor networks. In: Embedded Computing (MECO), 2012 Mediterranean conference, pp 196–199
5. Hochbaum DS, Maass W (1985) Approximation schemes for covering and packing problems in image processing and VLSI. J ACM (JACM) 32(1):130–136
6. Intelligent Systems Research Group, Department of Computer Engineering, Erciyes University, Turkiye (n.d.) Artificial Bee Colony (ABC) Algorithm. Retrieved 20 Oct 2017, from http://m f.erciyes.edu.tr/abc/
7. Karaboga D (2005) An idea based on honey bee swarm for numerical optimization (Vol. 200). Technical report-tr06, Erciyes university, engineering faculty, computer engineering department
8. Khedo KK, Perseedoss R, Mungur A (2010) A wireless sensor network air pollution monitoring system. Int J Wireless Mobile Netw 2(2)
9. Kulkarni RV, Venayagamoorthy GK (2011) Particle swarm optimization in wireless-sensor networks: a brief survey. IEEE Trans Syst Man Cybern Part C Appl Rev 41(2):262–267
10. Li S, Xu C, Pan W, Pan Y (2005) Sensor deployment optimization for detecting maneuvering targets. In: Information fusion, 2005 8th international conference on IEEE, vol 2, 7 p, July 2005
11. Lin FY, Chiu PL (2005) A near-optimal sensor placement algorithm to achieve complete coverage-discrimination in sensor networks. IEEE Commun Lett 9(1):43–45
12. Mini S, Udgata SK, Sabat SL (2014) Sensor deployment and scheduling for target coverage problem in wireless sensor networks. IEEE Sens J 14(3):636–644
13. Njoya AN, Abdou W, Dipanda A, Tonye E (2015). Evolutionary-based wireless sensor deployment for target coverage. In Signal-image technology and internet-based systems (SITIS), 2015 11th international conference on IEEE, pp 739–745, Nov 2015
14. Njoya AN, Abdou W, Dipanda A, Tonye E (2016) Optimization of sensor deployment using multi-objective evolutionary algorithms. J Reliab Intell Environ 2(4):209–220
15. Njoya AN, Thron C, Barry J, Abdou W, Tonye E, Konje NSL, Dipanda A (2017) Efficient scalable sensor node placement algorithm for fixed target coverage applications of wireless sensor networks. IET Wirel Sens Syst 7(2):44–54
16. Ozturk C, Karaboga D, Gorkemli B (2011) Probabilistic dynamic deployment of wireless sensor networks by artificial bee colony algorithm. Sensors 11(6):6056–6065
17. Sun S, Rappaport TS, Rangan S, Thomas TA, Ghosh A, Kovacs IZ, Rodriguez I, Koymen O, Partyka A, Jarvelainen J (2016) Propagation path loss models for 5G urban micro-and macro-cellular scenarios. In: Vehicular technology conference (VTC Spring), 2016 IEEE 83rd IEEE, pp 1–6, May 2016
18. Udgata SK, Sabat SL, Mini S (2009) Sensor deployment in irregular terrain using artificial bee colony algorithm. In: Nature and biologically inspired computing. NaBIC 2009. World congress on IEEE, pp 1309–1314, Dec 2009
19. Welzl E (1991) Smallest enclosing disks (balls and ellipsoids). In: Maurer H (ed) New results and new trends in computer science. Springer, Graz, pp 359–370
20. Xu Y, Yao X (2006) A GA approach to the optimal placement of sensors in wireless sensor networks with obstacles and preferences. In: Consumer communications and networking conference CCNC 2006, vol 1, 3rd IEEE, pp 127–131, Jan 2006

21. Zou Y, Chakrabarty K (2003) Sensor deployment and target localization based on virtual forces. In: INFOCOM 2003. Twenty-second annual joint conference of the IEEE computer and communications, vol 2. IEEE Societies, pp 1293–1303, Mar 2003

A Hybrid Bio—Inspired Algorithm for Protein Domain Problems

Manish Kumar and Hari Om

Abstract Multiple sequence alignment (MSA) is an important step for alignment, prediction and classification of protein sequences and their structural and behavior study. In this chapter, we have presented a novel approach for alignment of multiple protein sequences. Two different approaches such as the genetic algorithm and the biogeography based optimization technique were modified and merged to produce a hybrid algorithm (GA-BBO) for resolving multiple alignment problems of protein sequences. The results obtained by the proposed hybrid method are compared with some of the new alignment algorithm e.g., MO-strE, GAPAM, BBOMP,QBBOMSA and MOMSA-W etc. concluding that the new presented approach brings a remarkable accuracy when compared with existing methods over standard BALiBASE datasets.

Keywords Bioinformatics · Multiple sequence alignment · Genetic algorithm Biogeography based optimization

1 Introduction

The technique generally used to reveal and visualize the structure, feature or the evolutionary relationship between the biological molecules is known as sequence alignment. Sequences alignment technique is generally being used for drug designing as well as to improve the secondary and tertiary structure of RNA and protein sequences. The possible alignment and arrangement of biological molecules such as the Protein, DNA or RNA is known as multiple sequence alignment (MSA) [1].

M. Kumar (✉) · H. Om
Department of Computer Science and Technology, Madanapalle Institute of Science and Technology (UGC Autonomous), Post Box No: 14, Kadiri Road, Angallu (Village), Chittoor District, Madanapalle 517325, Andhra Pradesh, India
e-mail: manishkumar@cse.ism.ac.in

H. Om
e-mail: hariom4india@gmail.com

© Springer International Publishing AG, part of Springer Nature 2019
S. K. Shandilya et al. (eds.), *Advances in Nature-Inspired Computing and Applications*, EAI/Springer Innovations in Communication and Computing, https://doi.org/10.1007/978-3-319-96451-5_13

Over the last 25 years, the multiple sequence alignment (MSA) problem has gained a wide attention across the world because of its application in areas such as homology searches, genomic annotation, protein structure prediction, gene regulation networks, or functional genomics [2]. The main motive of multiple sequence alignment is to align sequences, based on their evolutionary relationship. Therefore, the development of an efficient and reliable MSA program is required.

While going through the MSA analysis, we have encountered a term know as sequence similarity which is also known a subset of sequence analysis or the sequence alignment. Multiple Sequence Alignment (MSA) belongs to a class of hard optimization problem and is considered to be the most challenging tasks in Bioinformatics [3]. In the subsequent analyses of protein families and their sequences, multiple sequence alignment acts as a pre-processing tool. In this study, we have considered the problems associated with the MSA of protein sequences. While going through the literature studies and the experimental analysis, we have seen that the protein databases contain many unstructured protein sequences [4]. These unstructured sequences often creates problem in proper aligning of sequences. Furthermore with increase of research in the area of sequencing of proteins, it is not worth to perform a detailed experimental analysis on protein sequences. As the process of doing sequencing experiments are very costly and time consuming. Due to which the structure of most of the currently available protein sequences remains unknown [5].

In order to know the structural and behavior nature of protein sequences and to overcome the challenges faced by protein in their databases, we have made possible efforts to resolve the problems related to alignment of multiple protein sequences through the hybrid combination of genetic algorithm and biogeography based optimization techniques (GA-BBO) [6].

In our approach, we have tested the proposed hybrid algorithm (GA-BBO) with 20 test cases (protein sequences from standard BALiBASE dataset) and compared the performances with some of the new and existing well know methods such as the MO-strE [7], GAPAM [8], BBOMP [9], QBBOMSA [10] and MOMSA-W [11]. After comparing with these methods and by observing the test results, we have concluded that the proposed approach is far better in aligning the protein sequences and for resolving MSA problem. However, in some test cases methods such as MO-strE and QBBOMSA has performed far better than us. Furthermore, in order to test the superiority of the proposed GA-BBO approach we have gone through a Wilcoxon signed rank test [12]. This is the hypothesis based test method generally being used for comparing different methods over protein sequences. This test also governed the superiority of the proposed hybrid method with respect to others.

The rest of the chapter is organized as follow. The next section describes the relevant literature study and related works required to understand and handle MSA problem. Section 3, describes the hybrid GA-BBO approach. Section 4, provides a detail experimental results over standard datasets. Finally, the concluding section presents our final consideration.

2 Related Work

In this section, we present the related studies based on soft computing approaches, existing methods and problem associated with multiple protein sequence alignments.

In general for MSA problems, local [13] or the global scheme are most preferred. These methods includes approaches like the progressive [14], iterative [15] and the classical. Global approach is well presented by the Needleman-Wunsch algorithm [16] and the Smith–Waterman algorithm [17], where as the local approach is defined in the dynamic method approach [18]. In the literature study, we have seen that in 1987 authors named Feng and Doolittle [19], applied the progressive method of Needleman and Wunsch [20] for predicting the relationship between sequences. But, fails to get the optimal result as the algorithm often suffers from the problem of early convergence or the local optima. In order to avoid such type of problems, it has been suggested in various literature studies [15] to use either iterative procedure or the stochastic approach.

Methods listed in [12, 16–25] have suggested that the existing approaches are insufficient in giving accurate alignment score with all the datasets. It has also been suggested to use genetic computation technique [26], because with this approach we can easily overcome the early convergence problem and can bring some important and positive improvements in getting optimal alignment results. Considering the above facts, strategies like HMM [27], genetic algorithm (GA) [26], and many other iterative based methods were developed and implemented for alignment problems [28].

Because of the drawbacks mentioned in [12, 16–25], it has become difficult for methods to construct a reliable alignment with optimal results. Progressive method [8, 29] is considered to be very fast and efficient for the alignments related problems. But, one common problem faced by progressive method is the alignment error. Whenever an error occurs in any alignment and if it somehow gets propagated to all the alignment then it becomes very hard to be removed. On the other hand, iterative methods [9] are slow in response and are used at place where quality of alignment is of prime importance and not the time taken to measure it.

The genetic computation or the evolutionary computation [26], which works on natural selection scheme is generally being used for implementing iterative methods. The iterative function is governed by using a fitness function and because of this feature it's being used by a number of scientists around the world.

Considering all the above discussions, we have presented a hybrid implementation of Genetic Algorithm (GA) [26] as well as of Biogeography Based Optimization (BBO) [30] approach to solve multiple protein sequence alignment problem. We have considered GA and BBO for our experimental study because GA does not require any source of algorithm to solve a given problem. And as discussed in earlier paragraphs, the only and foremost requirement for GA is the fitness function. Furthermore, GA does not suffer from any type of early or premature convergence and hence can easily be utilized for aligning multiple sequences of higher lengths. The advantage that BBO offers over other algorithms such as the GA and PSO is that, it is easy to control and handle and requires very less amount of parameters for its operation [31]. In order to improve the exploration ability of BBO, we have manipulated and distributed the multi-domain populations quite evenly across the habitats.

The study of species distribution over a biological ecosystem is known as Bio-geography [30]. The species or the organism in a biological community depends on isolation, habitat area, latitude, and elevation. The branch of biogeography that studies the distribution of plants is known as Phytogeography and which deals with animals is known as Zoogeography. Habitat or the island is defined as a place that suits the species or the solution set based on some features of the habitat or the island. The feature of the habitat may vary from problem to problem. Depending on these features, the immigration and emigration of species takes place. The immigration and emigration rate is the rate by which a species can leave or join a habitat based on certain features [31].

It has been suggested in many literature studies [32], that in coming years protein will play a crucial role in the development of drugs and medicines. Therefore, con-sidering the importance of protein in near future we have considered the alignment of multiple protein sequences for our experimental study. Recent developments in protein suggest that, there are various tools and techniques that have been developed for protein sequence analysis. One can expect a better and improved performance of alignment of protein sequences by proper utilizing the phylogenetic relation-ships among sequences [33]. Furthermore, there are many factors which make the alignment of protein sequences a challenging task. While going through the litera-ture studies [34–39] we have seen many challenges that exists in aligning protein sequences. The challenges or the problems for proteins sequences generally lie with their databases. It has been seen that, most of the available protein database contains huge alignment errors. Also, these databases are mostly seen as misaligned or less aligned regions within the sequences.

Methods like SAGA [28], MSA-GA [40] and RBT-GA [41] are based on genetic computation. Notredame and Higgins developed SAGA method in which they have implemented sequence alignment with genetic algorithm. This is considered as one of the finest work in the field of sequence alignment. This work involves different type of genetic operators applied to a group of population in order to improve the fitness of the alignments. RBT is also seen as a GA based iterative approach which is based on Dynamic Programming (DP) table. Methods which are evolutionary in nature provides an alternative approach to heuristic MSA. Furthermore, iterative or evolutionary approaches provides accurate alignment quality at the expense of runtime. Techniques such as the simulated annealing and evolutionary computation are seen as the successful measure to encounter problems like MSA.

Later, Karadimitriou and Kraft [42] introduced a program called MSA (not the multiple sequence alignment). In their approach, the alignments were first considered to be without gap. The chromosomes are designed only to encode the number of gaps. But, this approach is not considered to be meaningful because the alignments pro-duced by this method are lack of biological importance in real life. In the bit matrix, the positions of 1 represents the gaps and 0 corresponds to a nucleotide or residue. The concept of such a representation is much similar to that mentioned in [43]. An encoding scheme of alignments as bit matrix was further introduced by Isokawa et al. [44] and Wayama et al. [45]. In this scheme, 0 represents nucleotides and 1 corresponds to gap. But, the alignment produced by this method is of poor quality and was not considered for further research. Authors in [46], used the combination of GA-ACO to overcome the problems related to premature convergence. In this approach, GA is made to run with a randomly generated initial population and later, ant colony optimization (ACO) technique was applied on the alignments produced by the GA approach. But, as stated by the authors [46] this combination of GA-ACO is not efficient enough to overcome the problem of premature convergence. In order to avoid premature or early convergence problem the concept of reserve selection was introduced in [47]. The reserve selection is a new methodology which governs the selection of poor quality solutions for the coming generations. The authors in [47] has given the concept that even the poor quality solution may contains some potential building block which can contribute to the future evolutions and can easily tackle premature convergence problem.

We have also gone through the study of Particle Swarm Optimization (PSO) for handling MSA problem [48]. In [48], the PSO algorithm was merged with biological sequences with some local search operators for getting a better and fit solution sets. In [49] it has been demonstrated that, how CPSO attempts to avoid the speed convergence problem of PSO.

As we have considered some datasets of small lengths, therefore it is important to review methods which has been used for aligning small length sequences. CLUSTALW [50] is a progressive based method and can be utilized for small range of problems. Because of its accuracy, this method is mostly used around the world. It has been concluded from the literature that the two most widely accepted and used method for MSA are CLUSTAL X [34] and CLUSTAL W. These methods can easily be used for handling sequences of medium lengths as well as for small lengths. Another method which uses a local alignment technique is the DIANLIGN [38]. This method uses a segment to segment comparison rather than residue to residue comparison for construction of alignments. PIMA [51] is also a local dynamic programming approach to align the most conserved motifs. HMMT [27] is based on simulated annealing method whereas, the PRRP [28] is a robust and a global approach based on both progressive and iterative approach.

Here, in this chapter we have compared our proposed approach with some of the newly developed methods for MSA problem. The first method which we encountered from the literature study is the GAPAM [8, 29]. GAPAM is a progressive method which is based on production of initial population with the help of guide tree. This method is influenced by Muscle method [23] and are considered to be very useful for problems like MSA. BBO is a new evolutionary approach to find accurate solutions. For research based on BBO, a special attention is required for framing the problem (MSA) in accordance with this new technique. Comparing and using BBO technique for MSA problem is the need of the hour. Therefore, we have used BBO technique and compared it with some of the other new and existing BBO methods. Some biogeography based evolutionary algorithms which are considered in this chapter are BBOMP [9] and QBBO [10]. In BBOMP, the authors introduced the concept of changing and manipulating the initial populations for increasing the exploration capability of BBO. The improvement in exploring the ability of BBO is further introduced by QBBO. In QBBO, a new method using hill climbing algorithm is proposed to generate initial alignments. Then, with the help of a quantum representation technique the population are so initialized that the new generations will have a good balance between exploration and exploitation mechanisms. This method is bit different in initializing the population than other methods discussed earlier in the paragraph.

MOMSA [11] is another new method for MSA and introduces a new concept for population initialization with the help of a new mutation operator. This algorithm has shown some innovative improvement in overall score of the alignment when compared to GAPAM. Another new method developed in the recent years is MO-strE [7]. This method is generally based on three features mainly Structural information, non-gaps percentage and totally conserved columns. In our comparison analysis, MOMSA [11] method has show better results in most of the test cases as compared to other methods.

In our test analysis, we have encountered sequences of varying lengths (see Table 1). Some sequences are of larger length whereas some sequences are of medium and shorter lengths. To handle sequences of higher lengths we have gone through a detailed study of T- Coffee [52] and MAFFT [25] method. Both these methods are quick and can easily handle sequences of higher length. With all the above discussions, we conclude our literature survey and the section that follows will detail the presented approach to tackle MSA problem.

3   Proposed Approach

This section details about the presented GA-BBO approach.

3.1   Chromosomes Presentation and Population Initialization

The representation of initial population is described as follows:

In our (GA-BBO) scheme of population initialization, the population is generated at first. Then from the population of sequences, a sequence of largest length is recognized and based on this length gaps are inserted in all other sequences. Gaps are placed in such a way that the total size of the gap in particular sequences does not exceed to 15% of the total size of the largest length (recognized) sequence. Here, we have also tested the initialization of population with 25, 30 and 45% of gaps but were unable to find an optimal result which we found with 15% gap insertion [8]. The main reason behind considering low gap percentage is the time factor. Insertion of more number of gaps will take more time for initialization and will definitely affect the quality of alignments. Furthermore, insertion of large number of gaps will

Table 1 Experimental results with Ref. 1 to 5 of BALiBASE dataset

Ref. (1 to 5)	Methods									
	SN	ASL	Datasets	GAPAM	MOMSA-W	MO-SAStrE	BBOMP	QBBOMSA	GA-BBO
Ref. 1	3	374	1ped	0.498	0.738	0.716	0.746	0.758	**0.793**
	4	220	1uky	0.402	0.514	0.403	0.53	0.574	**0.878**
	4	474	2myr	0.317	0.437	0.544	0.291	0.427	**0.746**
	5	276	Kinase	0.487	0.849	0.808	0.618	0.703	**0.894**
Ref. 2	23	473	1lvl	0.781	**0.946**	0.825	n/a	n/a	0.850
	18	511	1pamA	0.860	**0.958**	0.913	0.946	0.824	0.856
	15	60	1ubi	0.767	0.921	0.911	n/a	n/a	**0.988**
	20	106	1wit	0.851	**0.920**	0.917	n/a	n/a	0.788
	16	294	2pia	0.828	**0.973**	0.879	0.827	0.897	0.912
	15	237	3grs	0.746	0.849	0.864	n/a	n/a	**0.877**
Ref. 3	23	287	Kinase	0.828	0.891	0.918	0.701	0.795	**0.929**
	19	427	4enl	0.800	0.815	0.862	n/a	n/a	**0.924**
	28	396	1ajsA	0.311	0.542	0.586	n/a	n/a	**0.655**
	22	97	1ubi	0.386	0.660	0.590	n/a	n/a	**0.768**
	19	511	1pamA	0.835	0.639	n/a	0.844	**0.856**	0.820
	24	220	1uky	0.468	0.923	0.673	n/a	n/a	**0.944**
Ref. 4	18	468	Kinase2	0.384	**1.000**	0.865	0.296	0.629	0.943
	6	848	1dynA	0.033	**0.800**	n/a	0.04	0.237	0.765
Ref. 5	8	328	2cba	0.852	**0.987**	n/a	0.798	0.871	0.859
	15	301	S51	0.835	0.981	n/a	0.915	0.869	**0.983**
Average Score				0.613	0.817	0.767	0.629	0.703	**0.858**

SN: Sequence Number
ASL.: Average Sequence Length

cost high gap penalty and will reduce the alignment quality on longer runs. After the initialization process, the obtained alignment is subjected for further operation as defined hereunder.

3.2 Score Evaluation

In this section, a formal definition of the sum-of-pairs of multiple sequence alignment is introduced which is used as a tool to calculate fitness.

As we have seen in Sect. 2 [32], all the genes and proteins having similar sequences characteristics will perform same function. Proteins are known to be the building blocks for the cells while the DNA keeps all genetic related information. A protein can be defined as the linear chain of amino acid. There are total 20 amino acids, denoted by G, H, V, L, K, M, F, P, S, T, W, Y A, R, N, D, C, Q, E, and I. Whereas, DNA are the chains of nucleotides. There are four different types of nucleotides, denoted by A, T, G, C. Therefore, we can simply represent both proteins and DNA molecules as strings of letters [33] (as represented above).

In our experimental analysis, we have calculated the sum of pair score for each of the protein sequences based on their fitness values. We have adopted a scoring matrix SM (a, b) to score an alignment between any two characters x and y [53].

In the experiment analysis, the penalty for the gaps is taken as:

$$J = \{G, H, V, L, K, M, F, P, S, T, W, Y A, R, N, D, C, Q, E, I\}$$

$$M(a, b) = \begin{cases} 7 \text{ if } a \in J \text{ and } b = - \\ 5 \text{ if } a = - \text{ and } b \in J \\ 2 \text{ if } a = - \text{ and } b = - \end{cases} \tag{1}$$

Equation (1) suggests that

If $a \in J$ and $b = -$ then the gap penalty is taken as 7.

If $a = -$ and $b \in J$ then the gap penalty is taken as 5.

And if, $a = -$ and $b = -$ then the gap penalty will be taken as 2.

If $a \in J$ and $b \in J$ then use Point Accepted Mutation (PAM) 120 matrix. In case of match occurs refer to PAM 120 matrix available online [54].

Here, the gap penalty mentioned in Eq. 1 is user defined and will remain same for a complete set of experiment. In our case, the penalty for gap opening and extension is not same.

3.3 Fitness Evaluation

To compare different alignments, a fitness function is defined based on the number of matching symbols and the number and size of gaps. For scoring purpose, PAM 120 Matrix [53] has been used to calculate score between different alignments.

In the experiment the fitness is calculated as:

$$\text{Fitness} = \sum_{i=1}^{n-1} \sum_{j=i+1}^{n} scoring\ matrix\left(l_i, l_j\right) \tag{2}$$

where,

$n = $ number of sequences,
$l_i = $ first sequence,
$l_j = $ second sequence.

The score for each column in an alignment is calculated by adding the score of each pair of symbols. After that, the overall alignment score is calculated considering Eqs. 1 and 2.

3.4 Crossover Operation

The crossover model followed in this experiment is hereunder:

In the proposed scheme, we first sorted the individuals (parents) according to their fitness scores and then among them the two fittest individuals were selected for crossover operation. The selection of parents is done in the mating pool.

As we all know that, the crossover operation is performed in between two strings or sequences. Therefore, as demonstrated in Fig. 1 we have selected two parents (sequences) and from a random selected point we made the crossover operation by simply cutting and swapping the sequences. The probability for crossover is 0.6%.

Figure 1 represents the operation performed by us for crossover between two parents. In the presented approach, we first randomly chosen a column in the parent alignment and define a cut point there. Then, by mutually swapping and replacing from the cut point we got two new offspring's (Child 1 and Child 2).

We have also tested multiple point crossover operation for our research. The results which we got using multiple point crossover is of least quality in comparison to what we got with the single point crossover operator. The selection of genetic operators can vary with respect to selection of parameters, size of datasets and representation of child population etc. In our case, result with single point crossover is optimal and differs a lot when compared with multiple point crossover operators. In multiple point crossover operator, the Childs (alignments) so produced consists of large amount of gaps (-) and therefore the quality of Childs keeps on decreasing when operated for

```
DLKFN F GDEV- LLDTR        HTKLDS ERCV FTSDEV
DETNN- E E −V-GGES L       S TK−D- ETDV - - TJH E
DKTNG− E E −AHLGEYG        R −V - - H TDDRTYCDDS
ESA AGT REEYG FCRTY        RTGH VB BBG - J − J F - S
```
         1st Parent   2nd Parent

```
HTKLDS GDEV- LLDTR         DLKFN F ERCV FT SDEV
S TK−D− E E − V-GGES L     DETNN- ET DV - - TJ HE
R −V - - H E E −AHLGEYG    DKTNG   TDD RTYCDDS
RTGH VBR EEYG FCRTY        ESA AGT BBG - J − J F - S
```
         1st Child   2nd Child

Fig. 1 Crossover operation

100 iterations. Due to all the above reasons, we opted single point crossover operator in place multiple point crossover operator.

3.5 Habitat

After the crossover, we move to biogeography operation. In this operation the resultant offspring produced by the crossover is being distributed in between four habitats named as H1, H2, H3 and H4 in the ratio of 40, 30, 20 and 10% of the total size of the offspring. The distribution is based on suitability index variable (SIV). SIV for the habitats described here is fitness score. The distribution is made according to the score of the particular column of the sequences (Childs). The first 40% columns having highest column score will move to habitat H1, the next 30% column having fitness less than habitat H1 will be put in H2 and the remaining columns are distributed among H3 and H4 according to the column scores. Here the habitat suitability index (HSI) is depends upon the fitness scores of the columns. After being evenly distributed, the chromosomes of the offspring are immigrated (λ) and emigrated (μ) in order to further improve the offspring in each of the habitats. The immigration and emigration rate for our experimental study is set as 12 and 5%. Immigration and emigration allows the movement and distribution of species (in our case chromosomes) in between the island or the habitat in order to maintain equilibrium and to modify a habitat by providing some species of higher fitness score (good species).

As we have seen in many evolutionary methods, that a global recombination strategy is used for producing a new offspring but in BBO migration strategy is used to produce or change the solution. Furthermore, evolutionary strategies are reproductive process while the BBO are the adaptive process. In our approach, the resultant alignment so produced after the migration operation is further subjected to calculate fitness score. And based on this the quality of alignments are evaluated. However, an alignment can further be improved, if it is subjected for mutation operation. We have made our alignment to go through mutation operators for further improvement in its quality [30].

The phenomena of habitat modification $\Omega(\lambda, \mu) : H^n \rightarrow H$ is known as a probabilistic operator that helps in adjusting the habitat H based on the nature of ecosystem H^n. The probability of a habitat which being modified, is totally depends upon immigration rate λ. The probability by which a habitat or island is being modified is depends upon μ.

The modification of habitat is detailed as:
Select H_i with probability $\propto \lambda_i$
If H_i is selected for operation
For j = 1 to n

   Select H_j with probability $\propto \mu_i$

If H_j is selected

   Randomly select an SIV σ from H_j
   Replace a random SIV in H_i with σ

End
End
End

3.6 Mutation Operation

After migration operation [31], our alignments are further subjected for mutation (permutation) operation. Mutation will help our hybrid algorithm (GA-BBO) to get rid of premature convergence and provide an optimal result.

The rate of mutation value is defined as:

$$m_i = m_{max}\left\{1 - \frac{P_i}{P_{max}}\right\}$$

where, m_{max} defines the mutation rate which usually ranges from 0 to 1(in our case mutation probability is 0.01%). P_i is the species count probability of habitat I whereas the P_{max} is the maximum of the species count for the habitat. (P_{max} depends on the size of the datasets considered in this chapter).

4 Test Set up

This section gives an overview of the parameters and the systems components used for the experiment.

4.1 Parameters Requirement for the Test Analysis

In the presented approach, the population is initially generated at first and the maximum number of generations (iteration) was set to 100 with a crossover probability of 0.6% and mutation rate of 0.01%. For the habitats, the immigration rate and the emigration rate is set to 12 and 5% respectively. The scoring matrix used for the experiment is PAM 120 for each Protein sequences. Here, for each of the given population size the algorithm will run for 100 iterations and after being operated with different operators it will produce a best child population (MSA).

4.2 System Requirement

The main objective of this research work is to observe the role of proposed hybrid bio-inspired algorithm (GA-BBO) for solving MSA problem of protein sequences in terms of quality of the sequence so aligned. Here, a sequence is judge to be the best or of high quality if it attains the highest fitness score. The experiments for the proposed approach has been performed using the hybrid combination of GA and BBO with JAVA programming on an Intel Core 2 Duo processor having 2.53 GHz CPU with 2 GB RAM.

4.3 Results and Evaluation

In this section, the methodology adopted for the experimental analysis is detailed. Moreover, results obtained on various datasets are presented and explained.

In order to evaluate our proposed hybrid algorithm (GA-BBO) with respect to other well known existing methods, we carried out the experiments with different datasets (ref. 1, ref. 2, ref. 3, ref. 4 and ref. 5) of different lengths from the BAliBASE version 2.0 standard database [55]. Due to the good performance of these datasets with other related and existing algorithms, forced us to choose them for our experimental analysis. The performance related information's were gained by referring to various literature studies which are listed in [28, 30, 38–40].

The proposed hybrid algorithm (GA-BBO) was executed for 100 independent runs (iterations) for 20 datasets (some of all datasets in Table 1). The fitness score is directly proportional to the level of similarity among the residue in the sequences and therefore, the scores of the alignments may be positive or negative. If the residues among two sequences are same, then a small numbers of gaps ("-") are needed to make the sequences aligned properly. On the other hand, if a number residues within the sequences are dissimilar then a large number of gaps are required for successful alignment of sequences. Gaps are generally allowed and permitted within the sequences so that identical or similar characters are aligned in successive columns. A user defined mechanism is also used for penalizing these gaps.

To know the reliability and quality of solutions obtained by the GA-BBO algorithm, we have considered a BAliscore, which is known to be an open source program of the BAliBASE benchmark. A BAliBASE dataset is known to be the best and standard dataset for alignment of biological sequences. It consists of 218 different lengths protein sequences. In BAliBASE a SP score (sum of pair) and TC Score (total column) is used to evaluate the quality of the obtained alignment.

In general, BAliBASE scores a solution (multiple sequence alignment) in the range between 0.0 and 1.0. A score of 1.0 implies that the solution so produced by the presented approach is very much similar to that of the reference alignment. Method such as the MOMSA-W is able to score a solution with a score 1 (refer Table 1). However, with the proposed methodology we were unable to construct an alignment whose score is equals to 1 (see Table 1). A score of 0 tells that nothing matches with the reference alignment, whereas a score in between 0 and 1 indicates that some part matches with the reference alignment. The scores which are nearer to 1, gives a better alignment result for a given dataset. A comparative test analysis of score with different methods is being compiled in Table 1. By visualizing Table 1, one can conclude that the hybrid method (GA-BBO) presented by us is much more efficient and reliable than other method it term of achieving high quality solutions.

In order to evaluate the overall performance of the presented approach, the average score of all test cases from ref. 1 to 5 were also calculated (bottom of Table 1). By seeing the outcome of average score solutions, one can concluded that the presented approach is much better (it terms of achieving high fitness score) in respect to other methods considered in this chapter. The scores are measured with reference to standard BAliBASE dataset [55]. Furthermore, the bold faced data's indicates the superiority of the presented approach among other methods.

We have tested GA-BBO approach with 20 datasets of BAliBASE and compared it with existing algorithms such as the MO-strE, GAPAM, BBOMP, QBBOMSA and MOMSA-W. Table 1 indicates that the presented approach is well enough to successfully align multiple protein sequences when tested over ref. 1 to 5 of BAliBASE datasets. Methods like the MOMSA-W and QBBOMSA performed better than GA-BBO method for eight test cases. But, in overall average score calculation the GA-BBO method outperformed all the other methods considered in this chapter.

MOMSA-W method [11] uses a progressive alignment technique and therefore able to produce better alignment results over different datasets. As presented and described in [11], the mutation operator used by MOMSA-W is efficient enough to produce better scores. Furthermore, performance of methods generally depends upon various factors. Such factors are the selection of parameters, experimental approach,

Table 2 Wilcoxon signed rank test results for the proposed hybrid (GA-BBO) method

Algorithms	Teat comparison with the proposed hybrid method			
	W+	W−	Whether proposed method is better or not	Hypothesis test decision
GAPAM	29	5	✓	Rejected
MOMSA-W	18	3	×	**Retained**
MO-SAStrE	32	7	✓	Rejected
BBOMP	29	1	✓	Rejected
QBBOMSA	22	5	✓	Rejected

evolutionary approach and the time taken to complete the experimental analysis. Because of all these factors, the said methods (MOMSA-W and QBBOMSA) have performed betters than the presented hybrid approach (GA-BBO) for few test cases.

In Table 1, the notation n/a represents the unavailability of results in literatures for the particular datasets/methods. For example, in ref. 2 for methods like BBOMP and QBBOMSA we don't have any related results for dataset 1lvl. Furthermore, just like above methods unavailability of results is also seen with MO-SAStrE. This scarcity of data is almost for all the datasets ranging from ref. 2 to 5. Because of these factors, we were unable to make a comparative result analysis with all methods listed in Table 1 (Fig. 2).

Figures 3, 4, 5, 6 and 7 represents the bar graph comparison of different methods listed in Table 1. By refereeing to these figures, we can easily analyze the performance of different methods over different datasets. Figure 8 gives the clear indication about the superiority of the presented approach (GA-BBO) with respect to other methods discussed in this chapter.

4.4 Statistical Evaluation of the GA-BBO Approach

In order to study and analyze the difference between any two existing algorithms, a test based on statistical performance has been done. A well known method for this purpose is the Wilcoxon signed rank test [12]. We have opted Wilcoxon signed rank method for our test analysis because this method has a unique feature that it does not requires normal distribution of population. Furthermore, this statistical method helps us to know the difference between paired scores when it becomes difficult to make or predict any assumption required by the paired-samples t test. The result obtained with our proposed method is presented in Table 2, where W (= W+ or W−) is the sum of ranks based on the absolute value of the difference between two test variables. The sign of positive(+) and negative (−) are used to classify cases into one of two samples i.e. either differences above zero (positive rank W+) or below zero (negative rank W−). For a null hypothesis, it is assumed that there is no significant

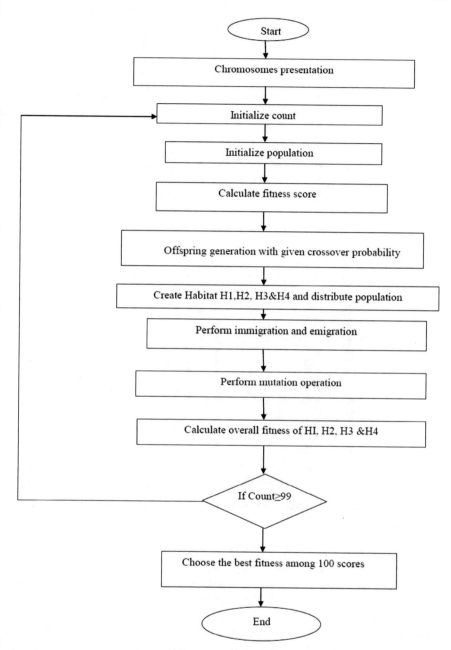

Fig. 2 Diagram for the GA-BBO approach

difference between two samples. However, if a null hypothesis is being rejected by
the hypothesis test, then we can assume that there is a significant difference, but if it

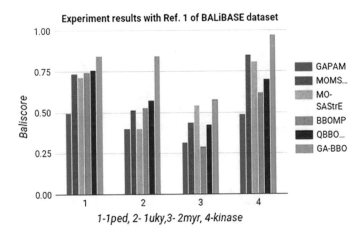

Fig. 3 Bar graph comparison result of scores between proposed and other methods over ref. 1

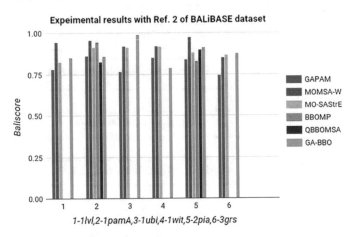

Fig. 4 Bar graph comparison result of scores between proposed and other methods over ref. 2

retains the null hypothesis then it means that there is no significant difference. On the basis of test results/rankings, we assumed two signs $\sqrt{}$ and × to make a comparison between any two algorithms. Sign "$\sqrt{}$" indicates that the proposed hybrid (GA-BBO) algorithm is much more efficient than the second, and "×" implies that there is no significant difference between the two algorithms. The total number of test case (datasets) are taken as 20 and the significance level is $\alpha = 0.05$. Hence, if $P < \alpha$, the null hypothesis is rejected, else the null hypothesis is retained. All tests were conducted with reference to the corresponding BAliscore (on 20 datasets) so as to get optimal results in comparison to other presented methods. To analyze the effectiveness of our hybrid (GA-BBO) method, we have taken 20 datasets from BALiBASE dataset version 2.0 and made a comparative analysis with respect to other methods such as the GAPAM, MOMSA-W, MO-SAStrE, BBOMP and QBBOMSA. The obtained

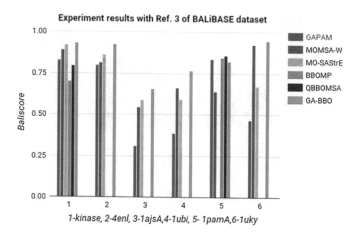

Fig. 5 Bar graph comparison result of scores between proposed and other methods over ref. 3

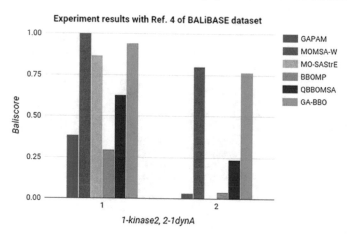

Fig. 6 Bar graph comparison result of scores between proposed and other methods over ref. 4

results gives a clear indication that the proposed method is far more better than other methods mentioned in this chapter. The statistical non parametric test also tells that the proposed method is better in achieving high alignment quality when subjected to different datasets. It can be concluded that the presented method is well enough to deal with the problem of aligning multiple protein sequences.

5 Conclusion

In this chapter, we have presented a new bio-inspired hybrid method (GA-BBO) with the combination of genetic computation and biogeography technique to solve

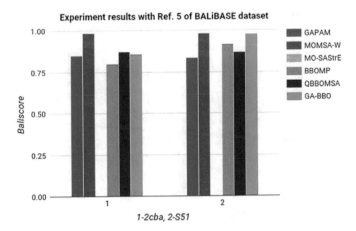

Fig. 7 Bar graph comparison result of scores between proposed and other methods over ref. 5

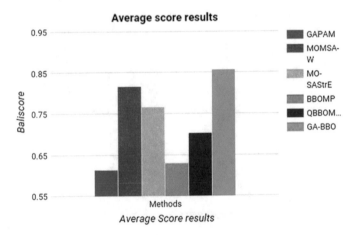

Fig. 8 Comparative results of average scores

the MSA problem of protein sequences with their structural and behavior study. In this approach, we have enhanced the genetic operators as well as the migration and mutation operators of BBO. The aim of the study reported in this chapter, is to propose an efficient algorithm that can easily handle MSA problem. When compared to other methods mentioned in [7–11] over standard datasets, the proposed method solutions outperformed other methods in respect to alignment quality. All in all, we can conclude that our presented hybrid approach have several qualities that not only enhance but also improve the alignment quality in various respect and can solve MSA problem quite efficiently. One common drawback that we faced with our proposed approach is that when we queries on large databases, the presented algorithm take a considerable amount of time to complete.

References

1. Hamidi S, Naghibzadeh M, Sadri J (2013) Protein multiple sequence alignment based on secondary structure similarity. In: International conference on advances in computing, communications and informatics, 1224–1229
2. Auyeung A, Melcher U (2005) Evaluations of protein sequence alignments using structural information. In: International conference on information technology: coding and computing, vol 2, 748–749
3. Pop M, Salzberg SL (2008) Bioinformatics challenges of new sequencing technology. Trends Gene. 24:142–149
4. Yonghua H, Bin M, Kaizhong Z (2004) SPIDER: software for protein identification from sequence tags with de novo sequencing error. In: Proceedings of computational systems bioinformatics conference, pp 206–215
5. Wen-W C, Tan-H T (1996) Statistical characterization of error sequences and its applications to error control. In: Proceedings of digital signal processing applications, vol 2, pp 625–629
6. Aniba MR, Poch O, Thompson JD (2010) Issues in bioinformatics benchmarking: the case study of multiple sequence alignment. Nucleic Acids Res 38:7353–7363
7. Ortuño FM et al (2013) Optimizing multiple sequence alignments using a genetic algorithm based on three objectives: structural information, nongaps percentage and totally conserved columns. Bioinformatics 29(17):2112–2121
8. Naznin F, Sarker R, Essam D (2012) Progressive alignment method using genetic algorithm for multiple sequence alignment. IEEE Trans Evol Comput 16(5):615–631
9. Zemali E, Boukra A (2015) Resolving the multiple sequence alignment problem using biogeography based optimization with multiple populations. J Bioinform Comput Biol 13:1–23
10. Zemali EA, Boukra A (2016) A new hybrid bio-inspired approach to resolve the multiple sequence alignment problem. In: 2016 international conference on control, decision and information technologies (CoDIT), St. Julian's, 108–113
11. Zhu H, He Z, Jia Y (2016) A novel approach to multiple sequence alignment using multiobjective evolutionary algorithm based on decomposition. IEEE J Biomed Health Inform 20(2):717–727
12. Corder GW, Foreman DI (2009) Nonparametric statistics for non statisticians: a step-by-step approach. New York
13. Changjin H, Tewfik AH (2009) Heuristic reusable dynamic programming: efficient updates of local sequence alignment. IEEE/ACM Trans Comput Biol Bioinf 6(4):570–582
14. Kupis P, Mandziuk J (2007) Evolutionary-progressive method for multiple sequence alignment. In: IEEE symposium on computational intelligence and bioinformatics and computational biology, 291–297
15. Mohsen B, Balaji P, Devavrat S, Mayank S (2007) Iterative scheduling algorithms. In: IEEE INFOCOM proceedings
16. Needleman SB, Wunsch CD (1970) A general method applicable to the search for similarities in the amino acid sequence of two proteins. J Mol Biol 48(3):443–453
17. Smith TF, Waterman MS (1981) Identification of common molecular subsequences. J Mol Biol 147(1):195–197
18. Zhimin Z h, Zhong w C (2013) Dynamic programming for protein sequence alignment. Int J BioSci Biotechnol 5(2)
19. Feng DF, Dolittle RF (1987) Progressive sequence alignment as a prerequisite to correct phylogenetic trees. J Mol Evol 25(4):351–360
20. Gotoh O (1982) An improved algorithm for matching biological sequences. J Mol Biol 162(3):705–708
21. Nguyen KD, Yi P (2011) An improved scoring method for protein residue conservation and multiple sequence alignment. IEEE Trans Nanobiosci 10(4):275–285
22. Pearson WR (2000) Flexible sequence similarity searching with the FASTA3 program package. Methods Mol Biol 132:185–219

23. Edgar RC (2004) MUSCLE: multiple sequence alignment with high accuracy and high through-put. Nucleic Acids Res 32(5):1792–1797

24. Li M, Ma B, Kisman D, Tromp J (2004) Pattern Hunter II: highly sensitive and fast homology search. J Bioinform Comput Biol 2(3):417–439

25. Katoh K, Kuma K, Toh H, Miyata T (2005) MAFFT version 5: improvement in accuracy of multiple sequence alignment. Nucleic Acids Res 33:511–518

26. Pengfei G, Xuezhi Wa, Yingshi H (2010) The enhanced genetic algorithms for the optimization design. In: 3rd international conference on biomedical engineering and informatics, vol 7, 2990–2994

27. Eddy S (1998) Profile hidden Markov models. Bioinformatics 14:755–763

28. Notredame C, Higgins DG (1996) SAGA: sequence alignment by genetic algorithm. Nucleic Acids Res 24(8):1515–1524

29. Naznin F, Sarker R, Essam D (2009) Iterative progressive alignment method (IPAM) for multiple sequence alignment. In: Computers & Industrial Engineering, 2009. CIE 2009. International Conference on, Troyes, 536–541

30. Simon D (2011) A probabilistic analysis of a simplified biogeography-based optimization algorithm. Evol Comput 19(2):167–188

31. Ekta, Kaur M (2015) Biogeography based optimization: a review. In: 2nd international conference on computing for sustainable global development (INDIACom), New Delhi, 831–833

32. Pei J, Grishin N (2007) PROMALS: towards accurate multiple sequence alignments of distantly related proteins. Bioinformatics 23:802–808

33. Thompson JD, Plewniak F, Poch O (1999) A comprehensive comparison of multiple sequence alignment programs. Nucleic Acids Res 27:2682–2690

34. Wong WC, Maurer-Stroh S, Eisenhaber F (2010) More than 1,001 problems with protein domain databases: transmembrane regions, signal peptides and the issue of sequence homology. PLoS Comput Biol 6

35. Taylor WR (2000) Protein structure comparison using SAP. Methods Mol Biol 143:19–32

36. A Razmara J, Deris SB, Parvizpour S (2009) Text-based protein structure modeling for structure comparison. In: International conference of soft computing and pattern recognition, 490–496

37. Mott R (2005) Alignment: statistical significance. In: Encyclopedia of life science

38. Morgenstern B, Dress A, Werner T (1996) Multiple DNA and protein sequence alignment based on segment-to-segment comparison. Proc Natl Acad Sci USA 93(22):12098–12103

39. Barton GJ, Sternberg MJE (1987) A strategy for the rapid multiple alignment of protein sequences. J Mol Biol 198(2):327–337

40. Gondro C, Kinghorn BP (2007) A simple genetic algorithm for multiple sequence alignment. Genet Mol Res 6(4):964–982

41. Taheri J, Zomaya AY (2009) RBT-GA: a novel metaheuristic for solving the multiple sequence alignment problem. BMC Genomics 10(1):S10, 1–11

42. Karadimitriou K, Kraft DH (1996) Genetic algorithms and the multiple sequence alignment program in biology. In: Tiersch TR, et al (eds) Proceedings of the second annual molecular biology and biotechnology conference, Baton Rough Area

43. Horng JT, Lin CM, Liu BJ, Lao CY (2001) Using genetic algorithm to solve multiple sequence alignment. In: Wingender E, et al (eds) Proceedings of German conference on bioinformatics, pp 883–890

44. Isokawa M, Wayama M, Shimizu T (1997) Multiple sequence alignment using a genetic algorithm. Genome Inform 7:176–177

45. Wayama W, Takahashi K, Shimizu T (1995) An approach to amino acid sequence alignment using a genetic algorithm. Genome Inform 6:122–123

46. Lee ZJ, Su SF, Chuang CC, Liu KH (2008) Genetic algorithm with ant colony optimization (GA-ACO) for multiple sequence alignment. Appl Soft Comput 8(1):55–78

47. Yang C, Jinglu H and Songnian Y (2008). Multiple sequence alignment based on genetic algorithms with reserve selection. In: IEEE international conference on networking, sensing and control, 1511–1516

48. Xu F, Chen Y (2009) A method for multiple sequence alignment based on particle swarm optimization. Springer 5755:965–973

49. Lei XJ, Sun JJ, Ma QZ (2009) Multiple sequence alignment based on chaotic PSO. In: Proceedings of the computational intelligence and intelligent systems, Springer, Vol 51, pp 351–360

50. Thompson JD, Higgins DG, Gibson TJ (1994) CLUSTAL W: improving the sensitivity of progressive multiple sequence alignment through sequence weighting, position-specific gap penalties and weight matrix choice. Nucleic Acids Res 22(22):4673–4680

51. Smith RF, Smith TF (1992) Pattern-induced multi-sequence alignment (PIMA) algorithm employing secondary structure-dependent gap penalties for use in comparative protein modeling. Protein Eng 5(1):35–41

52. Notredame C, Higgins DG, Heringa J (2000) T-coffee: a novel method for fast and accurate multiple sequence alignment. J Mol Biol 302(1):205–217

53. https://en.wikipedia.org/wiki/BLOSUM

54. Dayhoff MO, Schwartz RM, Orcutt BC (1978) A model of evolutionary change in proteins. Atlas Protein Seq Struct 5(3):345–351

55. Bahr A, Thompson JD, Thierry J-C, Poch O (2000) BALIBASE (benchmark alignment dataBASE): Enhancements for repeats, transmembrane sequences and circular permutation. Nucleic Acids Res 29(1):323–326

Modeling of Service Discovery Over Wireless Mesh Networks

Lungisani Ndlovu, Okuthe P. Kogeda and Manoj Lall

Abstract Wireless Mesh Networks (WMNs) have played a huge rule in networking environments by supporting seamless connectivity, Wide Area Networks (WANs) coverage, mobility features, etc. However, the rapid increase in the number of consumers on these networks brought an upsurge in competitions for available services and resources. This has led to link congestions, data collisions, and link interferences, which affects Quality of Service (QoS) . Therefore, the quick and timely discovery of the services and resources becomes an essential parameter in optimizing the performance of service discovery on these networks. In this study, we present Ndlovu Okuthe Manoj (NOM) model, a service discovery model that integrates the Particle Swarm Optimization (PSO) and Ant Colony Optimization (ACO) algorithms. The PSO is used to dynamically define and give different priorities to services on the network, based on varied workflow procedures. On the other hand, the ACO is used to effectively establish the most cost-effective path whenever each transmitter has to be searched to identify whether it possesses the requested service(s). Furthermore, we design and implement the Link Collision Reduction (LCR) algorithm. It's objective is to define the number of service receivers to be given access to the services simultaneously. We then simulate the proposed model in Network Simulator 2 (NS2), against Ant Colony based-multi constraints QoS-aware service selection (QSS) and FLEXIble Mesh Service Discovery (FLEXI-MSD) models. The results show an average service discovery throughput of 80%, service availability of 96%, service discovery delay of 1.8 s, and success probability of service selection of 89%.

Keywords Service discovery models · QoS · ACO · PSO · LCR · WMNs

L. Ndlovu (✉) · O. P. Kogeda · M. Lall
Department of Computer Science, Faculty of ICT, Tshwane University of Technology,
Pretoria, South Africa
e-mail: booakalu@yahoo.com

O. P. Kogeda
e-mail: kogedapo@tut.ac.za

M. Lall
e-mail: lallm@tut.ac.za

© Springer International Publishing AG, part of Springer Nature 2019
S. K. Shandilya et al. (eds.), *Advances in Nature-Inspired Computing
and Applications*, EAI/Springer Innovations in Communication and Computing,
https://doi.org/10.1007/978-3-319-96451-5_14

313

1 Introduction

Computer networks have been in existence for over past three decades worldwide. Different environments including offices, campuses, military bases, governmental institutions, etc., have had their electronic devices connected to each other to enable the communication and sharing of different services and resources. Consequently, Wireless Mesh Networks (WMNs) continue to play a crucial role in these environments by supporting seamless connectivity, Wide Area Networks (WANs) coverage, mobility features, etc. WMNs can be seen as an extension of multi-hop Mobile Ad hoc Networks (MANETS) wherein each node can communicate directly or indirectly with one or more peer nodes. However, the nodes in WMNs, have the ability to relay data on behalf of other nodes. Therefore, WMNs can be defined as decentralized network structures wherein nodes have the potential to relay messages on behalf of others. This facilitates the creation of a self-configuring system that extends its coverage range and increases its available bandwidth. Currently, there are two types of WMN topologies, namely; Full Mesh and Partial Mesh. In addition, WMNs can be subdivided into three different types of architectures, namely; infrastructure\backbone, client, and hybrid. In general, WMNs often comprise of Mesh Routers (MRs) and Mesh Clients (MCs). MRs have limited mobility and thus form a multi-hop wireless mesh backbone between MCs and Mesh Gateways (MGs) [13]. MRs have the potential to provide the routing capability for both gateway and bridging features and, thus, can be used as MGs. MCs comprise of end-user devices such as desktop computers, laptops, Personal Digital Assistants (PDAs), mobile phones, printers, etc. On the other hand, MGs facilitate the integration of WMNs with other wired/wireless networks such as the Internet, cellular, Wireless Sensor Networks (WSNs) , etc.

The attractive qualities of WMNs include low-cost deployment, robustness and the inheritance of useful characteristics from both ad hoc networking and the traditional wired infrastructure paradigms [22]. Furthermore, the services and resources are always available on these networks. However, discovering services and resources on these networks is still challenging [1]. For this reason, a lot of work has been done in the field of service discovery. Service discovery refers to the process whereby devices (mobile or stationary) are able to discover without any intervention, appropriate services requested by users. In most cases, this process consists of three entities, namely; services, service registries, and clients. A service is any tangible or intangible facility a device provides to other devices as long as it is useful [18]. Furthermore, services fall into two categories: software and hardware services. Software services include MP3 files, audio convention, electronic mailing (emailing), etc. On the other hand, hardware services include printers, scanners, Voice over Internet Protocol (VoIP) phones, etc. Moreover, service registries are entities that store information about available services and/or advertise information about available services to other service registries or service receivers. Clients represent both service receivers and service transmitters. Therefore, clients have the potential to provide services to or request services from other clients through service discovery models.

Moreover, there are three service discovery approaches for flooding the network with information: directory-based, non-directory-based, and hybrid approaches. The most suitable approach for WMNs is the directory-based approach, as it does not involve the broadcasting or multicasting of information. However, most recent service discovery models for WMNs perform broadcasting or multicasting when flooding the network with information, resulting to network overhead [1]. Apart from that, the number of users in WMNs continues to increase, leading to competitions for available services and resources. These competitions further lead to clogging as more users request services simultaneously [20].

In this study, therefore, we propose Ndlovu Okuthe Manoj (NOM) service discovery model that aids in optimizing the performance of service discovery in WMNs. In the proposed model, we integrate the Particle Swarm Optimization (PSO) and Ant Colony Optimization (ACO) algorithms. The PSO algorithm is used to define and give different priorities to the services on the network. The prioritization is based on varied workflow procedures. On the other hand, the ACO is used to determine the most cost-effective path between transmitters whenever each transmitter has to be searched to identify if it possesses the requested service(s). Furthermore, we design and implement the Link Collision Reduction (LCR) algorithm to define the number of service receivers to be granted access to the services simultaneously. We simulate the proposed model using Network Simulator 2 (NS2) against Ant Colony based-multi constraints QoS-aware service selection (QSS) model by Kumar et al. [16] and FLEXIble Mesh Service Discovery (FLEXI-MSD) model by Krebs [14]. The NS2 tool is a discrete-event network simulation tool that facilitates users to model the behavior of wired/wireless and satellite networks. Compared to those of QSS and FLEXI-MSD, the results of the proposed model show reduced link congestions and fewer data collisions and link interferences as well as an improved average service discovery throughput of 80%, service availability of 96%, service selection of 89%, and service discovery delay of 1.8 s.

The rest of this Chapter is organized as follows: In Sect. 2, we present a brief discussion of WMNs. In Sect. 3, we present an overview of the PSO and ACO. In Sect. 4, we discuss related works. The design and architecture of the proposed WMN are discussed in Sect. 5. In Sect. 6, we present the model, with simulation results and discussion, is presented in Sect. 7. Lastly, we discuss future works and conclude the Chapter in Sect. 8.

2 Wireless Mesh Networks

In this Section, we provide a deeper insight into WMNs. We discuss the topologies and different types of WMNs. We also discuss the characteristics of these networks. Furthermore, we look into the critical factors to be considered in designing these networks. We then discuss the application areas of these networks. Later, we present a brief overview of the PSO and ACO algorithms. Thereafter, we discuss service discovery, its components, as well as the steps involved in this process. Lastly, we provide an insight into existing service discovery models and the various gaps available in these models.

However, mesh networks are categorized into two types of topologies, namely; Full mesh and Partial mesh. We show the differences of these topologies in Figs. 1 and 2. As shown in Fig. 1, a Full mesh topology occurs when every node is connected to every other node on the network. This topology yields the greatest amount of redundancy, so if a node fails, network traffic can be directed to other nodes. Full mesh is usually reserved for backbone networks. While, on the other hand, with partial mesh, some nodes are organized in a full-mesh topology, but others are connected to only one or two nodes on the network. Looking at Fig. 2, Partial mesh topologies are less expensive to implement but yields less redundancy compared to Full mesh topologies.

2.1 Overview of Wireless Mesh Networks

In the last decade, computer networks have become a never-ending quest. The availability of services and resources has made WMNs be the most recognized with the growth of wireless fidelity and outdoor wireless networks. Furthermore, mobile

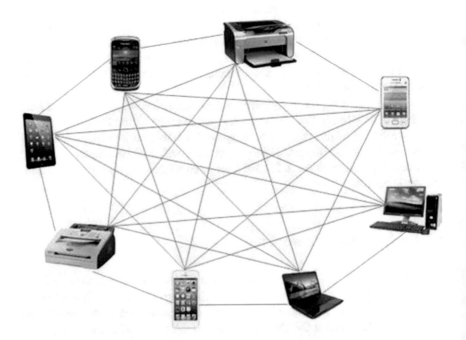

Fig. 1 Typical full mesh topology

Fig. 2 Typical partial-mesh topology

devices such as laptops, cell phones, Personal Digital Assistants (PDAs) , and tablets have become available at an affordable cost [19]. These networks provide services independent of both the network owners and equipment vendors and further enable users to introduce self-tailored services within their devices. These networks support services such as the Internet, e-commerce, audio streaming, Voice over Internet Protocol (VoIP), Video on Demand (VoD), file and printer sharing, etc. Ndlovu et al. [21] defined WMNs as heterogeneous and decentralized network structures whereby pervasive nodes are able to relay data over the network. On the other hand, Zeng et al. [29] defined WMNs as communication networks consisting of wireless devices organized in a mesh topology.

In its most general form, a WMN consist of MRs and MCs wherein MRs have limited mobility to form a multi-hop wireless mesh backbone between MCs and Mesh Gateways (MGs) . Furthermore, some MRs have the potential to provide routing capabilities for gateway/bridge functions simultaneously and, thus, can be utilized as MGs. The purpose is to enable the integration of WMNs with other networks such as the Internet, etc. MCs are desktop computers, laptops, PDAs, mobile phones, printers, etc., sharing services and resources on the network. Furthermore, there are three types of WMNs, namely; Infrastructure\Backbone, Client, and Hybrid. The differences between these types are briefly discussed in the next Section.

2.2   Types of Wireless Mesh Networks

There are three types of WMNs and are discussed as follows:

A.   Infrastructure/Backbone WMN

This type of architecture comprises of MRs forming a backbone for wired/wireless MCs. This architecture can also be built by integrating various radio technologies in addition to the IEEE 802.11 technologies used to interconnect nodes within WMNs. The benefit of this architecture is that MRs form a mesh of self-configuring as well as self-healing links among themselves [2]. As shown in Fig. 3, MRs are responsible for the communication between MCs as well as other networks such as the Internet through MGs functionalities. As mentioned in Sect. 2.1, MGs are responsible for integrating WMNs with other networks.

B.   Client WMN

This type of architecture comprises of MCs only, which are communicating with each other in a peer-to-peer technique. As shown in Fig. 4, in this type of architecture, client nodes constitute the actual network to perform routing and configuration functionalities and to provide end-user applications to customers [2]. However, this, in turn, increases the workload to client nodes leading to poor QoS when the network size increases.

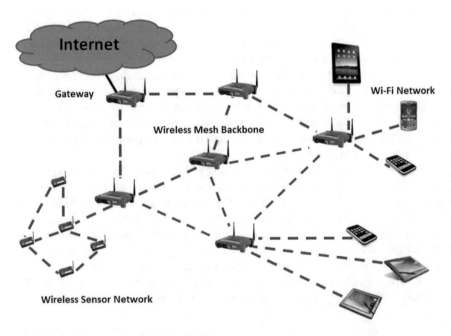

Fig. 3 Typical infrastructure/backbone WMN

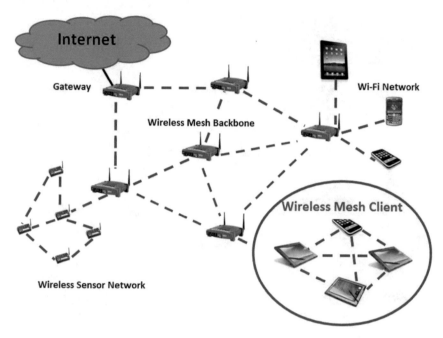

Fig. 4 Typical client WMN

C. **Hybrid WMN**

This is an integration of Infrastructure/Backbone and Client WMNs. For this reason, in this type of WMN, MCs can access the network through MRs or by directly meshing with other MCs [2]. As shown in Fig. 5, the Infrastructure/Backbone architecture allows the integration of WMNs and other networks. This architecture yields an improved QoS compared to Infrastructure and Client WMNs.

2.3 *Characteristics of Wireless Mesh Networks*

The research work by Akyildiz et al. [2], discussed the essential characteristics of WMNs as follows:

A. WMNs are multi-hop wireless networks.
    WMNs are considered as multi-hop networks which aid in achieving higher throughput without sacrificing effective radio range via shorter link distances, less interference between the nodes, and more efficient frequency reuse.
B. WMNs support for ad hoc networking, and capability of self-forming, self-healing, and self-organization.

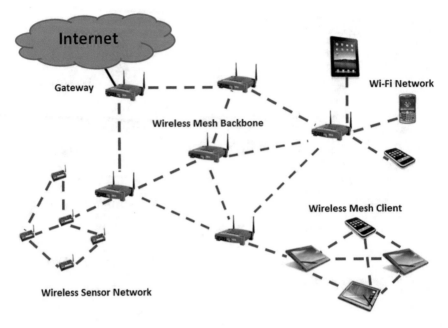

Fig. 5 Typical hybrid WMN

Introducing WMNs topology in any network enhances that networks' performance. This is because WMNs are flexible, easy to deploy and configure, fault
tolerance, and self-healing. By having such features, WMNs have low upfront
investment requirement, and the network can grow gradually as needed.

C. The mobility in WMNs depends on the type of mesh nodes.
   As mentioned in Sect. 2.1, the mobility depends on the type of node. Nodes such
   as MRs have limited mobility. However, MCs can either be stationary or mobile.

D. WMNs support multiple types of network access.
   WMNs permit both backhaul accesses to the Internet and client to client network
   communication. WMNs can also be integrated with other networks and that is
   achieved through MGs.

E. The power consumption constraint in WMNs depends on the type of mesh nodes.
   In most case, MRs do not have strict constraints on power consumption. However,
   due to the fact that MCs are mobile, and may, therefore, require power efficient
   routing protocols.

F. WMNs are compatible and interoperable with other existing networks.
   Since WMNs are built on IEEE 802.11 technologies, therefore, they must also
   be compatible with the IEEE 802.11 standards by supporting both the mesh
   capable and conventional Wi-Fi clients.

G. WMNs have wireless infrastructure/backbone.
   In most cases, WMNs have wireless infrastructure/backbone, which is composed
   of MRs providing large coverage, connectivity, as well as robustness on the
   network.

H. WMNs support the integration of WMNs and conventional networks.
   The integration of WMNs and other networks is easily supported in WMNs. This is because MRs provide the gateway and bridging functions, which enables users from a particular network to access services and resources located on another network.
I. The routing and configurations are dedicated to particular nodes in WMNs.
   In WMNs, MRs are responsible for routing and configurations, which reduces the load for MCs compared to other types of networks such as MANETs, whereby routing and configurations are performed by clients.
J. Multiple radios.
   In most cases, MRs are equipped with multiple radios to perform routing and access functionalities, which aids in the separation of two main types of traffic on the network.
K. Lastly, WMNs are able to self-heal.
   Whenever a fault is experienced within the entire network, the network is able to reconfigure itself.

2.4 Benefits of Wireless Mesh Networks

The benefits of WMNs include [5, 14, 22]:

A. WMNs are considered as highly flexible and scalable networks.
B. The routers in WMNs have limited mobility for the purpose of forming a multi-hop communication.
C. WMNs provide a large coverage area and lower costs of backhaul connections.
D. WMNs also include low-cost deployment, robustness and inherit the useful characteristics from both the ad hoc networking paradigm and traditionally wired infrastructure paradigms.
E. WMNs are the right solution for metropolitan area networks, providing last few miles connectivity.

2.5 Critical Design Factors in Wireless Mesh Networks

Pathak and Dutta [22] presented a survey of network design problems and joint design approaches in WMNs. The purpose was to discuss the critical design factors to be considered before WMNs can be designed, deployed, and operated. These factors are briefly discussed as follows:

A. Interference measurement and modeling—how to estimate and model the interferences in the dynamic wireless environment.
B. Power control—how to assign transmission power levels to nodes having transmission requirements.

C. Topology control—how to control the topology by choosing or avoiding certain links on the network.
D. Link Scheduling—how to schedule link transmissions to achieve feasible and conflict-free transmission schedule.
E. Channel/radio assignment—how to assign multiple channels to single or multiple radios at mesh nodes.
F. Routing—how to choose routing paths to satisfy the end-to-end traffic demands between mesh nodes.
G. Network planning and deployment—how to deal with the topological and deployment factors such as the placement of a gateway.
H. Performance modeling and capacity analysis—how to understand the best and worst case theoretical capacity.
I. Joint power control, topology control, link scheduling, routing or channel/radio assignment—how to deal with cross-layer optimization of more than one problem simultaneously.

2.6 Applications of Wireless Mesh Networks

WMNs have been seen as the cost-effective type of networks supporting seamless connectivity, WAN coverage, and mobility features. These benefits and the availability of services and resources have led WMNs to be the best solution for many applications. Some of the applications that motivate the research and development of WMNs include [25]:

A. Broadband internet access—many companies realize the potential of WMNs as an internet access solution and produced a broad range of related products. This is because WMNs offer considerable advantages as an internet broadband access technology.
B. Indoor WLAN coverage—many companies leveraged the multi-hop capabilities of WMNs to eliminate the need for cables. In such a deployment, at least one of the routers is connected to the external network and, hence, becomes a gateway. Other routers double up as Access Points (APs) in order to forward data from (and to) wireless clients to the gateway.
C. Mobile user access—most mobile users seeking connectivity outside the sparse coverage of WLAN hot spots have to settle for the slow and expensive 19.2 kbps cellular digital packet data (CDPD) or, more recently, for GPRS (usually 20–30 kbps—theoretical maximum 171.2 kbps). Properly designed WMNs can easily deliver higher bandwidth than the best 3 g technology.
D. Connectivity—Sometimes, providing network connectivity can be cumbersome, expensive, time-consuming or unsightly. network users should consider constructing WMNs specifically geared toward providing connectivity whereby routers form a wireless "cloud" that can be seen from outside as one big Ethernet switch.

3 Overview of Particle Swarm Optimization and Ant Colony Optimization Algorithms

In this Section, we discuss the PSO and ACO algorithms. These are algorithms used to eradicate the various gaps found in the conventional service discovery models for WMNs. Later, we then discuss the benefits and drawbacks of these two models.

3.1 Particle Swarm Optimization

Selvi and Umarani [23] stated that both the PSO and ACO algorithms are the most important and recent methods and optimization methods in networking. PSO algorithm was proposed for the optimization of continuous nonlinear functions [10]. The algorithm defines three behaviors of flocks in a swarm: cohesion—(sticking together), separation—(not coming too close) and alignment—(following the general heading of the flock). Its responsibility is to simulate a random search in the design space for the maximum value of the objective. Furthermore, it was implemented on two paradigms: one globally oriented (GBEST) and one locally oriented (LBEST).

It works more similar to the Genetic Algorithms (GAs). However, the information sharing in these two algorithms differs in the sense that any entity can share information in GAs while only GBEST and LBEST entities can share information with other entities in the PSO algorithm. PSO can be used to search for optimal solutions and to efficiently face the classification of multiclass databases as discussed by Selvi and Umarani [23], which is one reason why it is employed in this work.

As previously discussed, this algorithm is better than GAs as it is easily programmable, is faster in convergence, and provides effective solutions. However, in this study, we employ the PSO algorithm mainly for the classification and prioritization of services on the network.

3.2 Ant Colony Optimization

The ACO is a technique used to calculate an optimized path between two positions in a space. This algorithm was proposed by Dorigo in 1992 as a multi-agent approach to difficult combinatorial optimization problems such as the traveling salesman problem and the quadratic assignment problem [9]. The algorithm mimics the cooperative behavior of real-time ant colonies, which are able to find the most cost-effective path between the nests and food sources.

The basic idea of the ACO is taken from the real-time ants when they are searching for food. As they search for food ants deposit a chemical called pheromone, which is meant for marking the route from their colony to the food source. Other ants are attracted by the pheromone and start following its trails. However, as time goes on the pheromone evaporates making ants lose interest in following that route.

In any ACO technique, packets represent ant agents, traveling across the network to find the most cost-effective routes. The selection of the route is based on the concentration of pheromones. Unlike GAs, the ACO is able to run continuously and adapt to network changes in real time and this facilitates network routing and urban transportation systems [23]. In most cases, the process of ACO has random paths at the beginning that converge into a single most effective path to reduce time, energy, and collisions.

3.3 Benefits and Drawbacks

The benefits and drawbacks of PSO and ACO include [23]:

A. PSO is based on intelligence.
B. It can be applied in both scientific research and engineering use
C. It cannot work out problems of scattering and optimization.
D. ACO provides positive feedback accounts for the rapid discovery of good solutions.
E. The probability distribution changes after iteration.

4 Overview of Service Discovery

In this Section, we present an overview of service discovery by discussing services and their different types. We also briefly discuss the process of service discovery, the components of service discovery. Lastly, we discuss the steps involved in the process of service discovery.

4.1 Services

A service is an act or process whereby a device is serving other devices. A service can be anything such as printing, emailing, scanning, user account management, etc. Consequently, Mian et al. [18] defined a service as any tangible or intangible facility a device provides to other devices that are useful. On the other hand, Kogeda [12] discussed a service as more of an abstraction of the network-element-oriented or equipment-oriented view. Furthermore, identical services can be provided by different network elements, and different services can be provided by the same network element [18]. There are two types of services known as software and hardware services. Software services can be anything such as MP3 files while hardware services can be printers, etc. [18].

However, some services rely on other services to perform their duties and this is known as Network Service Dependency. Kogeda [12] discussed Network Service Dependency as the process whereby one component requires a service performed by another component to execute its functions. For instance, a printing service depends on the network and power supply; otherwise, it cannot perform its duty. However, in this Chapter, we do not dwell much on Network Service Dependency.

4.2 Service Discovery

Service discovery is the process whereby devices whether mobile or stationary are able to discover appropriate services as requested by users without any intervention. It involves the ability to discover and recognize the available services in a network [4]. The purpose of service discovery is to reduce human effort such as configuring and maintaining hundreds of devices in large networks, especially WMNs as MCs join and leave the network anytime.

In order to benefit from the process of service discovery, service discovery models enable these capabilities. Their responsibility is to autoconfigure the network to discover services without any user intervention. Therefore, if there is no service discovery model then services cannot be discovered without user involvement and their existence remains unknown to service receivers [8].

There are two ways to discover services; (a) discovery-advertisement whereby service transmitters advertise their services to service receivers. (b) discovery-search whereby service receivers advertise their requests to service transmitters [6].

Furthermore, there are three types of service discovery approaches for flooding information to the network and are known as directory-based, non-directory-based, and hybrid approaches. The most suitable approach for WMNs is a directory-based approach as there are service directories (servers) wherein service providers can locate or register their services to them. The advantage is that there is no frequently disconnection while accessing services on the network. While non-directory-based approaches apply broadcasting or multicasting of messages, which harm the network performance.

Nevertheless, recent approaches perform broadcasting or multicasting for flooding messages on the network, which leads to network overhead [1]. Consequently, we have based the proposed model on a directory-based approach.

4.3 Entities of Service Discovery

The process of service discovery includes entities such as clients, service registries, and services, which are discussed as follows:

A. Clients are entities representing both service receivers and service transmitters. These entities either provide services or expect to discover services from other entities with the help of service discovery models.

B. Registries are entities that store information about available services advertised by service transmitters within the network. Its responsibility is to advertise services to service receivers so that they can be aware of services that are available on the network.

C. Services refer to support provided by entities to other entities within the network in order to perform their tasks. There are two types of services in networking and are known as software and hardware services.

4.4 The Steps Involved in Service Discovery

Ahmad and Khalid [1] stated that service discovery involves steps such as bootstrapping, service registration, service discovery, and network resilience. However, we have considered the first three steps and provide a discussion as follows:

A. Bootstrapping refers to the process performed by nodes, which aids in the identification of neighbor nodes and available services. The bootstrapping node uses unicast and multicast communication techniques to declare its existence on the network and to advertise the available services on the network.

B. Service registration which refers to the process which enables the service transmitters to register their services to one of the nearest servers. The purpose of storing the information about available services is to advertise this information via unicast or multicast to service receivers.

C. Service discovery which is defined as the process performed by end-user devices to discover appropriate services requested by a user and without manually configuring devices and programs. The purpose of this process is to reduce human effort and precious time spent during the network configuration especially when looking for available services.

D. Network resilience which deals with the ability to provide and maintain an acceptable level of service on the network especially in the face of faults and challenges.

4.5 The Challenges in Service Discovery

In the past years, a lot of work has been proposed in the field of service discovery in computer networks, which resulted in the development of several service discovery models. These models include Sun's Jini technology, Microsoft's Universal Plug and Play (UPnP), Salutation, Service Location Protocol (SLP) [11], Apple's Bonjour,

Bluetooth Service Discovery Protocol (SDP), etc. The most dominant ones included UPnP, Bluetooth SDP, Apple's Bonjour, and Sun's Jini [6].

However, most of these models were developed mainly for small office, and home or enterprise networks [1, 14, 15, 31]. For this reason, these models are not scalable enough to be employed in WMNs. Furthermore, the work by Villaverde et al. [26] discussed that these protocols were not designed to be power efficient and consequently introduce significant overhead when adapted to wireless networks composed of devices with limited power supply.

With the aim of trying to diminish the drawbacks mentioned above, a Mesh-enhanced Service Location Protocol (mSLP) was presented by Zhao et al. [30]. However, this model too does not scale well in WMNs as service registration must be replicated between all the servers [1, 15].

On the other hand, other models were employed for MANETs whereby energy consumption and message dissemination were the two major concerns. Consequently, these models were found not scalable enough for WMNs [14, 15]. The primary reason is that these models apply flooding techniques as there are no dedicated servers on MANETs, which leads to network overhead [14]. Additionally, flooding of query messages consumes a lot of bandwidth, computation resources, and battery resources [28]. The work by Deepa and Swamynathan [7], Zakarya and Rahman [28], further discussed node mobility as another challenge experienced in MANETs. The primary reason is that it affects service availability because of frequent disconnection between service transmitters and receivers [3, 7].

In this study, therefore, we proposed NOM model in order to solve the mentioned drawbacks and optimize the performance of service discovery in WMNs. The proposed model provides an improved service discovery throughput, service availability, service selection probability, and reduced service discovery delay.

5   Related Work

Several types of research have been ongoing currently in various ways to optimize the performance of service discovery in WMNs. In this Section, we discuss briefly some of these researches.

5.1   QoS Aware Service Selection (QSS) Model

Kumar et al. [16] developed a QoS-aware service selection (QSS) model in order to provide QoS in the presence of constraints such as delay, jitter, and service availability. The research discovered that most of the existing models were mainly concerned about energy consumption which should not be the concern in WMNs as mesh nodes have limited mobility. In the model, services were discovered based on a defined cost-effective (CE) metric which guides ants regarding the path to choose between the source and destination. Compared to other models, ants in the model are not launched randomly but based on the Guided Search Evaluation (GSE) criterion. The GSE criterions also aid in decision-making when two paths have the same CE metric. The authors claim that the model provided better performance than other conventional models with respect to convergence, end-to-end delay (jitter), and service availability. However, the authors did not consider having a definite number of users given access to services simultaneously which yields to link congestions, data collisions, interference, etc. Additionally, in the model, different services are not given priorities, which affect QoS. Similarly, NOM model uses ACO to choose the most cost-effective path whenever each transmitter has to be searched to identify if it possesses the requested service.

5.2   FLEXI-Mesh Service Discovery (MSD) Model

Krebs [14] developed a cross-layer service discovery model called the FLEXI-Mesh Service Discovery (MSD) model. The proposed model was employed to ensure that wireless MRs dynamically established a virtual backbone. In the model, nodes automatically reorganize themselves in order to adapt to network changes in terms of structure and stability. The model uses the Optimized Link State Routing (OLSR) protocol and Domain Name System Service Discovery (DNS-SD). The OLSR protocol aids in finding most cost-effective routes. Meanwhile, the DNS-SD is used to offer clients an opportunity to discover, using standard DNS-messages, a list of named instances of the requested service. The authors claim that an adaptive (hybrid) service discovery mechanism which switches between ad hoc and supernode-based backbone systems outperforms a static operation mode system. This is because WMNs vary in network stability, device capabilities, and different mobility patterns. However, the problem with OLSR is that it requires a lot more bandwidth and Central Processing Unit (CPU) power to compute optimal paths on the network than does the ACO. Additionally, DNS-SD limits message forwarding to a maximum of hops, which makes it unfit for WMNs. Moreover, authors did not take into account factors such as clogging, which could be caused by many users given access to services at the same time. In this study, we have carefully considered such bottlenecks. We have also given different priorities to services to improve QoS.

5.3   A Model to Assure a Certain Level of QoS

Ahmad and Khalid [1] presented a service discovery model that assured certain levels of QoS. The model was proposed to avoid flooding and broadcasting of packets, which leads to discovery overhead. The aim was to reduce the routing update overhead by using different exchange periods for different entries in the routing table. In the model, a combination of Fisheye State Routing (FSR) and OLSR are used to minimize the flooding and broadcasting of packets. This new routing protocol aided in dividing the network into different scope levels for service advertisement and discovery. Meanwhile, the Normalized Link Failure Frequency (NLFF) metric is used to randomly select backbone nodes to be service directories based on the networks' stability constraints. The authors claim that the model would achieve a reduced network load; reasonable mean time delay by the requests initiated by the clients, a great average hit ratio of successful attempts, and reduced battery power consumption. However, this model did not take into account factors such as service response and availability, nor did the research did consider defining the number of service receivers to be granted access to services simultaneously. Additionally, the authors did not provide any experimental evaluations. On the other hand, in our proposed model, we give different priorities to the available services, which assure a certain level performance during service discovery.

5.4 A Scalable Two-Layered Approach for Distributed Service Discovery

Krebs and Krempels [15] developed a scalable two-layered approach for distributed service discovery in WMNs. The authors proposed the model after realizing that other models were optimized for highly dynamic environments. The model uses OLSR to apply anycast functionality. This aids in increasing service cache reliability, client transparency, and locality improvements, which further enables clients to communicate with the best service cache to register or query for services. This model integrated two replicate techniques that are exponential backoff and sends control in order to prevent severe duplicate information dissemination. The authors claim that this technique reduced delay and increases successful anycast queries. However, in this model service records are not consistent within service caches at any time. Additionally, the model does not have a definite number of receivers to be given access to service at the same time, which leads to clogging. Moreover, services are not given different priorities. On the other hand, we use ACO rather than OLSR.

5.5 Distributed-Hash Table Localized Service Discovery (DLSD)

Wirtz et al. [27] developed a Distributed-Hash Table Localized Service Discovery (DLSD) in order to maintain high transmission performance and minimize interference in WMNs. The study discovered that services and data discovery using undirected broadcast or multicast messages significantly degrade network performance because of link interferences and data collisions. In the model, Distributed-Hash Tables (DHTs) are used to offer consistent mapping of services and data, which permits directed unicast discovery and access. This aids in minimizing multi-hop transmission and ensured global reachability. The authors' claim that by searching in local sub-address spaces, their model reduces the number of overlays and underlay hops compared to Mesh-DHT and Chord. However, Mesh-DHT experiences delay in the event whereby the requested item is not found within the most local provider because it has to be searched within other providers through broadcasting or multi-casting, which leads in traffic congestions and data collisions. Additionally, we argue that the authors did not consider limiting the number of receivers to be given access to services simultaneously. We further considered choosing the most cost-effective path whenever transmitters are to be searched for the requested services.

5.6 Existing Service Discovery Routing Protocols

A number of routing protocols have been proposed. However, routing protocols have been designed for specific network sizes. The aim has been to establish the most cost-effective path between the source and destination, which is achieved by allowing the exchange of routing tables or known networks between routers. Additionally, there are two types of routing protocols: static and dynamic routing protocols. However, the work by Manvi et al. [17] presented a performance analysis of an Ad hoc On-Demand Distance Vector (AODV), Dynamic Source Routing (DSR), and swarm intelligence routing protocols. Their objective was to evaluate, using extensive simulation experiments, the routing performance of these routing protocols for Vehicular Ad Hoc Networks (VANETs). For the movement of vehicles, they used the random way-point model. In order to analyze the performance of these protocols, the following performance measures were considered: Throughput, Latency, Delivery ratio, and Delivery cost. Through experiments, the authors discovered that swarm routing protocols provide more promising results than do AODV and DSR. Additionally, Shruthi and Hemanth [24] concurred that swarm intelligence routing protocols outperform the standard MANET routing protocols like AODV, DSDV, DSR. Consequently, we have used ACO, which is a swarm-based routing protocol and aided in choosing the most cost-effective path whenever transmitters are to be searched for the requested services.

Unfortunately, various existing service discovery models have been more concerned about energy consumption and message dissemination. Some of these models applied flooding and broadcasting of query messages, which led in link congestion, data collisions, link interferences, etc. In these models, services were not given different priorities based on their varied workflow procedures, which does not assure a certain level of performance during service discovery. Additionally, these models did not consider having a definite number of receivers to be given access to services simultaneously. These gaps yielded to poor QoS. In this study, we propose NOM service discovery model in order to optimize the performance of service discovery in WMNs by prioritizing services, defining a number of receivers to be given access to services simultaneously, as well as choosing the cost-effective path between transmitters when each transmitter has to be searched to identify if it possesses the requested services.

6 System Architecture

In setting up the proposed WMN architecture various resources are pooled as shown in Fig. 6. These components include Layer 3 switches, secured MGs, servers, Linksys wireless router/Access Points (APs) representing MRs, and end-user devices such as PDAs representing MCs. The network consists of 3 MRs connecting clients to the network. Recent Linksys wireless routers include the Linksys EA9500 Max-Stream™ AC5400 MU-MIMO Gigabit Router, Linksys RE6500 AC1200 Dual-Band Wireless Range Extender, Linksys EA6900 AC1900 Dual-Band Smart Wi-Fi Wireless Router, Linksys RE4100W N600 Dual-Band Wireless Range Extender, and Linksys EA6350 AC1200+ Dual-Band Smart Wi-Fi Wireless Router. These MRs connect to the Layer 3 switch configured as the Transmitter, since it has the capacity to handle large amounts of traffic, possess different interface ports including Gigabit interface that connects to the secure server.

The recent Layer 3 switches now on the market include the Cisco SG300 series, Cisco 3650, 4500, and 6500, NETGEAR M4100, SG500-28P, and others. The transmitter provides centralized network management and configurations. Its responsi-

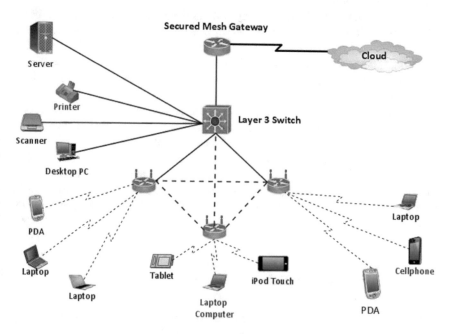

Fig. 6 Typical WMN proposed by Ndlovu et al. [20]

bility is to facilitate the communication between service transmitters and receivers. Furthermore, the advantages also include: the combination of routing and bridging features simultaneously, multiple Ethernet interfaces, ease of configuration, and lower cost-effective than purchasing and installing a router. Additionally, it supports advanced QoS features such as the prioritizing, classifying, policing, marking, queuing, and scheduling of packets. Consequently, in this study, it aids in the prioritization of services according to their varied workflow procedures.

The secured gateway is statistically configured with an Internet Protocol (IP) address. It facilitates the connection to remote sites and\or other networks. The server is configured with a static IP address used as a Domain Name System (DNS) address. Its purpose is to host files and databases as well as to deal with configurations and security issues. Consequently, it serves to store information about available services. The MRs are statistically configured with unique IP addresses. Lastly, MCs are configured with dynamic IP addresses through the Dynamic Host Configuration Protocol (DHCP).

In the proposed NOM model, we have mathematically modeled the network by a directed graph, which is represented by $G = (V, E)$ whereby $V = (v_1, v_2, \ldots, v_n)$ represents the number of mesh nodes on the network and $E = (e_1, e_2, \ldots, e_n)$ represents the number of links for both internal and external communications such as connections to the Internet. As mentioned in Sect. 1, WMNs are comprised of MRs and MCs. Any MR has the potential to provide routing capability for gateway/bridge functions. Therefore, among V nodes, J may represent the gateway, so $|V| - J$ represents ordinary MRs.

As shown in Fig. 6, on WMNs, Linksys wireless router or APs, have the potential to relay messages to and/or from mesh clients in a single hop or multi-hop manner. These APs further relay messages to and from the Internet through MGs. Clients (MCs) are end-user devices such as desktop computers, laptops, PDAs, mobile phones, etc., connected to MRs using wireless or wired links. Additionally, MCs represent both service transmitters and receivers.

In this study, therefore, we propose a model for facilitating service receivers to discover services without any user intervention. Consequently, any receiver joining the network expects the network to have n services available. Thus $s = (s_1, s_2, \ldots, s_n)$ wherein s represents the set of services and s_i represents each service on the network. Consequently, $s_i \in s$ and $i = 1, 2, 3, \ldots, \infty$. Since services have different workflow procedures, int this study, all available services are given different priorities in order to guarantee a certain level of performance. These priorities are represented by $s_n > s_{n-1} > s_{n-2} > \cdots > s_1$.

7 Modelling and Simulation

In this Section, we present the proposed NOM service discovery model and simulations. We discuss in depth the implementation stages involved.

7.1 Model Implementation

Based on Fig. 6, we use the PSO algorithm to dynamically define and give different priorities to services on the network. Its purpose is to guarantee a certain level of performance to data flow during service discovery. On the other hand, we use the ACO algorithm to effectively establish the most cost-effective path between transmitters whenever each transmitter has to be searched to identify whether it possesses the requested service(s). Its purpose is to reduce end-to-end delay and minimize packets loss during transmission. In the model, every transmitter maintains its routing table whereby ants are launched from one Transmitter to another in search for the requested service(s). Furthermore, we design and implement the LCR algorithm to define the number of service receivers to be granted access to services simultaneously. The purpose is to have a definite number of receivers to be granted access to service in order to solve congestion problems.

In the model, we begin by determining and defining the numbers of Transmitters, T_x and Receivers, R_x in the wireless network as given by Eq. (1):

$$R_x \geq T_x \tag{1}$$

We then define and prioritize service receivers, r, and transmitters, t, on the network using the PSO algorithm as given by Eq. (2):

$$X_r^l \leq X_r \leq X_r^u \quad \forall r$$
$$X_t^l \leq X_t \leq X_t^u \quad \forall t \tag{2}$$

where: X represents the priority given to each service transmitter and receiver. But L and U represent the least and highest priority for each service transmitters and receivers.

Therefore, each receiver must be a member of a given transmitter, and this is denoted by Eq. (3):

$$r_n \in t \tag{3}$$

Furthermore, we define and prioritize services supported by the network using the PSO algorithm as given by Eq. (4).

$$X_s^l \leq X_s \leq X_s^u \quad \forall s \tag{4}$$

where: X represents the priority given to each service. But L and U represent the lowest and highest priority given to each service.

Each transmitter and receiver must be able to receive or send a service request or acknowledgment. This is represented by Eq. (5):

$$s_n \exists t \ \& \ s_n \exists r \tag{5}$$

No particular receiver can request more services than are on the network and that the t_n can offer and provide.

Furthermore, the ACO is employed to determine the most cost-effective path whenever each transmitter has to be searched to identify whether it possesses the requested service. This is represented by the following equation:

$$P_{t_1 t_n}^{(r)} = \begin{cases} \dfrac{\tau_{t_1 t_n}^{\propto}}{\sum_{t_n \in N_{t_1}^{(r)}} \tau_{t_1 t_n}^{\propto}} & \text{if } t_n \in N_{t_1}^{(r)} \\ \\ 0 & if\ t_n \nexists N_{t_1}^{(r)} \end{cases} \tag{6}$$

Any receiver/requester, r, located at transmitter t_1, use pheromone trail $\tau_{t_1 t_n}^{\propto}$ to compute the probability, P, of choosing t_n as the next transmitter. $\propto$ denotes the degree of importance of the pheromones and $N_{t_1}^{(r)}$ indicates the set of neighborhood transmitters of the receiver when located at the transmitter t_1. The neighborhood of transmitter t_1 contains all the transmitters directly connected to the transmitter t_1 except for the predecessor transmitter. This limits the receiver from returning to the same transmitter visited immediately before transmitter t_1. A given receiver travels from one transmitter to the next in search of services.

Before returning to the home transmitter, the nth receiver deposits $\Delta \tau^r$ of pheromone on arcs, it has visited. The pheromone value $\tau_{t_1 t_n}$ on the arc (t_1, t_n) traversed is updated using Eq. (7), which gives us the total delay to get the service:

$$\tau_{t_1 t_n} \leftarrow \tau_{t_1 t_n} + \Delta \tau^{(r)} \tag{7}$$

Therefore, because of the increase in pheromone, the probability of this arc being selected by the forthcoming receivers on the network increases.

Also, we have to consider that whenever a receiver moves to the next transmitter in search of services, the pheromone evaporates from all the arcs according to the following relation:

$$\tau_{t_1 t_n} \leftarrow (1 - p)\tau_{t_1 t_n}; \forall (t_1, t_n) \in A \tag{8}$$

where: $p \in (0, 1)$ is a parameter and A denotes the segments/arcs traveled by the receiver r in its path from home to destination. A decrease in pheromone intensity favors the exploration of different paths during the search process. This, in return, favors the elimination of poor choices made in the path selection. Iteration, therefore, is a complete cycle involving receiver movement, pheromone evaporation, and pheromone deposit.

After all the receivers have returned to the home transmitter, the pheromone information is updated using the relation given in Eq. (9):

$$\tau_{t_1 t_n} = (1 - p)\tau_{t_1 t_n} + \sum_{r=1}^{N} \Delta \tau_{t_1 t_n}^{(r)} \tag{9}$$

where: $p \in (0, 1)$ represents the evaporation rate (pheromone decay factor) and $\Delta\tau_{t_1 t_n}^{(r)}$ is the amount of pheromone deposited on arc $t_1 t_n$ by the best receiver.

The major goal of pheromone update is to increase the pheromone value associated with good or promising paths. The pheromone deposited on arc $t_1 t_n$ by the best receiver is given by Eq. (10):

$$\Delta\tau_{t_1 t_n}^{(r)} = \frac{Q}{L_r} \tag{10}$$

where: Q represents a constant and L_r is the length of the path traveled by the receivers.

However, for every service request, a maximum of five receivers are allowed to request services simultaneously and then leave the media for the next receivers. This is because none of the existing models specified or defined the number of receivers to be granted access to service simultaneously causing congestion problems. Consequently, in the proposed model, we apply the LCR to eliminate such bottlenecks.

Algorithm: Link Congestion Reduction (LCR)

Input: Number of service receivers (r)
Output:
While number of waiting for service receivers > 0 **Do**
  If number of waiting for service receivers > 4 **Then**
    For service receivers to be given access = 1 to 5 **Do**
      Discover the requested services on the network.
      Then leave the media for the next service receivers.
    End For
  Else
    Discover the requested services on the network.
  End If
End While

In discovering any service by the r_n, the 1st five receivers in priority order can request the available services on the network. The above iteration is implemented for the remaining receivers on the network until all get to receive the requested services.

Lastly, a maximum of ten services can be requested at any particular time in the model, and those ten services must be in priority order. This improves the QoS, which further improves service discovery throughput, service availability, service selection, and service discovery delay.

7.2 Simulation Set up

The simulations were carried out using NS2 Version 2.35 (NS-2.35) with a developed model of IEEE 802_11b/g. We employed this tool because it is freely available on the Internet; it works on different operating systems, and as it has many users, so help from others is readily available. The simulation scenario of the topology used to carry out the simulations is shown in Fig. 7. This topology represents the proposed WMN to carry out the simulations. The setup consists of 43 nodes. There are 30 mobile nodes (service receivers) randomly placed on a network and numbered 13–43. For the mobility, a random waypoint is used wherein each node moves—around until its mobility-time ends.

We use Omni-directional Antenna for wireless transmission and reception of data. We also use Link Layer (LL) to process and transmit the messages. Furthermore, we use Queue/DropTail/PriQueue to give different priorities to service requests based on the priorities assigned to service transmitters by the PSO algorithm, thus improving QoS. The bandwidth is 512 Kb with the delay of 5 ms, and message length of 64 bytes. We have 3 transmitters and each labeled "Transmitter" with 600 meters coverage. The transmitters are placed at (413, 517), (1045, 700), and (1345, 174). Furthermore, we have 10 services labeled with service names such as Voice-on-Demand, etc. The services are randomly placed nearby each transmitter. The proposed model is implemented on the distribution (transmitter) layer of the network.

We then run the simulations 100 s for each evaluation. The evaluations were performed a minimum of 5 times in order to obtain convincing results. The primary reason is that NS2 is not scalable, so simulation results might not be reliable. The results are recorded in trace files and used to calculate throughput, delay, service availability, as well as service selection. The R Programming is used to graphically display the analyzed results obtained. However, in order to calculate the average

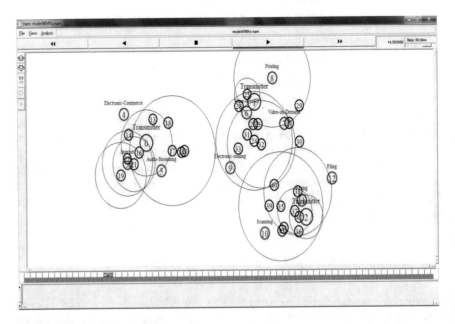

Fig. 7 WMN simulation scenario

discovery throughput, service available, service selection, and discovery delay, we have to code, compile and run AWK programs. We observe packet losses in the simulations. Lastly, we evaluate NOM model against QSS and FLEXI-MSD.

As mentioned in Sect. 1, in the model, we integrate the PSO, ACO, and LCR algorithm. The PSO algorithm is used to define and prioritize services on the network. The purpose is to give different priorities to different services in order to assure a certain level of performance during service discovery in WMNs. ACO is used to find and choose the cost-effective path between transmitters. The purpose is to determine the shortest path between transmitters whenever each has to be searched to identify if it possesses the requested service. ACO is an adaptive nature inspired routing technique applied to routing protocols in wired/wireless networks. Its responsibility is to find the shortest path between the source and destination nodes.

When compared to other routing protocols such as Routing Information Protocol (RIP) and Open Shortest Path First (OSPF), this technique can be attached to data, supports the frequent transmission of ant agents, reduces routing overhead, and updates an entry in a pheromone table independently.

There is quite a number of routing protocols that apply the ACO technique. However, these ACO-based routing algorithms are designed for different network sizes. In this work, we employ AntHocNet, a routing protocol proposed for wireless networks such as MANETs, with less or more than 50 nodes compared to other protocols. Furthermore, AntHocNet is more efficient and scalable to deliver packets than other MANETs routing protocols such as AODV, OLSR, etc. In the proposed model, the goal of ants is to travel from one transmitter to another using the shortest path in search for the requested service(s) while depositing pheromones and updating the routing tables. Once one of the ants finds the requested service, all other ants are notified about that path through pheromone trails. Additionally, rather than OLSR, AntHocNet does not rely on bandwidth and energy which reduces bandwidth usage and energy consumption.

NS2 which is an object-oriented discrete-event network simulator targeted at networking research. The purpose is to evaluate the behavior and performance of communication networks. The proposed model has generated service discovery throughput, service discovery delay, service availability and success probability of service selection results as shown and discussed in Sect. 8.

8 Results and Discussions

In order to test and evaluate the performance of the proposed service discovery model, we had set up a framework that evaluates the factors of QoS in NS2. Simulations have been conducted for the NOM, QSS, and FLEXI-MSD models. The two models were chosen mainly because they were also proposed to improve discovery rate, discovery delay, and success probability of service selections. The results are recorded to ascertain the objectives of this study.

Fig. 8 Service discovery
throughput

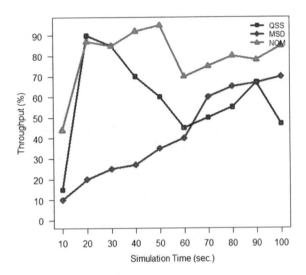

8.1 Service Discovery Throughput

We observe an incline in service discovery throughput between 10 and 20 s in the proposed model as shown in Fig. 8. More and steady service discovery throughput is observed between 20 and 50 s. However, the service discovery throughput drops a bit between 60 and 100 s. The model has achieved an average service discovery throughput of 80% thereby reducing link congestion, data collisions, and link interferences while improving QoS.

The proposed NOM model improved service discovery throughput compared to QSS and FLEXI-MSD models as shown in Fig. 8. This is because we defined a maximum number of service receivers to be granted access to the services simultaneously and then leave the media for the next service receivers. This reduces link congestion, data collisions, and link interferences. We also permit a maximum number of services, in priority order, to be accessed by the defined number of service receivers simultaneously, which improves QoS.

8.2 Service Availability

More services become unavailable at 10 and 20 s and this is because the simulation is at the initialization state wherein nodes (whether mobile or stationary) are busy locating themselves available in the entire network. However, as time goes on, the service availability becomes improved in the NOM model as shown in Fig. 9. That is why between 30 and 60 s, the service availability elevates to 85, 95, 94 and 93% respectively, realizing an average service availability of 92%. This shows the need

Fig. 9 Service availability

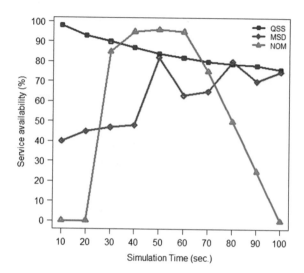

and importance of considering mobility issues for future service discovery models. The service availability is further affected by the number of requests processed simultaneously.

In the proposed model, the service availability is improved compared to QSS and FLEXI-MSD models. However, looking at Fig. 9, the service availability at the start of the simulation is very low. This is because of the mobility of nodes while joining the network. The service availability then rises up and drops after sometimes and that is because, at some point, all the services become unavailable as a result of being in use by service receivers. This means as the services become available, the service availability rises up. Furthermore, the service availability is also affected by the mobility of nodes on the network.

8.3 Service Selection

NOM model has an improved success probability of service selection being witnessed compared to QSS and FLEXI-MSD models. Through Fig. 10, we can observe an incline in service selection for the proposed model. This observation is attributed to having a definite number of service receivers to be given access to services simultaneously, which ensures that all service receivers get to access available services while reducing packet losses caused by link congestions, data collisions, as well as link interferences.

Here we present the success probability of service selection obtained by NOM model compared to QSS and FLEXI-MSD models. Through Fig. 10, we observe a promising success probability of service selection in NOM compared to QSS and FLEXI-MSD. This is because the proposed model ensures that all service receivers

Fig. 10 Service selection

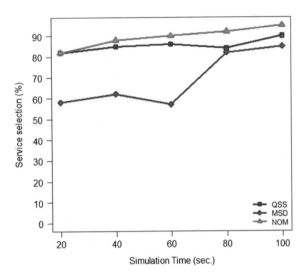

get to be given access to the services. This is achieved by having a definite number of service receivers given access to services simultaneously, which aids in reducing packet losses that could be caused by link congestions, data collisions, as well as link interferences.

8.4 Service Discovery Delay

NOM model has the least service discovery delay of 1.8 s being witnessed compared to QSS and FLEXI-MSD models, that witness more service discovery delay of 3.5 and 6.8 s respectively, as shown in Fig. 11. A maximum of 2.3 s of service discovery delay is witnessed in the proposed model. The observation is attributed to reduced link congestions, data collisions, and link interferences.

Through Fig. 11, we present the service discovery delay obtained throughout simulations by NOM model compared to QSS and FLEXI-MSD models. It is clear that NOM provides a reduced service discovery delay. These promising results are achieved as link congestion, data collisions, and link interferences are reduced. We reduced such by defining a number of service receivers to be given access to services simultaneously, which also minimizes bandwidth usage.

9 Conclusion and Future Work

In most of the existing service discovery models, energy consumption and message dissemination seem to be two major concerns that were most taken into consid-

Fig. 11 Service discovery delay

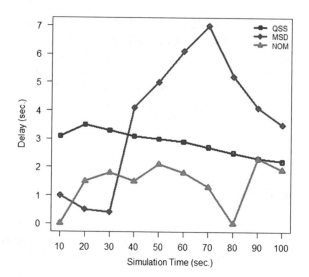

eration. Furthermore, various existing models perform flooding and broadcasting of messages, leading to link congestions, data collisions, and link interferences in WMNs. Apart from that, these networks continue to experience the increasing number of consumers competing for available services and resources. These competitions degrade QoS and form a major contribution to the poor performance of service discovery on these networks. Consequently, the quick and timely discovery of services and resources becomes an essential parameter in optimizing the performance of service discovery on these networks.

In this Chapter, we introduced and presented various factors affecting the performance of service discovery in WMNs. An overview, benefits, critical design factors and application areas of WMNs were also presented. Furthermore, we discussed existing service discovery model and various gaps found in these models. Thereafter, we presented a detailed implementation, simulation, testing, and evaluation of the proposed model. On the other hand, we analyzed and discussed the results obtained through simulations.

In the model, we integrated the PSO, ACO, and LCR algorithms to optimize the performance of service discovery in WMNs. We used the PSO algorithm to dynamically define and prioritize services on the network in order to guarantee a certain level of performance to data flow during service discovery. We used ACO algorithm to choose the cost-effective path whenever each transmitter has to be searched to identify if it possesses the requested service(s) or resource(s). Furthermore, we designed and implemented the LCR algorithm to define the number of service receivers that must be granted access to the available services simultaneously.

We demonstrated by Network Simulator 2 (NS2) that the proposed model outperforms QSS and FLEXI-MSD models. NOM improves service discovery throughput, service availability, service selection success probability, and reduces service discovery delay compared to QSS and FLEXI-MSD. The model yielded an average service discovery throughput of 80%, service availability of 96%, service selection of 89, and service discovery delay of 1.8 s from the simulation and testing results in NS2.

Furthermore, research could focus on applying the improved PSO known as Accelerated Particle Swarm Optimization (APSO). Considering the advances of the Internet of Things (IoT) , security is also an important area of research to be considered in the field of service discovery in WMNs. Lastly, we recommend these finding for enterprise telecommunication and networking industries.

References

1. Ahmad F, Khalid S (2012) Scalable design of service discovery mechanism for an ad-hoc network using wireless mesh network. Int J Smart Sens Ad-Hoc Netw 1(4):95–99
2. Akyildiz IF, Wang X, Wang W (2005) Wireless mesh networks: a survey. Comput Netw 47(4):445–487
3. Al Mallah R, Quintero A (2009) A light-weight service discovery protocol for ad hoc networks. J Comput Sci 5(4):330–337
4. Bashah NSK, Jørstad I (2010) Service discovery in future open mobile environments. In: Fourth international conference on proceedings of digital society, 2010. ICDS'10, held in Netherlands Antilles on 10–16 Feb 2010, St. Maarten
5. Benyamina D, Hafid A, Gendreau M (2012) Wireless mesh networks design—a survey. IEEE Commun Surv Tutorials 14(2):299–310
6. Bruda SD, Salehi F, Malik Y, Abdulrazak B (2012) A peer-to-peer architecture for remote service discovery. Procedia Comput Sci 10:976–983
7. Deepa R, Swamynathan S (2010) A service discovery model for mobile ad hoc networks. In: 2010 international conference on trends in information, telecommunication and computing (ITC), held in India on 12–13 Mar 2010
8. Dittrich A, Salfner F (2010) Experimental responsiveness evaluation of decentralized service discovery. In: Proceedings of parallel & distributed processing, workshops and Phd forum (IPDPSW), held in USA on 19–23 Apr 2010, Atlanta, Georgia
9. Dorigo M, di Caro G, Gambardella LM (1999) Ant algorithms for discrete optimization. Artif Life 5(2):137–172
10. Eberhart RC, Kennedy J (1995) A new optimizer using particle swarm theory. In: Proceedings of the sixth international symposium on micro machine and human science, held in Japan on 4–6 Oct 1995, Nagoya
11. Guttman E (1999) Service location protocol: Automatic discovery of IP network services. IEEE Internet Comput 3(4):71–80
12. Kogeda OP (2012) Cellular network service dependency modeling for network faults prediction. In: Proceedings of the World Congress on engineering and computer science, held in USA on 24–26 Oct 2012, San Francisco
13. Komba GM, Kogeda OP, Zuva T (2014) A new gateway location protocol for mesh networks. In: Proceedings of the World Congress on engineering and computer science, held in USA on 22–24 Oct 2014, San Francisco
14. Krebs M (2009) Dynamic virtual backbone management for service discovery in wireless mesh networks. In: 2009 IEEE wireless communications and networking conference, held in Hungary on 24–26 Sept 2009, Budapest
15. Krebs M, Krempels K-H (2009) Optimistic on-demand cache replication for service discovery in wireless mesh networks. In: 2009 5th IEEE consumer communications and networking conference, held in USA on 10–13 Jan 2009, Las Vegas, Nevada

16. Kumar N, Iqbal R, Chilamkurti N, James A (2011) An ant based multi-constraints QoS-aware service selection algorithm in wireless mesh networks. Simul Model Pract Theory 19(9):1933–1945
17. Manvi S, Kakkasageri M, Mahapurush C (2009) Performance analysis of AODV, DSR, and swarm intelligence routing protocols in vehicular ad hoc network environment. In: International conference on proceedings of future computer and communication, 2009. ICFCC 2009, held in Malaysia on 3–5 Apr 2009, Kuala Lumpur
18. Mian AN, Baldoni R, Beraldi R (2009) A survey of service discovery protocols in multihop mobile ad hoc networks. IEEE Pervasive Comput 8(1):66–74
19. Mitsugi J, Sato Y, Ozawa M, Suzuki S (2014) An integrated device and service discovery with UPnP and ONS to facilitate the composition of smart home applications. In: 2014 IEEE world forum on proceedings of internet of things (WF-IoT), held in South Korea on 06–08 Mar 2014, Seoul
20. Ndlovu L, Lall M, Kogeda OP (2016) An improved ant-based service discovery model for wireless mesh networks. In: Proceedings of Southern Africa telecommunication networks and applications conference (SATNAC), held in South Africa on 4–7 Sept 2016, George, Western Cape
21. Ndlovu L, Lall M, Kogeda OP (2018) Enhanced service discovery model for wireless mesh networks. J Adv Comput Intell Intell Inf 22(1):44–53
22. Pathak PH, Dutta R (2011) A survey of network design problems and joint design approaches in wireless mesh networks. IEEE Commun Surv Tutorials 13(3):396–428
23. Selvi V, Umarani DR (2010) Comparative analysis of ant colony and particle swarm optimization techniques. Int J Comput Appl (0975–8887), 5(4):1–6
24. Shruthi VR, Hemanth SR (2015) Simulation of ACO technique using NS2 simulator. Int J Eng Trends Technol 23(8):403–406
25. Sichitiu ML (2005) Wireless mesh networks: opportunities and challenges. In: Proceedings of World Wireless Congress, held in USA, on 22–25 May 2005, Atlanta, GA
26. Villaverde BC, Alberola RDP, Jara AJ, Fedor S, Das SK, Pesch D (2014) Service discovery protocols for constrained machine-to-machine communications. IEEE Commun Surv Tutorials 16(1):41–60
27. Wirtz H, Heer T, Serror M, Wehrle K (2012) DHT-based localized service discovery in wireless mesh networks. In: 2012 IEEE 9th international conference on mobile Ad-Hoc and sensor systems (MASS 2012), held in USA on 8–11 Oct 2012, Las Vegas, Nevada
28. Zakarya M, UR RAHMAN I (2013) A short overview of service discovery protocols for MANETs. VAWKUM Trans Comput Sci 1(2):1–6
29. Zeng G, Wang B, Mutka M, Xiao L, Torng E (2012) Efficient link-heterogeneous multicast for wireless mesh networks. Wireless Netw 18(6):605–620
30. Zhao W, Schulzrinne H, Guttman E (2000) mSLP-mesh-enhanced service location protocol. In: Ninth international conference on proceedings of computer communications and networks, 2000, held in USA on 18–18 Oct 2000
31. Zhu F, Mutka MW, Ni LM (2005) Service discovery in pervasive computing environments. IEEE Pervasive Comput 4(4):81–90

Index

© Springer International Publishing AG, part of Springer Nature 2019
S. K. Shandilya et al. (eds.), *Advances in Nature-Inspired Computing and Applications*, EAI/Springer Innovations in Communication and Computing,
https://doi.org/10.1007/978-3-319-96451-5

Printed in the United States
By Bookmasters